No-tillage Seeding in Conservation Agriculture

Second Edition

This book is dedicated to the scientists and students whose work is reviewed, together with their long-suffering families. Such people were driven by a desire to make no-tillage as sustainable and risk-free as possible, and in the process to make food production itself sustainable for the first time in history. The odds were great but the results have been significant and will have far-reaching consequences.

No-tillage Seeding in Conservation Agriculture

Second Edition

C.J. Baker, K.E. Saxton, W.R. Ritchie, W.C.T. Chamen,
D.C. Reicosky, M.F.S. Ribeiro, S.E. Justice and P.R. Hobbs

Edited by

C.J. Baker and K.E. Saxton

Published by
Food and Agriculture Organization of the United Nations

www.cabi.org

CABI is a trading name of CAB International

CABI Head Office
Nosworthy Way
Wallingford
Oxfordshire OX10 8DE
UK

Tel: +44 (0)1491 832111
Fax: +44 (0)1491 833508
E-mail: cabi@cabi.org
Website: www.cabi.org

CABI North American Office
875 Massachusetts Avenue
7th Floor
Cambridge, MA 02139
USA

Tel: +1 617 395 4056
Fax: +1 617 354 6875
E-mail: cabi-nao@cabi.org

A catalogue record for this book is available from the British Library, London, UK.

Library of Congress Cataloging-in-Publication Data

No-tillage seeding in conservation agriculture/C.J. Baker . . . [et al.] edited by C.J. Baker and K.E. Saxton.-- 2nd ed.
p. cm.
Rev. ed. of: No-tillage seeding/C.J. Baker. 1996.
Includes bibliographical references and index.
ISBN 1-84593-116-5 (alk. paper)
1. No-tillage. I. Baker, C.J. (C. John) II. Saxton, Keith E., 1937- III. Baker, C.J. (C. John). No-tillage seeding. IV. Food and Agriculture Organization of the United Nations. V. Title.

S604.B36 2006
631.5'31--dc22

2005035401

Published jointly by CAB International and FAO.
Food and Agriculture Organization of the United Nations (FAO)
Viale delle Terme di Caracalla, 00100 Rome, Italy
Website: www.fao.org

ISBN-10: 1-84593-116-5 (CABI)
ISBN-13: 978-1-84593-116-2 (CABI)
ISBN: 92-5-105389-8 (FAO)

Typeset by AMA DataSet Ltd, Preston, UK.
Printed and bound by CPI Group (UK) Ltd, Croydon, CR0 4YY

Contents

Contributors

C.J. Baker, *Centre for International No-tillage Research and Engineering (CINTRE), Feilding, New Zealand*

W.C.T. Chamen, *4Ceasons Agriculture and Environment, Maulden, Bedfordshire, UK*

P.R. Hobbs, *Department of Crops and Soil Science, Cornell University, Ithaca, New York, USA*

S.E. Justice, *National Agriculture and Environment Forum (NAEF), Kathmandu, Nepal*

D.C. Reicosky, *United States Department of Agriculture, Agricultural Research Service, Morris, Minnesota, USA*

M.F.S. Ribeiro, *Instituto Agronômico do Paraná (IAPAR), Ponta Grossa, Parana, Brazil*

W.R. Ritchie, *Centre for International No-tillage Research and Engineering (CINTRE), Feilding, New Zealand*

K.E. Saxton, *Retired, formerly United States Department of Agriculture, Agricultural Research Service, Pullman, Washington, USA*

Foreword to the Second Edition

The Food and Agriculture Organization (FAO) has a history of supporting the development and extension of conservation agriculture cropping systems. No-tillage seeding is one of the key operations of conservation agriculture; no-till seeding, together with the principles of cover crops and crop rotation, constitute conservation agriculture. The availability of suitable technology and equipment is a necessary precondition for making conservation agriculture work. Special equipment is required not only for direct seeding and planting, but also for the management of crop residues and cover crops.

The earlier book, entitled *No-tillage Seeding: Science and Practice*, by Baker, Saxton and Ritchie, was, at the time of its publication, one of the most comprehensive publications covering the engineering aspects of no-tillage seeding as well as the agronomic and environmental background for no-tillage farming. It has been valuable as a reference for scientists and students, and also as a guide for practitioners. A case was reported where a farmer after reading this book bought a no-till planter and converted his farm to no-till.

This new book, *No-tillage Seeding in Conservation Agriculture*, provides a broader picture of the equipment used in conservation agriculture cropping systems. It includes chapters on material not previously covered, for example, the management of crop residues and cover crops, preparation for the no-tillage seeding operation, and controlled-traffic farming as a complementary technology. There are also new chapters describing no-tillage seeding technologies for small-scale farmers. Technology developments from South America and South Asia are described, including manual equipment, draught-animal equipment and equipment for power tillers. The subject of greenhouse gases as driving forces for climate change is also discussed in a chapter on carbon sequestration under no-tillage farming systems.

We hope that this book contributes to a better understanding of the engineering components of conservation agriculture. It is also our wish that it helps with the introduction and expanded application of this technology. Conservation agriculture is a valuable approach to cropping that can lead to more productive, competitive and sustainable agricultural systems with parallel benefits to the environment and to farmers and their families.

Shivaji Pandey
Director
Theodor Friedrich
Senior Agricultural Engineer
Agricultural Support Systems Division
FAO
Rome, November 2005

Preface

And he gave for his opinion, that whoever could make two ears of corn or two blades of grass to grow upon a spot of ground where only one grew before, would deserve better of mankind, and do more essential service to his country than the whole race of politicians put together.

Jonathan Swift, *Gulliver's Travels* (1726)
'A Voyage to Brobdingnag'

The authors of this book describe and analyse no-tillage technologies, particularly those related to no-tillage seed drilling, from a variety of accumulated experiences over the past 40 years. Most of us set out to discover why no-tillage did not always work and how to overcome these obstacles. The more we learned the more appealing no-tillage farming became. The understanding and system science have now been acquired and tested to the point where we are ever more confident it represents the future of farming.

Some of the reported research started from knowledge that none of the traditional drills, planters or opener technologies used for tillage farming then provided a fail-safe methodology for untilled, residue-covered soils. Inevitably that resulted in new machine designs and evaluations, and combined associated technologies. The guiding premise was that every functional part of any new design had to have a verifiable scientific reason and performance, which often resulted in a long evolution.

No functional assumptions were made. All commonly held ideas about what seeds required were challenged or discarded and new experiments set up to determine their requirements specifically in untilled soils. This new knowledge was combined with whatever existing knowledge proved still to be applicable. In other cases the rules for tilled soils simply did not apply, or were proved wrong, when applied to untilled soils. Undisturbed soils were found to provide different resources and challenges from those of tilled soils, thus requiring different approaches to seed sowing.

Other authors report what happened to soil when ploughing ceases. Everyone by now knows that no-tillage is good and ploughing is bad for the soil, but what are the causal mechanisms and can the improvements or damage be quantified? Can the gains be further improved by techniques such as controlled-traffic farming? Still other authors studied available equipment and management methods and relate these to no-tillage systems and applications, large and small. Only when the capabilities of modern no-tillage equipment are understood and fully integrated into a crop production enterprise can it be fully quantified and realistic local recommendations made.

Collectively these authors have provided a comprehensive overview of what makes a successful no-tillage enterprise work. This includes machinery design and operating principles, the interactions of machines with the soil, the importance of parallel inputs, such as herbicides, pesticides and controlled traffic, and the management of the system as a whole, including quantifying the importance of soil carbon and tracking carbon dioxide emissions as a function of soil disturbance. They have also provided a guide to experimental procedures for evaluation of variables.

The book is not intended to be a blueprint on how to design any one style of no-tillage machine, component or system. It is a record of the comparative performances of several different machine design options and management practices, tested under controlled scientific conditions, and how these have been found to integrate into a whole no-tillage system. Much of the information is about the biological performance of machines and soils, since both primarily perform biological functions. But mechanical performance is not ignored either. The interface between the two is particularly important.

The reader is invited to place his or her own value on the relevance of the data presented. The relevance some of the authors placed on the data led to the design of the disc version of a winged opener, called Cross Slot®. Others will see different things in the data. However, independent research and field experience have increasingly shown that the data and the conclusions drawn from them have been remarkably accurate and prophetic.

The relevance of the book is that it illustrates that there are now ways and means to make no-tillage more fail-safe than tillage and to obtain crop yields not only equal to those from tillage but, in many cases, superior. Untilled soils contain greater potential to germinate, establish and grow plants than tilled soils ever did. And, of course, they are much more environmentally friendly. The problem for humankind has been to learn and understand how to harness that potential. We hope this book goes some way towards achieving that objective.

The book expands on the first edition, entitled *No-tillage Seeding: Science and Practice* (Baker, Saxton and Ritchie, ISBN 0 85199 103 3, first published by CAB International in 1996 and reprinted in 2002).

1 The 'What' and 'Why' of No-tillage Farming

C. John Baker and Keith E. Saxton

No farming technique yet devised by humankind has been anywhere near as effective as no-tillage at halting soil erosion and making food production truly sustainable.

Since the early 1960s farmers have been urged to adopt some form of conservation tillage to save the planet's soil, to reduce the amount of fossil fuels burnt in growing food, to reduce runoff pollution of our waterways, to reduce wind erosion and air quality degradation and a host of other noble and genuine causes. Charles Little in *Green Fields Forever* (1987) epitomized the genuine enthusiasm most conservationists have for the technique. But early farmer experience, especially with no-tillage, suggested that adopting such techniques would result in greater short-term risk of reduced seedling emergence, crop yield or, worse, crop failure, which they were being asked to accept for the long-term gains outlined above.

Farmers of today were unlikely to see many short-term benefits of their conservation practices. Leaving a legacy of better land for future generations was one thing, but the short-term reality of feeding the present generation and making a living was quite another. Not unreasonably, short-term expediency often took priority. Although some countries already produce 50% or more of their food by no-tillage (e.g. Brazil, Argentina and Paraguay), it is estimated that, worldwide, no-tillage currently accounts for only some 5–10% of food production. We still have a long way to go. Certainly there have been good, and even excellent, no-tillage crops, but there have also been failures. And it is the failures that take prime position in the minds of all but the most forward-looking or innovative farmers.

Tillage has been fundamental to crop production for centuries to clear and soften seedbeds and control weeds. So now we are changing history, not always totally omitting tillage (although that is certainly a laudable objective) but significantly altering the reasons and processes involved. Most people understand tillage to be a process of physically manipulating the soil to achieve weed control, fineness of tilth, smoothness, aeration, artificial porosity, friability and optimum moisture content so as to facilitate the subsequent sowing and covering of the seed. In the process, the undisturbed soil is cut, accelerated, impacted, inverted, squeezed, burst and thrown, in an effort to break the soil physically and bury weeds, expose their roots to drying or to physically destroy them by cutting. The objective of tillage is to create a weed-free, smooth, friable soil material through which relatively unsophisticated seed drill openers can travel freely.

During no-tillage, few, if any, of the processes listed above take place. Under no-tillage, other weed-control measures, e.g. chemicals, must substitute for the physical disturbance during tillage to dislodge, bury or expose existing weeds. But part of the tillage objective is also to stimulate new weed seed germination so that fresh weeds get an 'even start' and can therefore be easily killed in their juvenile stages by a single subsequent tillage operation. No-tillage, therefore, must either find another way of stimulating an 'even start' for new weeds, which would then require a subsequent application of herbicide or avoid stimulating new weed growth in the first place.

In his keynote address to the 1994 World Congress of Soil Science, Nobel Prize-winner Norman Borlaug estimated that world cereal production (which accounts for 69% of world food supply) would need to be raised by 24% by the year 2000 and doubled by the year 2025. More importantly, Borlaug estimated that grain yields would need to increase by 80% over the same time span because creating new arable land is severely limited throughout the world. Until now, yield increases have come largely from increased fertilizer and pesticide use and genetic improvement to the species grown. The challenge is for no-tillage to contribute to future increases, while simultaneously achieving resource preservation and environmental goals. But this is only going to happen if no-tillage is practised at advanced technology levels.

The notion of sowing seeds into untilled soils is very old. The ancient Egyptians practised it by creating a hole in untilled soil with a stick, dropping seeds into the hole and then closing it again by pressing the sides together with their feet. But it was not until the 1960s, when the herbicides paraquat and diquat were released by the then Imperial Chemical Industries Ltd (now Syngenta) in England, that the modern concept of no-tillage was born because now weeds could be effectively controlled without tillage.

For the preceding decade it had been recognized that, for no-tillage to be viable, weeds had to be controlled by some other method than tillage. But the range of agricultural chemicals then available was limited because of their residual effects in the soil. A delay of several weeks was necessary after spraying before the new crop could be safely sown, which partly negated saving of time, one of the more noteworthy advantages of no-tillage compared with tillage. Paraquat and diquat are almost instantly deactivated upon contact with soil. When sprayed onto susceptible living weeds, the soil beneath is almost instantly ready to accept new seeds, without the risk of injury.

This breakthrough in chemical weed control spawned the birth of true no-tillage. Since then, there have been other broader-spectrum translocated non-residual chemicals, such as glyphosate, which was first introduced as Roundup by Monsanto. Other generic compounds, such as glyphosate trimesium (Touchdown) and glufosinate ammonium (Buster), were later marketed by other companies, which have expanded the concept even further.

In other circumstances non-chemical weed control measures have been used. These include flame weeding, steam weeding, knife rolling and mechanical hand weeding. None of the alternative measures has yet proved as effective as spraying with a translocated non-residual herbicide. These chemicals are translocated to the roots of the plant thereby affecting a total kill of the plant. Killing the aerial parts alone often allows regeneration of non-affected plant parts.

The application of any chemicals within agricultural food production correctly raises the question of human and biological safety. Indeed, many chemicals must be very carefully applied under very specific conditions for specific results, just like any of the modern pharmaceuticals that assist in cures and controls. Through careful science, and perhaps some good fortune, glyphosate has been found to be non-toxic to any biological species other than green plants and has been safely used for many years with virtually no known effects other than the control of undesired plants.

An even more recent development using genetic modification of the crops themselves has made selected plant varieties immune to very specific herbicides such as glyphosate. This unique trait permits planting the crop without weed concerns until the crop is well established and then spraying both the crop and the weeds with a single pass. The susceptible weeds are eliminated and the immune crop thrives, making a full canopy that competes with any subsequent weed growth, usually through to harvest. Only selected crops such as maize and soybean are currently commonly used in this fashion, but they have already attained a very significant percentage of the world's acreage. With this success, other important food and fibre crops are being modified for this capability.

What is No-tillage?

As soon as the modern concept of no-tillage based on non-residual (and mostly translocated) herbicides was recognized, everyone, it seems, invented a new name to describe the process. 'No-tillage', 'direct drilling' or 'direct seeding' are all terms describing the sowing of seeds into soil that has not been previously tilled in any way to form a 'seedbed'. 'Direct drilling' was the first term used, mainly in England, where the modern concept of the technique originated in the 1960s. The term 'no-tillage' began in North America soon after, but there has been recent support for the term 'direct seeding' because of the apparent ambiguity that a negative word like 'no' causes when it is used to describe a positive process. The terms are used synonymously in most parts of the world, as we do in this book.

Some of these names are listed below with their rationales, some only for historical interest. After all, it's the process, not the name, that's important.

Chemical fallow, or *chem-fallow*, describes a field currently not cropped in which the weeds have been suppressed by chemical means.

Chemical ploughing attempted to indicate that the weed control function usually attributed to ploughing was being done by chemicals. The anti-chemical lobby soon de-popularized such a restrictive name, which is little used today.

Conservation tillage and *conservation agriculture* are the collective umbrella terms commonly given to no-tillage, minimum tillage and/or ridge tillage, to denote that the inclusive practices have a conservation goal of some nature. Usually, the retention of at least 30% ground cover by residues after seeding characterizes the lower limit of classification for conservation tillage or conservation agriculture, but other conservation objectives include conservation of money, labour, time, fuel, earthworms, soil water, soil structure and nutrients. Thus, residue levels alone do not adequately describe all conservation tillage or conservation agricultural practices and benefits.

Disc-drilling reflects the early perception that no-tillage or direct drilling could only be achieved with disc drills (a perception that proved to be erroneous); thus some started referring to the practice as disc-drilling. Fortunately the term has not persisted. Besides, disc drills are also used in tilled soils.

Drillage was a play on words that suggested that under no-tillage the seed drill was in fact tilling the soil and drilling the seed at the same time. It is not commonly used.

Minimum tillage, *min-till* and *reduced tillage* all describe the practice of restricting the amount of general tillage of the soil to the minimum possible to establish a new crop and/or effect weed control or fertilization. The practice lies somewhere between no-tillage and conventional tillage. Modern practice emphasizes the amount of surface residue retention as an important aim of minimum or reduced tillage.

No-till is a shortening of no-tillage and is not encouraged by purists, for grammatical reasons.

Residue farming describes conservation tillage practices in which residue retention is the primary objective, even though many of the 'conservation tillage' benefits previously mentioned may also accrue.

Ridge tillage, or *ridge-till*, describes the practice of forming ridges from tilled soil into which widely spaced row crops are drilled. Such ridges may remain in place for several seasons while successive crops are no-tilled into the ridges, or they might be re-formed annually.

Sod-seeding, *undersowing*, *oversowing*, *overdrilling* and *underdrilling* all refer to the specific no-tillage practice of drilling new pasture seeds into existing pasture swards, collectively referred to as pasture renovation. The correct use of the term oversowing does not involve drilling at all, but rather is the broadcasting of seed on to the surface of the ground. Each of the other listed terms involves drilling of the seed.

Stale seedbed describes an untilled seedbed that has undergone a period of fallow, usually (but not exclusively) with periodic chemical weed control.

Strip tillage, or *zone tillage*, refers to the practice of tilling a narrow strip ahead of (or with) the drill openers, so the seed is sown into a strip of tilled soil but the soil between the sown rows remains undisturbed. 'Strip tillage' also refers to the general tilling of much wider strips of land (100 or more metres wide) on the contour, separated by wide fallowed strips, as an erosion-control measure based on tillage.

Sustainable farming is the end product of applying no-tillage practices continuously. Continuous cropping based on tillage is now considered to be unsustainable because of resource degradation and farming inefficiencies, while continuous cropping based on no-tillage is much more likely to be sustainable on a long-term basis under most agricultural conditions. Some discussions of 'sustainability' include broader considerations beyond the preservation of natural resources and food production, such as economics, energy and quality of life.

Zero-tillage was synonymous with no-tillage and is still used to a limited extent today.

The most commonly identified feature of no-tillage is that as much as possible of the surface residue from the previous crop is left intact on the surface of the ground, whether this be the flattened or standing stubble of an arable crop that has been harvested or a sprayed dense sward of grass. In the USA, where the broad category of conservation tillage is generally practised as an erosion-control measure, the accepted minimum amount of surface covered by residue after passage of the drill is 30%. Most practitioners of the more demanding option of no-tillage or direct seeding aim for residue-coverage levels of at least 70%.

Of course, some crops, such as cotton, soybean and lupin, leave so little residue after harvest that less than 70% of the ground is likely to be covered by residue even before drilling. Such a soil, however, can be equally well direct drilled as a fully residue-covered soil in the course of establishing the next crop. Thus it is also regarded as true no-tillage. What is no-tillage to one observer may not be no-tillage to another, depending upon the terms of reference and expectations of each observer.

The most fundamental criterion common to all no-tillage is not the amount of residue remaining on the soil after drilling, but whether or not that soil has been disturbed in any way prior to drilling. Even then, during drilling, as will be explained later, such a seemingly unambiguous definition becomes confused when you consider the actions of different drills and openers in the soil. Some literally till a strip as they go, while others leave all of the soil almost undisturbed. So the untilled soil prior to drilling might well become something quite different after drilling.

This book is focused on the subject of 'no-tillage' in which no prior disturbance or manipulation of the soil has occurred other than possibly minimal disturbance by operations such as shallow weed control, fertilization or loosening of subsurface compacted layers. Such objectives are entirely compatible with true no-tillage. Any disturbance before seeding is expected to have had very minimal surface disturbance of soil or residues.

Depending on the field cropping history and the available seeding machine capability, it may be necessary to perform one or more very minimal-disturbance functions for best crop performance. The most common of these needs is the application of fertilizer when that function can not be made part of the seeding operation. Early no-tillage seeding trials often simply broadcast the fertilizer over the soil surface expecting it to be carried into the soil profile by precipitation, but two things became readily apparent. First, only the nitrogen component was moved by water, leaving the remaining forms, such as phosphorus and potassium, on or near the soil surface. And even then preferential flow of soluble nitrogen down earthworm and old root channels often meant that much of it bypassed the juvenile roots of the newly sown crop (see Chapter 9).

Secondly, emerging weeds between the crop plants readily helped themselves as the first consumers of this fertilizer and 'outgrew' the crop. Subsurface placement is now the only recommended procedure, often banded near the seeding furrow or emerging crop row.

Where herbicides are less available, it may prove more economical to perform a weeding pass prior to seeding to reduce the weed pressures on the emerging crop. If used in conservation agriculture, this operation must be very shallow and leave the soil surface and residues nearly intact ready for the seeding operation. Typical implements that can achieve this quality of weed control are shallow-running V-shaped chisels or careful hand hoeing.

Historical compaction arising from many years of repetitive tillage often cannot be undone 'overnight' by switching to no-tillage. While soil microbes are rebuilding their numbers and improving soil structure, a process that may take several years even in the most favourable of climates, historical compaction may still exist. Temporary relief can often be achieved by using a subsoiling machine that cracks and bursts subsurface zones while causing only minor disturbance at the surface.

But sometimes overly aggressive subsoilers cause so much surface disturbance that full tillage is then required to smooth the surface again. This seemingly endless negative spiral must be broken if the benefits of no-tillage are to be gained. All that is required is a less aggressive or shallow-acting subsoiler that allows no-tillage to take place after its passage without any further 'working' of the soil surface layer.

Another effective method is to sow a grass or pasture species in the compacted field and either graze this with light stocking or leave it ungrazed as a 'set-aside' area for a number of years before embarking on a no-tillage programme thereafter without tillage. A rule of thumb for how many years of pasture are required to restore soil organic carbon (SOC) and ultimately the structural damage done by tillage was established by Shepherd *et al.* (2006) for a gley soil (Kairanga silty clay loam) under maize in New Zealand soils as:

> Where tillage has been undertaken for up to 4 consecutive years, it takes approximately 1½ years of pasture to restore SOC levels for each year of tillage.
>
> Where tillage has been undertaken for more than 4 consecutive years, it takes up to 3 years of pasture to restore SOC levels for each year of tillage.

The rate of recovery of soil structure lags behind the recovery rate of SOC. The more degraded the soil, the greater the lag time.

Why No-tillage?

It is not the purpose of this book to explore in detail the advantages and disadvantages of either no-tillage or conservation tillage. Numerous authors have undertaken this task since Edward Faulkner and Alsiter Bevin questioned the wisdom of ploughing in *Ploughman's Folly* (Faulkner, 1943) and *The Awakening* (Bevin, 1944). Although neither of these authors actually advocated no-tillage, it is interesting to note that Faulkner made the now prophetic observation that 'no one has ever advanced a scientific reason for ploughing'. In fact, long before Faulkner's and Bevin's time, the ancient Peruvians, Scots, North American Indians and Pacific Polynesians are all reported to have practised a form of conservation tillage (Graves, 1994).

None the less, to realistically focus on the methods and mechanization of no-tillage technologies, it is useful to compare the advantages and disadvantages of the technique in general as measured against commonly practised tillage farming. The more common of these are summarized below with no particular order or priority. Those followed by an * can be either an advantage or a disadvantage in differing circumstances.

In Chapter 2 we shall expand on the advantages (benefits) of no-tillage, particularly those derived either directly or indirectly from enhancement of SOC levels, and in Chapter 3 we shall examine the risks of no-tillage in more detail.

Advantages

Fuel conservation. Up to 80% of fuel used to establish a crop is conserved by converting from tillage to no-tillage.

Time conservation. The one to three trips over a field with no-tillage (spraying, drilling and perhaps subsoiling) results in a huge saving in time to establish a crop compared with the five to ten trips for tillage plus fallow periods during the tillage process.

Labour conservation. Up to 60% fewer person-hours are used per hectare compared with tillage.

Time flexibility. No-tillage allows late decisions to be made about growing crops in a given field and/or season.

Increased soil organic matter. By leaving the previous crop residues on the soil surface to decay, soil organic matter near the surface is increased, which in turn provides food for the soil microbes that are the builders of soil structure. Tillage oxidizes organic matter, resulting in a cumulative reduction, often more than is gained from incorporation.

Increased soil nitrogen. All tillage mineralizes soil nitrogen, which may provide a short-term boost to plant growth, but such nitrogen is 'mined' from the soil organic matter, further reducing total soil organic matter levels.

Preservation of soil structure. All tillage destroys natural soil structure while no-tillage minimizes structural breakdown and increases organic matter and humus to begin the rebuilding process.

Preservation of earthworms and other soil fauna. As with soil structure, tillage destroys humans' most valuable soil-borne ally, earthworms, while no-tillage encourages their multiplication.

Improved aeration. Contrary to early predictions, the improvement in earthworm numbers, organic matter and soil structure usually result in improved soil aeration and porosity over time. Soils do not become progressively harder and more compact. Quite the reverse occurs, usually after 2–4 years of no-tillage.

Improved infiltration. The same factors that aerate the soil result in improved infiltration into the soil. Plus residues reduce surface sealing by raindrop impact and slow down the velocity of runoff water.

Preventing soil erosion. The sum of preserving soil structure, earthworms and organic matter, together with leaving the surface residues to protect the soil surface and increase infiltration, is to reduce wind and water soil erosion more than any other crop-production technique yet devised by humans.

Soil moisture conservation. Every physical disturbance of the soil exposes it to drying, whereas no-tillage and surface residues greatly reduce drying. In addition, accumulation of soil organic matter greatly improves the water-holding capacity of soils.

Reduced irrigation requirements. Improved water-holding capacity and reduced evaporation from soils lessen the need for irrigation, especially at early stages of growth when irrigation efficiency is at its lowest.

*Moderating soil temperatures.** Under no-tillage soil temperatures in summer stay lower than under tillage. Winter temperatures are higher where snow retention by residue is a factor, but spring temperatures may rise more slowly.

Reduced germination of weeds. The absence of physical soil disturbance under no-tillage reduces stimulation of new weed seed germination, but the in-row effect of this factor is highly dependent on the amount of disturbance caused by the no-tillage openers themselves.

Improved internal drainage. Improved structure, organic matter, aeration and earthworm activity increase natural drainage within most soils.

Reduced pollution of waterways. The decreased runoff of water from soil and the chemicals it transports reduces pollution of streams and rivers.

Improved trafficability. Untilled soils are capable of withstanding vehicle and animal traffic with less compaction and structural damage than tilled soils.

Lower costs. The total capital and/or operating costs of all machinery required to establish tillage crops are reduced by up to 50% when no-tillage substitutes for tillage.

*Longer replacement intervals for machinery.** Because of reduced hours per hectare per year, tractors and advanced no-tillage drills are replaced less often and reduce capital costs over time. Some lighter no-tillage drills, however, may wear out more quickly than their tillage counterparts because of the greater stresses involved in operating them in untilled soils.

*Reduced skills level.** While achieving successful no-tillage is a skilful task in itself, the total range of skills required is smaller than the many sequential tasks needed to complete successful tillage.

Natural mixing of soil potassium and phosphorus. Earthworms mix large quantities of soil potassium and phosphorus in the root zone, which favours no-tillage because it sustains earthworm numbers and increases plant nutrient availability.

Less damage of new pastures. The more stable soil structure of untilled soils allows quicker utilization of new pastures by stock with less plant disruption during early grazing than where tillage has been employed.

More recreation and management time. The time otherwise devoted to tillage can be used to advantage for further management inputs (including the farming of more land) or for family and recreation.

Increased crop yields. All of the above factors are capable of improving crop yields to levels well above those attained by tillage – but only if the no-tillage system and processes are fully practised without short cuts or deficiencies.

Future improvements expected. Modern advanced no-tillage systems and equipment have removed earlier expectations of depressed crop yields in the short term to gain the longer-term benefits of no-tillage. Ongoing research and experience have developed systems that eliminate short-term depressed yields while at the same time raising the expectation and magnitudes of yield increases in the medium to longer term.

Disadvantages

*Risk of crop failure.** Where inappropriate no-tillage tools and weed- or pest-control measures are used, there will be a greater risk of crop yield reductions or failure than for tillage. But where more sophisticated no-tillage tools and correct weed- and pest-control measures are used, the risks will be less than for tillage.

*Larger tractors required.** Although the total energy input is significantly reduced by changing to no-tillage, most of that input is applied in one single operation, drilling, which may require a larger tractor or more animal power, or conversely a narrower drill.

New machinery required. Because no-tillage is a relatively new technique, new and different equipment has to be purchased, leased or hired.

*New pest and disease problems.** The absence of physical disturbance and retention of surface residues encourages some pests and diseases and changes the habitats of others. But such conditions also encourage their predators. To date, no pest or disease problems have proved to be insurmountable or untreatable in long-term no-tillage systems.

Fields are not smoothed. The absence of physical disturbance prevents soil movement by machines for smoothing and levelling purposes. This puts pressure on no-tillage drill designers to create machines that can cope with uneven soil surfaces. Some do this better than others.

Soil strength may vary across fields. Tillage serves to create a consistently low soil strength across each field. Long-term no-tillage requires machines to be capable of adjusting to natural variations in soil strength that occur across every field. Since soil strength dictates the penetration forces required to be applied to each no-tillage opener, variable soil strength places particular demands on drill designs if consistent seeding depths and seed coverage are to be attained.

*Fertilizers are more difficult to incorporate.** General incorporation of fertilizers is more difficult in the absence of physical burial by machines, but specific incorporation at the time of drilling is possible and desirable, using special designs of no-tillage openers.

Pesticides are more difficult to incorporate. As with fertilizers, general incorporation of pesticides (especially those that require pre-plant soil incorporation) is not readily possible with no-tillage, requiring different pest-control strategies and formulations.

*Altered root systems.** The root systems of no-tillage crops may occupy smaller volumes of soil than under tillage, but the total biomass and function of the roots are seldom different and anchorage may in fact be improved.

*Altered availability of nitrogen.** There are three factors that affect nitrogen availability during early plant development under no-tillage:

The decomposition of organic matter by soil microbes often temporarily 'locks up' nitrogen, making it less plant-available under no-tillage.

No-tillage reduces mineralization of soil organic nitrogen that tillage otherwise releases.

The development of bio-channels in the soil from earthworms and roots causes preferential flow of surface-applied nitrogenous fertilizers into the soil, which may bypass shallow, young crop roots.

Each (or all) of these factors may create a nitrogen deficiency for seedlings, which encourages placing nitrogen with drilling. Fortunately some advanced no-tillage drills have separate nitrogen banding capabilities that overcome this problem.

*Use of agricultural chemicals.** The reliance of no-tillage on herbicides for weed control is a cost and environmental negative but is offset by the reduction in surface runoff of other chemical pollutants (including surface-applied fertilizers) and the fact that most of the primary chemicals used in no-tillage are 'environmentally friendly'. Small-scale agriculture may require more hand weeding, but with greater ease than with tilled soils.

*Shift in dominant weed species.** Chemical weed control tends to be selective towards weeds that are resistant to the range of available formulations, requiring more diligent use of crop rotations by farmers and commitment by the agricultural chemical industry to researching new formulations.

*Restricted distribution of soil phosphorus.** Relatively immobile soil phosphorus tends to become distributed in a narrower band within the upper soil layers under no-tillage because of the absence of physical mixing. Improved earthworm populations help reduce this effect and also cycle nutrient sources situated below normal tillage levels.

*New skills are required.** No-tillage is a more exacting farming method, requiring the learning and implementation of new skills, and these are not always compatible with existing tillage-related skills or attitudes.

Increased management and machine performance. There is only one opportunity with each crop to 'get it right' under a no-tillage regime. Because no-tillage drilling is literally a once-over operation, there is less room for error compared with the sequential operations involved in tillage. This places emphasis on the tolerance of no-tillage drills to varying operator skill levels and their ability to function effectively in suboptimal conditions.

*No-tillage drill selection is critical.** Few farmers can afford to own several different no-tillage drills awaiting the most suitable conditions before selecting which one to use. Fortunately more advanced no-tillage drills are capable of functioning consistently in a wider range of conditions than most tillage tools, making reliance on a single no-tillage drill for widely varying conditions both feasible and a practical reality.

Availability of expertise. Until the many specific requirements of successful no-tillage are fully understood by 'experts', the quality of advice to practitioners from consultants will remain, at best, variable. Local, successful no-tillage farmers often become the best advisers.

*Untidy field appearance.** Farmers who have become used to the appearance of neat, 'clean', tilled seedbeds often find the retention of surface residues ('trash') 'untidy'. But, as they come to appreciate the economic advantages of true no-tillage, many such farmers gradually come to see residues as an important resource rather than 'trash' requiring disposal.

*Elimination of 'recreational tillage'.** Some farmers find driving big tractors and tilling on a large scale to be recreational. Others regard it as a chore and health-damaging. Farmers in developing countries regard tillage as burdensome or impossible.

Figure 1.1 shows some of the likely short- and long-term trends that might arise as a result of converting from tillage to no-tillage.

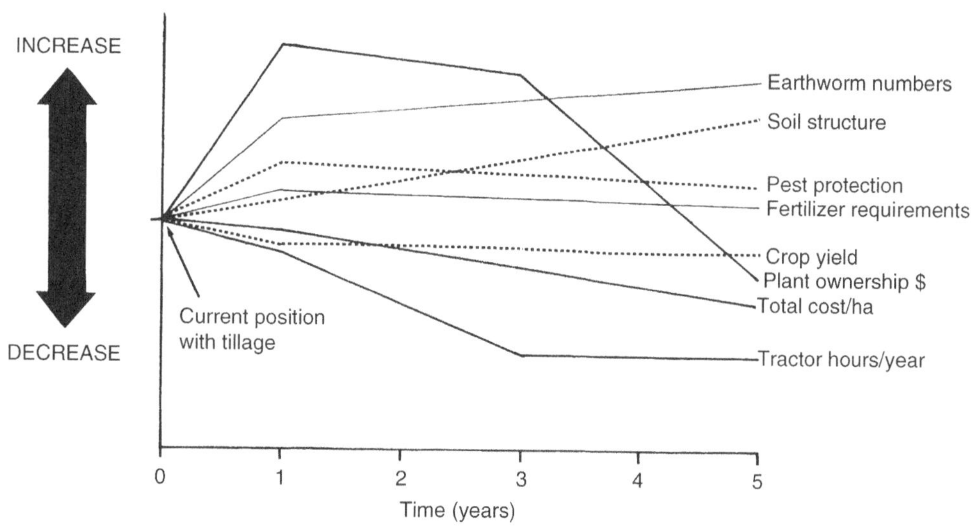

Fig. 1.1. The likely short- and long-term trends that might arise as a result of converting from tillage to no-tillage (from Carter, 1994).

Each identified item or process progresses over the years from stopping tillage as the effects of no-tillage take precedent. The realization is that the effects of no-tillage are developed as the soil and its physical and biological characteristics change. The result of these combined processes has been observed and documented in nearly every soil and climate worldwide, to the point of becoming common knowledge. It is in this transition stage that many who convert to no-tillage farming become disillusioned and sceptical that the benefits will in fact occur.

Summary of the 'What' and 'Why' of No-tillage

No-tillage farming is a significant methodology shift in production farming as performed over the past 100 years of mechanized agriculture. It intuitively requires new thinking by the producers of the 'what' and 'why' to change the processes. Only by encompassing the full scope of 'why' we should change from an enormously successful food production system shall we move forward with confidence to develop 'what' a modern no-tillage farming system should incorporate. The short-term advantages far outweigh the disadvantages, and in the longer term it involves no less than making world food production sustainable for the first time in history.

2 The Benefits of No-tillage

Don C. Reicosky and Keith E. Saxton

Intensive tillage farming reduces soil organic matter and degrades soil quality – no-tillage farming enhances soil quality and sustains long-term agriculture.

Introduction

Sustainable food and fibre production of any given field and region requires that the farming methods be economically competitive and environmentally friendly. To achieve this result requires adopting a farming technology that not only benefits production but provides an environmental benefit to the long-term maintenance of the soil and water resources upon which it is based. We must reduce pollution and use our resources in line with the earth's carrying capacity for sustainable production of food and fibre.

The responsibility of sustainable agriculture lies on the shoulders of farmers to maintain a delicate balance between the economic implications of farming practices and the environmental consequences of using the wrong practices. This responsibility entails producing food and fibre to meet the increasing population while maintaining the environment for a sustained high quality of life. The social value of an agricultural community is not just in production, but in producing in harmony with nature for improved soil, water and air quality and biological diversity.

Sustainable agriculture is a broad concept that requires interpretation at the regional and local level. The principles are captured in the definition reported by El-Swaify (1999) as: 'Sustainable agriculture involves the successful management of resources for agriculture to satisfy changing human needs, while maintaining or enhancing the quality of the environment and conserving natural resources.'

Conservation agriculture, especially no-tillage (direct seeding), has been proved to provide sustainable farming in many agricultural environments virtually around the world. The conditions and farming scales vary from humid to arid and vegetable plots to large prairie enterprises. All employ and adapt very similar principles but with a wide variety of machines, methods and economics.

The benefits of performing crop production with a no-tillage farming system are manyfold. Broad subjects discussed here only begin to provide the science and results learned over recent decades of exploring and developing this farming method. In addition to improved production and soil and water resource protection, many other benefits accrue. For example, it saves time and money, improves timing of planting and harvesting, increases the potential for double cropping, conserves soil water through decreased evaporation and increased infiltration, reduces fuel, labour and machinery requirements and enhances the global environment.

Principles of Conservation Agriculture

Conservation agriculture requires implementing three principles, or pillars, as illustrated in Fig. 2.1. These are: (i) minimum soil tillage disturbance; (ii) diverse crop rotations and cover crops; and (iii) continuous plant residue cover. The main direct benefit of conservation agriculture and direct seeding is increased soil organic matter and its impact on the many processes that determine soil quality. The foundation underlying the three principles is their contribution and interactions with soil carbon, the primary determinant of long-term sustainable soil quality and crop production.

Conservation tillage includes the concepts of no-tillage, zero-tillage and direct seeding as the ultimate form of conservation agriculture. These terms are often used interchangeably to denote minimum soil disturbance. Reduced tillage methods, sometimes referred to as conservation tillage, such as strip tillage, ridge tillage and mulch tillage, disturb a small volume of soil and partially mix the residue with the soil and are intermediate in their soil quality effects. These terms define the tillage equipment and operation characteristics as they relate to the soil volume disturbed and the degree of soil–residue mixing. Intensive inversion tillage, such as that from mouldboard-ploughing, disc-harrowing and certain types of powered rotary tillage, is not a form of conservation tillage. No-tillage and direct seeding are the primary methods of conservation tillage to apply the three pillars of conservation agriculture for enhanced soil carbon and its associated environmental benefits.

True soil conservation is largely related to organic matter, i.e. carbon, management.

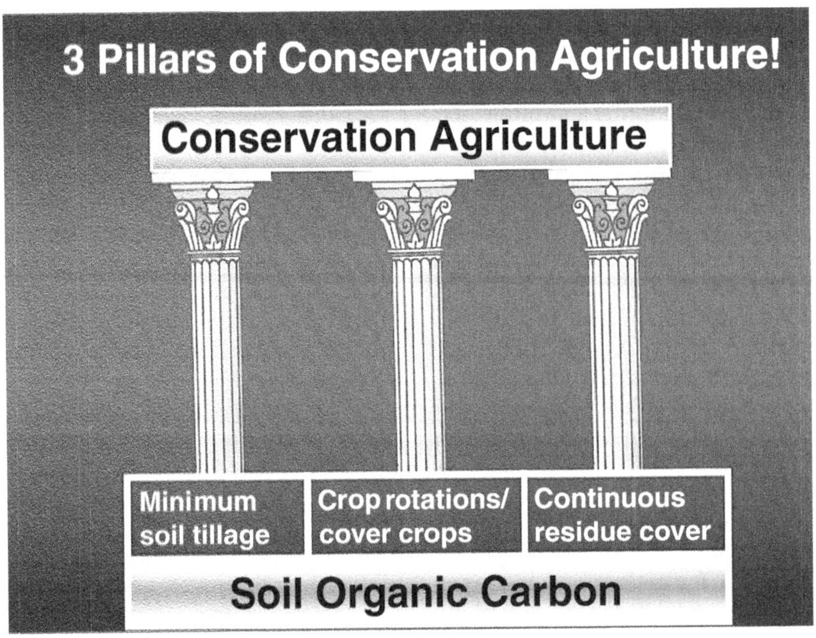

Fig. 2.1. Schematic representation of the three pillars or principles of conservation agriculture supported by a foundation of soil carbon.

By nothing more than properly managing the carbon in our agricultural ecosystems, we can have less erosion, less pollution, clean water, fresh air, healthy soil, natural fertility, higher productivity, carbon credits, beautiful landscapes and sustainability. Dynamic soil quality encompasses those properties that can change over relatively short time periods, such as soil organic matter, soil structure and macroporosity. These can readily be influenced by the actions of human use and management within the chosen agronomic practices. Soil organic matter is particularly dynamic, with inputs of plant materials and losses by decomposition.

Crop Production Benefits

Producing a crop and making an economic profit are universal goals of global farming. Production by applying no-tillage methods is no different in these goals, but there are definite benefits for the achievement, which we outline in this chapter. But these benefits only occur with fully successful no-tillage farming. There are certainly obstacles and risks in moving from traditional tillage farming, which has been the foundation technology for centuries, as outlined in Chapter 3.

Acceptable crop production requires an adequate plant stand, good nutrition and moisture with proper protection from weed, insect or disease competition. Achieving the plant stand in untilled, residue-covered soils is the first major obstacle, a particular challenge in modern mechanized agriculture, but certainly surmountable, as explained in the core of this text. Providing adequate nutrition and water for full crop potentials is readily achieved with the benefits of no-tillage, as discussed below.

Weed-control methods, by necessity, shift to dependence on chemicals, flame-weeding, mechanical crushing or hand picking for full no-tillage farming to stay within the goal of minimum soil disturbance. Chemical developments in recent decades have made great strides in their effectiveness, environmental friendliness and economic feasibility. Supplemental techniques of mowing, rolling and crushing without soil disturbance are showing significant promise to reduce weed presence and increase the benefit of cover crops and residues. Experience has shown that controlling insects and diseases has generally been less of a problem with no-tillage, even though there are often dire predictions about the potential impact of surface residues harbouring undesirables. As with weeds, crop health and pest problems are not likely to be avoided but may well shift to new varieties and species with the change in the field environment.

As a result of these developments and skilled applications, it has been repeatedly shown that crop production can be equalled and exceeded by no-tillage farming compared with traditional tillage methods. Because many soils have been tilled for many years, it is not uncommon to experience some yield reduction in the first few no-tillage years, largely because, as discussed later, it takes time for the soil to rebuild into a higher quality. This ‘transition period reduction’ can often be overcome or even averted with increased fertility, strategic fertilizer banding with drill openers and careful crop selection.

The full benefit of no-tillage comes in the reduced inputs. Most notable are the reduced inputs by minimizing labour and machine hours spent establishing and maintaining the crop. Reduced machine costs alone are significant, since all tillage equipment is dispensable. True no-tillage farming requires only an effective chemical sprayer, seeding–fertilizing drill and harvester.

With no seedbed preparation of the soil by tillage, seed drilling has become the major limitation to many efforts to successfully change to no-tillage farming. Modifying drills used in tillage farming has generally not been very successful, resulting in undesirable crop stands for optimum production. Many were not equipped to provide simultaneous fertilizer banding; thus it had to be provided by a supplemental minimum-tillage machine or, in the worst case, surface-applied, where it was very ineffective and stimulated weed growth.

Fortunately, drill development has progressed to now provide acceptable seeding in many cases, but, as described in later chapters, many still do not fully meet all desirable attributes, especially in relation to the amount of soil disturbance they create.

As a result of science and technique developments of recent years, no-tillage crop production now not only is feasible but has significant economic benefits. Combining and multiplying this result by the further benefits of soil and environmental qualities make no-tillage farming a highly desirable method of crop production. Further, many are now finding personal and social benefits from the reduced labour inputs, which remove much of the demanded time and drudgery often associated with traditional farm life. A common remark by successful no-tillage farmers is 'It has brought back the fun of farming.'

Increased organic matter

Understanding the role of soil organic matter and biodiversity in agricultural ecosystems has highlighted the value and importance of a range of processes that maintain and fulfil human needs. Soil organic matter is so valuable for its influence on soil organisms and properties that it can be referred to as 'black gold' because of its vital role in physical, chemical and biological properties and processes within the soil system.

The changes of these basic soil properties, called 'ecosystem services', are the processes by which the environment produces resources that sustain life and which we often take for granted. An ecosystem is a community of animals and plants interacting with one another within their physical environment. Ecosystems include physical, chemical and biological components such as soils, water and nutrients that support the biological organisms living within them, including people. Agricultural ecosystem services include production of food, fibre and biological fuels, provision of clean air and water, natural fertilization, nutrient cycling in soils and many other fundamental life support services. These services may be enhanced by increasing the amount of carbon stored in soils.

Conservation agriculture through its impact on soil carbon is the best way to enhance ecosystem services. Recent analyses have estimated national and global economic benefits from ecosystem services of soil formation, nitrogen fixation, organic matter decomposition, pest biocontrol, pollination and many others. Intensive agricultural management practices cause damage or loss of ecosystem services, by changing such processes as nutrient cycling, productivity and species diversity (Smith *et al.*, 2000). Soil carbon plays a critical role in the harmony of our ecosystems providing these services.

Soil carbon is a principal factor in maintaining a balance between economic and environmental factors. Its importance can be represented by the central hub of a wagon wheel, a symbol of strength, unity and progress (Reicosky, 2001a). The 'spokes' of this wheel in Fig. 2.2 represent incremental links to soil carbon that lead to the environmental improvement that supports total soil resource sustainability. Many spokes make a strong wheel. Each of the secondary benefits that emanate from soil carbon contributes to environmental enhancement through improved soil carbon management. Soane (1990) discussed several practical aspects of soil carbon important in soil management. Some of the 'spokes' of the environmental sustainability wheel are described in the following paragraphs.

Based on soil carbon losses with intensive agriculture, reversing the decreasing soil carbon trend with less tillage intensity benefits a sustainable agriculture and the global population by gaining better control of the global carbon balance. The literature holds considerable evidence that intensive tillage decreases soil carbon and supports increased adoption of new and improved forms of no-tillage to preserve or increase storage of soil organic matter (Paustian *et al.*, 1997a, b; Lal *et al.*, 1998). The environmental and economic benefits of conservation agriculture and no-tillage demand their consideration in the development of improved soil carbon storage practices for sustainable production.

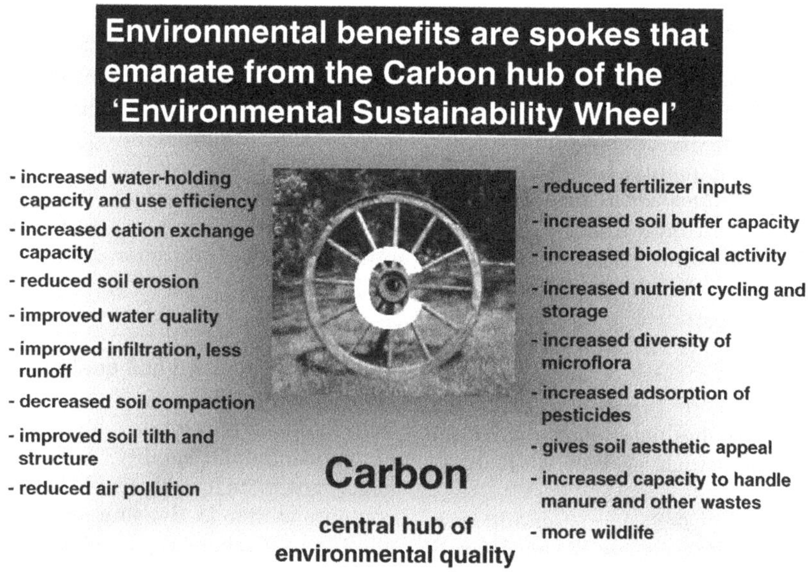

Fig. 2.2. Environmental sustainability wheel with benefits emanating from the soil carbon hub.

Increased available soil water

Increased soil organic matter has a significant effect on soil water management because of increased infiltration and water-holding capacity. Enhanced soil water-holding capacity is a result of increased soil organic matter, which more readily absorbs water and releases it slowly over the season to minimize the impacts of short-term drought. Hudson (1994) showed that, for some soil textures, for each 1% weight increase in soil organic matter, the available water-holding capacity in the soil increased by 3.7% volume. Other factors being equal, soils containing more organic matter can retain more water from each rainfall event and make more of it available to plants. This factor and the increased infiltration with higher organic matter and the decreased evaporation with crop residues on the soil surface all contribute to improved water use efficiency.

Increased organic matter is known to increase soil infiltration and water-holding capacity, which significantly affect soil water management. Under these situations, crop residues slow runoff water and increase infiltration by earthworm channels, macropores and plant root holes (Edwards *et al.*, 1988). Water infiltration is two to ten times faster in soils with earthworms than in soils without earthworms (Lee, 1985).

Soil organic matter contributes to soil particle aggregation, which makes it easier for water to move through the soil and enables plants to use less energy to establish root systems (Chaney and Swift, 1984). Intensive tillage breaks up soil structure and results in a dense soil, making it more difficult for plants to fully access the nutrients and water required for their growth and production. No-tillage and minimum-tillage farming allows the soil to restructure and accumulate organic matter for improved plant water and nutrient availability.

Reduced soil erosion

Crop residue management practices have included many agricultural practices to reduce soil erosion runoff and off-site sedimentation. Soils relatively high in C, particularly with crop residues on the soil surface, very effectively increase soil organic matter and reduce soil erosion loss. The primary role of soil organic matter to reduce soil erodibility is to stabilize the surface aggregates through reduced crust formation and surface sealing, resulting in less runoff (Le Bissonnais, 1990). Reducing or eliminating runoff that carries sediment from fields to rivers and streams is a major enhancement of environmental quality. Under these situations, crop residues act as tiny dams that slow down water runoff from fields, allowing the water more time to soak into the soil.

Crop residues on the surface not only help hold soil particles in place but keep associated nutrients and pesticides on the field. The surface layer of organic matter minimizes herbicide runoff and, with conservation tillage, herbicide leaching can be reduced by as much as half (Braverman *et al.*, 1990).

Increased soil organic matter and crop residues on the surface will significantly reduce wind erosion (Skidmore *et al.*, 1979). Depending on the amount of crop residues left on the soil surface, soil erosion can be reduced to near zero as compared with that from an unprotected, intensively tilled field. Wind or water soil erosion causes soil degradation and variability to the extent of a resulting crop yield decline.

Papendick *et al.* (1983) reported that the original topsoil on most hilltops had been removed by tillage erosion in the Palouse region of the Pacific Northwest of the USA. Mouldboard ploughs were identified as the primary cause, but all tillage implements will contribute to this problem (Groves *et al.*, 1994; Lobb and Kachanoski, 1999). Soil translocation from mouldboard plough-based tillage can be greater than soil loss tolerance levels (Lindstrom *et al.*, 1992; Groves *et al.*, 1994; Lobb *et al.*, 1995, 2000; Poesen *et al.*, 1997). Soil is not directly lost from the fields by tillage translocation; rather, it is moved away from the convex slopes and deposited on concave slope positions.

Lindstrom *et al.* (1992) showed that soil movement on a convex slope in southwestern Minnesota, USA, could result in a sustained soil loss level of approximately 30 t/ha/year from annual mouldboard-ploughing. Lobb *et al.* (1995) estimated soil loss in southwestern Ontario, Canada, from a shoulder position to be 54 t/ha/year from a tillage sequence of mouldboard-ploughing, tandem-discing and C-tine cultivating. In this case, tillage erosion, as estimated through resident caesium-137, accounted for at least 70% of the total soil loss. The net effect of soil translocation from the combined effects of tillage and water erosion is an increase in spatial variability of crop yield and a likely decline in soil carbon, related to lower soil productivity (Schumacher *et al.*, 1999).

Enhanced soil quality

Soil quality is the fundamental foundation of environmental quality. Soil quality is largely governed by soil organic matter (SOM) content, which is dynamic and responds effectively to changes in soil management, tillage and plant production. Maintaining soil quality can reduce the problems of land degradation, decreasing soil fertility and rapidly declining production levels that occur in large parts of the world needing the basic principles of good farming practice.

Soil compaction in conservation tillage farming is significantly reduced by the reduction of traffic and increased SOM (Angers and Simard, 1986; Avnimelech and Cohen, 1988). Soane (1990) presented several mechanisms by which soil 'compactibility' can be affected by SOM:

1. Improved internal and external binding of soil aggregates.
2. Increased soil elasticity and rebounding capabilities.
3. Reduced bulk density due to mixing organic residues with the soil matrix.
4. Temporary or permanent existence of root networks.
5. Localized change of electrical charge of soil particle surfaces.
6. Change in soil internal friction.

While most soil compaction occurs during the first vehicle trip over the tilled field, reduced weight and horsepower requirements associated with no-tillage can also help minimize compaction. Additional field traffic required by intensive tillage compounds the problem by breaking down soil structure. Maintenance of SOM contributes to the formation and stabilization of soil structure. The combined physical and biological benefits of SOM can minimize the effect of traffic compaction and result in improved soil tilth.

While it is commonly known that tillage produces a well-fractured soil, sometimes requiring several tillage passes, it is a misconception that this is a well-aggregated, healthy soil. These soils never fare well when judged against modern knowledge of high 'soil quality'. A tilled soil is poorly structured, is void of many microorganisms and has poor water characteristics, just to name a few characteristics. As soils are farmed without tillage and supplied with residues, they naturally improve in overall quality, again support many microorganisms and become 'mellow' to the point of being easily penetrated by roots and earthworms. This transition takes several years to accomplish but invariably occurs given the opportunity.

Many traditional experienced farmers will often ask, 'How many years of no-tillage are possible before the soil becomes so compact as to require tillage?' No-tillage experience has shown exactly the opposite effect: once a no-tilled soil has regained its quality, it will continue to resist compaction and any subsequent tillage will cause undue damage. Most soils will continue to build organic matter and improve in quality criteria for years into the practice of no-tillage farming if the sequence is not broken by the thunderous effect of tillage.

Improved nutrient cycles

Improved soil tilth, structure and aggregate stability enhance the gas exchange and aeration required for nutrient cycling (Chaney and Swift, 1984). Critical management of soil airflow, with improved soil tilth and structure, is required for optimum plant function. It is the combination of many factors that results in comprehensive environmental benefits from SOM management. The many attributes suggest new concepts on how we should manage the soil for long-term aggregate stability and sustainability.

Ion adsorption or exchange is one of the most significant nutrient cycling functions of soils. Cation exchange capacity (CEC) is the quantity of exchange sites that can absorb and release nutrient cations. SOM can increase this capacity of the soil from 20 to 70% over that of the clay minerals and metal oxides present. In fact, Crovetto (1996) showed that the contribution of organic matter to the cation exchange capacity exceeded that of the kaolinite clay mineral in the surface 5 cm of his soils. Robert (1996) showed that there was a strong linear relationship between organic carbon and the cation exchange capacity of his experimental soil. The capacity was increased fourfold with an organic carbon increase from 1 to 4%. The toxicity of other elements can be inhibited by SOM, which has the ability to adsorb soluble chemicals. Adsorption by clay minerals and SOM is an important means by which plant nutrients are retained in crop rooting zones.

Increased infiltration and concerns over the use of nitrogen in no-tillage agriculture require an understanding of the biological, chemical and physical factors controlling nitrogen losses and the relative impacts of contrasting crop production practices on nitrate leaching from agroecosystems. Domínguez *et al.* (2004) evaluated the leaching of water and nitrogen in plots with varying earthworm populations in a maize system. They found that the total flux of nitrogen in soil leachates was 2.5-fold greater in plots with increased earthworm populations than in those with lower populations. Their results are dependent on rainfall amounts, but do indicate that earthworms can increase the leaching of water and inorganic nitrogen to greater depths in the profile, potentially increasing nitrogen leaching from the system. Leaching losses were lower on the organically fertilized plots, attributed to higher immobilization potential.

Reduced energy requirements

Energy is required for all agricultural operations. Modern, intensive agriculture requires much more energy input than traditional farming methods since it relies on the use of fossil fuels for tillage, transportation, grain drying and the manufacture of fertilizers, pesticides and equipment used to apply agricultural inputs and for generating electricity used on farms (Frye, 1984). Reduced labour and machinery costs are economic considerations that are frequently given as additional reasons to use conservation tillage practices.

Practices that require lower energy inputs, such as no-tillage versus conventional tillage, generally result in lower inputs of fuel and a consequent decreases of CO_2-carbon emissions into the atmosphere per unit of land area under cultivation. Emissions of CO_2 from agriculture are generated from four primary sources: manufacture and use of machinery for cultivation, production and application of fertilizers and pesticides, the soil organic carbon that is oxidized following soil disturbance (which is largely dependent on tillage practices) and energy required for irrigation and grain drying.

A dynamic part of soil carbon cycling in conservation agriculture is directly related to the 'biological carbon' cycle, which is differentiated from the 'fossil carbon' cycle. Fossil carbon sequestration entails the capture and storage of fossil-fuel carbon prior to its release to the atmosphere. Biological carbon sequestration entails the capture of carbon from the atmosphere by plants. Fossil fuels (fossil carbon) are very old geologically, as much as 200 million years. Biofuels (bio-carbon) are very young geologically and can vary from 1 to 10 years in age and as a result can be effectively managed for improved carbon cycling. One example of biological carbon cycling is the agricultural production of biomass for fuel. The major strength of biofuels is the potential to reduce net CO_2 emissions to the atmosphere. Enhanced carbon management in conservation agriculture may make it possible to take CO_2 released from the fossil carbon cycle and transfer it to the biological carbon cycle to enhance food, fibre and biofuel production, for example, using natural gas fertilizer for plant production.

West and Marland (2002) conducted a carbon and energy analysis for agricultural inputs, resulting in estimates of net carbon flux for three crop types across three tillage intensities. The analysis included estimates of energy use and carbon emissions for primary fuels, electricity, fertilizers, lime, pesticides, irrigation, seed production and farm machinery. They estimated that net CO_2-carbon emissions for crop production with conservation, reduced and no-tillage practices were 72, 45 and 23 kg carbon/ha/year, respectively.

Total carbon emission values were used in conjunction with carbon sequestration estimates to model net carbon flux to the atmosphere over time. Based on US average crop inputs, no-tillage emitted less CO_2 from agricultural operations than did conventional tillage, with 137 and 168 kg of carbon/ha/year, respectively. The effect of changes in fossil-fuel use was the dominant factor 40 years after conversion to no-tillage.

This analysis of US data suggests that, on average, a change from conventional tillage to no-tillage will result in carbon sequestration in soil, plus a saving in CO_2 emissions from energy use in agriculture. While the enhanced carbon sequestration will continue for a finite time until a new equilibrium is reached, the reduction in net CO_2 flux to the atmosphere, caused by the reduced fossil-fuel use, can continue indefinitely, as long as the alternative practices are continued.

Lal (2004) recently provided a synthesis of energy use in farm operations and its conversion into carbon equivalents (CE). The principal advantage of expressing energy use in terms of carbon emission as kg CE lies in its direct relation to the rate of enrichment of atmospheric CO_2 concentration. The operations analysed were carbon-intensive agricultural practices that included tillage, spraying chemicals, seeding, harvesting, fertilizer nutrients, lime, pesticide manufacture and irrigation. The emissions for different tillage methods were 35.3, 7.9 and 5.8 kg CE/ha for conventional tillage, chisel tillage or minimum tillage and no-tillage methods of seedbed preparation, respectively.

Tillage and harvest operations account for the greatest proportion of fuel consumption within intensive agricultural systems.

Frye (1984) found fuel requirements using reduced tillage or no-tillage systems were 55 and 78%, respectively, of those used for conventional systems that included mouldboard-ploughing. On an area basis, savings of 23 kg/ha/year in energy carbon resulted from the conversion of conventional tillage to no-tillage. For the 186 million ha of cropland in the USA, this translates to a potential reduction in carbon emissions of 4.3 million metric tonnes carbon equivalent (MMTCE)/year.

These results further support the energy efficiencies and benefits of no-tillage. Conversion of ploughed tillage to no-tillage, using integrated nutrient management and pest management practices, and enhancing water use efficiency can save carbon emissions and at the same time increase the soil carbon pool. Thus, adopting conservation agriculture techniques is a holistic approach to management of soil and water resources. Conservation agriculture improves efficiency and enhances productivity per unit of carbon-based energy consumed and is a sustainable strategy.

Carbon Emissions and Sequestration

Tillage or soil preparation has been an integral part of traditional agricultural production. Tillage fragments the soil, triggers the release of soil nutrients for crop growth, kills weeds and modifies the circulation of water and air within the soil. Intensive tillage accelerates soil carbon loss and greenhouse gas emissions, which have an impact on environmental quality.

By minimizing soil tillage and its associated (CO_2) emissions, global increases of atmospheric carbon dioxide can be reduced while at the same time increasing soil carbon deposits (sequestration) and enhancing soil quality. The best soil management systems involve minimal soil disturbance and focus on residue management appropriate to the geographical location, given the economic and environmental considerations. Experiments and field trials are required for each region to develop proper knowledge and methods for optimum application of conservation agriculture.

Since CO_2 is the final decomposition product of SOM, intensive tillage, particularly the mouldboard plough, releases large amounts of CO_2 as a result of physical disruption and enhanced biological oxidation (Reicosky *et al.*, 1995). With conservation tillage, crop residues are left more naturally on the surface to protect the soil and control the conversion of plant carbon to SOM and humus. Intensive tillage releases soil carbon to the atmosphere as CO_2, where it can combine with other gases to contribute to the greenhouse effect.

Soils store carbon for long periods of time as stable organic matter. Natural systems reach an equilibrium carbon level determined by climate, soil texture and vegetation. When native soils are disturbed by agricultural tillage, fallow or residue burning, large amounts of carbon are oxidized and released as CO_2 (Allmaras *et al.*, 2000). Duxbury *et al.* (1993) estimated that agriculture has contributed 25% of the historical human-made emissions of CO_2 during the past two centuries. However, a significant portion of this carbon can be stored, or sequestered, by soils managed with no-tillage and other low-disturbance techniques. Increased plant production greater than that of native soil levels by the addition of fertilizers or irrigation can enhance carbon sequestration.

Carbon is a valuable environmental natural resource throughout the world's industrial applications of production and fossil energy consumption. Releasing carbon to the atmosphere by energy processes may be offset by capturing carbon with plant biomass and subsequently soil carbon sequestration in the form of organic matter. Energy consumers may at some time be required to compensate for their atmospheric carbon emissions by contracting with those who can sequester atmospheric carbon. Conservation agriculture may be able to provide this sequestration benefit and thus be compensated for its role in maintaining low net carbon emissions. While this 'carbon trading' mechanism is still in the discussion stage, it provides an important potential benefit.

A more detailed explanation of carbon dioxide emissions and sequestration is given in Chapter 17, together with comments on how these interact with nitrous oxide and methane emissions and the potential for carbon trading.

Summary of the Benefits of No-tillage

Conservation tillage, and particularly no-tillage, agriculture has universal appeal because of numerous benefits. Improved production with fewer inputs and reduced time and energy are often cited as the highlights. Conservation agriculture techniques benefit the farmers and the whole of society, and can be viewed as both 'feeding and greening the world' for global sustainability. Agricultural policies are needed to encourage farmers to improve soil quality by storing carbon as SOM, which will also lead to enhanced air quality, water quality and productivity and help to mitigate the greenhouse effect.

Some of the more important benefits of conservation tillage farming are:

1. Improved crop production economics.
2. Increased SOM.
3. Improved soil quality.
4. Reduced labour requirements.
5. Reduced machinery costs.
6. Reduced fossil-fuel inputs.
7. Less runoff and increased available plant water.
8. Reduced soil erosion.
9. Increased available plant nutrients.
10. Improved global environment.

3 The Nature of Risk in No-tillage

C. John Baker, W. (Bill) R. Ritchie and Keith E. Saxton

The ultimate decision to adopt a no-tillage system will have more to do with how farmers perceive it altering their business risks than anything else.

The risks associated with no-tillage are those that result in reduced income to the farmer through impaired crop performance and/or increased costs. To be a sustainable technique, the failure rate for no-tillage must be no more, and preferably less, than that for tillage (Baker, 1995).

While early sceptics of the no-tillage concept forecast many and varied problems that would ultimately lead to the downfall of the practice, experience has shown that there are no insurmountable obstacles in most circumstances. The fact remains, however, that many farmers are still reluctant to attempt the new technique, fearing that it may increase their risks of crop failure or reduced yield.

The perception of risk is probably the single biggest factor governing the rate of adoption of no-tillage, and it is likely to remain so for a long time. Only education and personal experiences will finally put risk into perspective. Recent results convincingly show that no-tillage is not inherently more risky than conventional tillage, even in the short term. Indeed, it can reduce the risk factor during crop establishment if it is undertaken and managed correctly. Of course, tillage is also subject to increased risk under poor management. It is therefore pertinent to explore the concept of risk during crop establishment and growth, and to explain how this is affected by sound no-tillage practices.

What is the Nature of Risk in No-tillage?

To plant and grow a crop with no-tillage, a farmer undertakes an economic risk that is affected by three functional risk categories: (i) biological; (ii) physical; and (iii) chemical. These risks are comparable between tillage and no-tillage systems because almost all of them are the everyday risks of cropping either way. Only their relative levels and remedies differ between the two techniques. The combined effects of the functional risks result in economic risks. The results and associated implications are sometimes surprising and are examined at the end of this chapter.

Biological risks

Biological risks arise from pests, toxins, diseases, seed vigour, seedling vigour, nutrient stress and, ultimately, crop yield.

The change to residue farming in general, which is the cornerstone of no-tillage, can have a marked effect on the incidence of diseases and pests, both positively and negatively. Seed placement and soil and residue disturbance by various drill or opener designs can influence all of these factors.

Pests

The change in earthworm and slug populations creates the most common pest problems in no-tillage. Slugs are particularly prone to proliferate in residue in high-humidity climates and must often be controlled by chemical means. Earthworms, on the other hand, can be either beneficial or damaging, depending on type. Earthworms generally provide positive effects that help aerate, drain and cycle nutrients. All of the effects of earthworms are not yet known but some of their benefits in wet soils are explained in detail in Chapter 7. While tillage destroys earthworms, no-tilled soil nearly always has a significant and important increase in populations, and they are a great 'indicator' organism for other beneficial biota developments. Other damaging worms, such as wireworms, are generally not different regarding crop risks.

Slugs (*Deroceras reticulatum*) (Follas, 1981, 1982) find shelter beneath the soil in many types of seed slots and feed on sown seeds and establishing seedlings. Clearly, slugs increase the biological risks of no-tillage. But they are relatively cheaply countered by the application of a suitable molluscicide.

Other pests can increase their damage risk because of increased surface residues or decreased physical destruction by tillage machines. But then so too do many of their predators.

An example of pest–drill interaction is that experienced with inverted T-shaped slots (see Chapter 4), which create subsurface soil-slot environments that are higher in soil humidity than either tilled soils or other no-tillage slots. Soil fauna that are sensitive to soil humidity, such as slugs and earthworms, tend to congregate in such slots. These may have both positive and negative effects for the sown crop (Carpenter *et al.*, 1978; Chaudhry, 1985; Baker *et al.*, 1987; Basker *et al.*, 1993).

Diseases

The most common soil disease that no-tillage appears to encourage is *Rhizoctonia*. Disturbance of the soil during tillage appears to partly destroy the fungal mycelia. Other fungal diseases are carried over in cereal residue and decaying organic matter in root channels, requiring diligent use of crop rotations or application of appropriate fungicides. On the other hand, the soil disease take-all (*Gaeumannomyces graminis*) appears to become more confined under no-tillage because of reduced soil movement.

A concept called 'green bridge' was identified by Cook and Veseth (1993), in which certain root bacteria from recent chemically killed plants can readily transfer to new seedlings if no-tillage seeding is undertaken within 14–21 days after the green crop begins dying. The specific pathogen has not yet been identified, but some delay after spraying and before no-tillage seeding appears to be an advantage where these bacteria exist, particularly in instances of continuous cereal cropping.

Toxins

The risks arising from toxins relate mainly to contact between seeds and decaying residue within the sown slot under persistently wet conditions (see Chapter 7). This risk, which is peculiar to no-tillage in cold wet soils, is eliminated by the use of no-tillage openers that effectively separate seed from the residues (Chaudhry, 1985) or the use of neutralizing agents sown with the seed (Lynch, 1977, 1978; Lynch *et al.*, 1980).

The most common occurrence of residue effects has been experienced with double-disc drills seeding into wet, soft soils with surface residues. The residues tend to be folded and 'tucked' or 'hairpinned' into the seed slot with the seed dropped in the same location, which results in both the seed and residue experiencing decaying conditions and poor plant stands.

Some explanations for early no-tillage failures assumed that allelopathic exudates from dying plants may have killed newly sown seeds. But later detailed explanations for the causes of seedling emergence failures pointed to other (largely physical) factors and it has been hard to find any confirmed cases of allelopathy having played any role at all.

Nutrient stress

Without soil tillage to stir and mix applied fertilizer applications, careful attention must be paid to placing the fertilizer in untilled soils to optimize crop uptake and yield. Bands of fertilizer to the side and below the seed have proved to be very effective, sometimes utilizing one fertilizer band for each pair of seed rows. While it is important to place fertilizers far enough away from seeds and seedlings to avoid toxicity problems (see 'Chemical Risks'), it also appears that separation distances can (and indeed should) be much closer than those commonly accepted for tilled soils (see Chapter 9). Fertilizer banding has been found to be optimally accomplished by simultaneously seeding and fertilizing with a combination direct seed drill and fertilizer dispenser, and which is now common practice.

Again, the risk under no-tillage increases only if inappropriate equipment is used. On the other hand, there is voluminous evidence to show that, when fertilizers are placed correctly, no-tillage crop yields may be greater than those obtained from tilled soils (see Chapter 9). Thus, while the risk of nutrient stress under no-tillage may increase with inappropriate equipment, it may decrease compared with tillage if improved designs of no-tillage drills and planters are utilized.

Physiological stress

It has been stated that untilled seedbeds are not as 'forgiving' as their tilled counterparts (Baker, 1976a). This is often true because seedlings have to emerge through covering material that is physically more resistant than friable tilled soils. If the seeds are sown into mellow soils that have been no-tilled for several years or with scientifically designed furrow openers, such as inverted-T-shaped slots, the micro-environment of the slots will actually place less physiological stress on the seedlings than will a tilled soil. Thus physiological stress at the time of seedling emergence need not increase the biological risks. It may actually decrease the risk (see Chapter 5). Figure 3.1 shows the difference in growth between seedlings established within contrasting no-tillage slots resulting from physiological stress.

Seed quality

International seed testing authorities throughout the world test mainly for purity and optimally wetted germination as the main indicators of seed quality. But there are also agreed voluntary tests that describe other aspects of seed quality. One such test, the 'accelerated ageing' or 'vigour' test, examines a seed's ability to germinate after experiencing a period of stress (usually high or low temperature). It is possible for a given seed line to record a high-percentage germination but a low-percentage vigour. Therefore final germinations counts give no real indication of the vigour of a seed line although interim counts might be helpful in this respect.

There is an important interaction between seed vigour and drill opener designs, which can have important impacts on biological risk, and operators need to understand this interaction. No-tillage openers that create inverted-T-shaped slots produce about as favourable a micro-environment as it is possible to create for seeds, in either tilled or untilled soils. The main attribute is the availability of both vapour-phase and liquid-phase water. This ensures that even low-vigour seeds will germinate, almost regardless of the soil conditions.

In contrast, seeds sown into tilled soils or less favourable no-tillage slots that only provide liquid-phase water for germination of seeds are less likely to germinate. Farmers usually attribute such failures to a variety of reasons, but seldom test the vigour of

Fig. 3.1. Growth responses of wheat seedlings as a result of physiological stress when sown by a winged opener (left) and double disc (right) no-tillage openers.

the seed they had sown. When germination of low-vigour seeds does occur in tilled soils and open no-tillage slots, emergence of the seedlings is seldom restricted because of the friable nature of tilled soils and the open nature of vertical no-tillage slots. But the ensuing crop is likely to perform poorly.

Extensive field experience with inverted-T-shaped no-tillage slots, where even low-vigour seeds will often germinate under unfavourable conditions, have shown that the seedlings often did not have the vigour to emerge and were instead found twisted, weak and un-emerged beneath the soil surface. Observers at first attributed such twisting to fertilizer burn, but it is now known that fertilizer burn causes shrivelling and premature death of seedlings, not twisting. When vigour tests were carried out over a 3-year period on some 40 lines of seeds that had shown symptoms of subsurface seedling twisting in inverted-T no-tillage slots, all seed lines were found to be of low vigour (some as low as 18%).

The question is: What can be done about the problem? The responsibility rests with both the seed industry and individual no-tillage farmers. The seed industry needs to improve the quality of the seeds it offers for sale or at least be prepared to disclose information on seed vigour to farmers. Some companies already do this. No-tillage farmers, for their part, need to seek information from the seed industry about the vigour of particular seed lines and to be prepared to pay more for high-vigour lines. Those drill manufacturers that market advanced no-tillage seed drills need to advise purchasers that the weakest part of the system may now be seed quality, whereas previously it had been drill quality.

Physical risks

Weather

Weather is likely to be the most variable and uncontrollable element in farming, and performing no-tillage won't change that. However, no-tillage does have the opportunity to significantly modify the impact by several means, some already mentioned or obvious. Increased available plant water is often the first noticeable effect, since residues and minimal soil disturbance reduce evaporation and increase infiltration.

Improved trafficability in wet soil is often a surprising no-tillage effect. With only one or two no-tillage crop years, the 'fabric' of the soil strengthens (mainly through improved soil structure) and animal or machine treading causes much less compaction with fewer surface depressions. It is common knowledge that no-tilled fields are accessible for seeding or spraying several days sooner following rainfall than tilled soils, with less damage by surface compaction. No-tilled soils are not more dense or compact than tilled soils; they just have more resistance to down pressures as a result of the increased organic matter and structure.

No-tillage also moderates excessive weather effects, such as extreme rainfalls and temperatures. With the surface residues protecting the surface against raindrop impact, runoff and erosion, rills and gullies don't form. Residues minimize the high wind profiles from having an impact on the soil surface and significantly reduce wind erosion. And very subtle dampening of soil temperature variations often prevents freezing of overwintering plants. No-tillage seeding into standing residues has allowed successful winter wheat crops in far more northerly climates in the northern hemisphere than previously possible, with increased yields compared with spring-seeded crops.

Young *et al.* (1994) showed how seasonal weather variations could affect the risk of altering the profitability of conservation tillage (which includes a component of no-tillage) compared with conventional tillage (Fig. 3.2). They pointed out that the period 1986 to 1988 was particularly dry in the Palouse area of Washington State, which favoured the profitability of conservation tillage. The 1990/91 winter was particularly cold, which also favoured conservation tillage. At other times (1989 and 1990) the weather did not favour either technique. In this manner the relative risks of changing profitability are clearly illustrated. Such risks cannot be predicted with any accuracy, but they can be minimized by selecting conservation tillage techniques and/or machines with the widest possible tolerance of changing weather patterns.

It is obvious that no-tillage machines cannot control the weather. But it has been repeatedly noted that when no-tillage is undertaken with appropriate residue manipulation and seeding machines designed with proper seeding slots, seeds and seedlings have considerably better protection from weather variations (e.g. too hot, cold, dry, windy or wet) than when that soil is either tilled or drilled with inappropriate

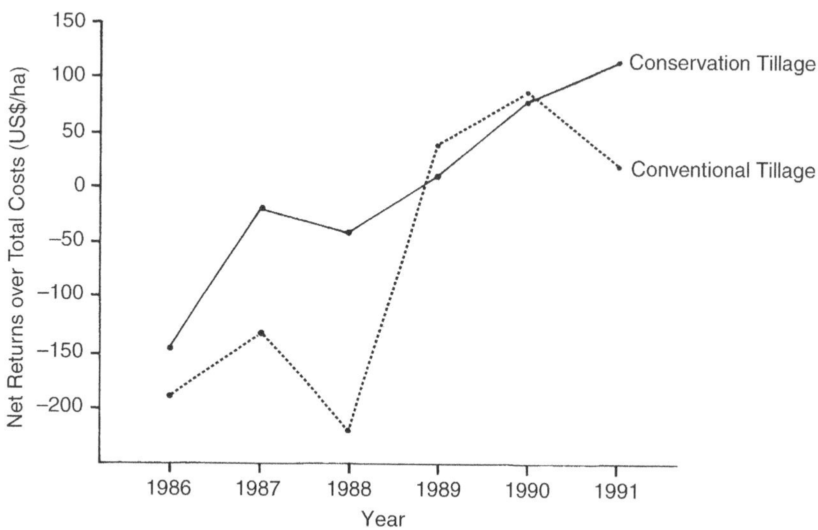

Fig. 3.2. The relative profitability of two crop establishment systems in Washington State over 5 years (from Young *et al.*, 1994).

no-tillage equipment. Thus, risks arising from inclement weather have the potential to be reduced under no-tillage if appropriate methods and equipment are used.

Machine function

Many of the physical risks arise from how well no-tillage machines perform their intended functions. The machine's designers must understand and incorporate the required capabilities to perform its intended functions in a wide variety of soil types, residues and weather conditions. These variations can change widely even within a single field or on a single day. There is much risk inserted into the farming system from a machine that operates at different levels of performance on different days in different parts of a field. A successful no-tillage drill must have a wide tolerance of changing, sometimes even hostile, conditions.

There are few more important physical functions than creating the correct micro-environment for the seeds within the soil. Different drill openers differ markedly in their abilities to do this (see Chapter 4) and this affects the level of risk associated with different machines. To reduce machine-related risks, the openers of no-tillage drills must follow ground surface variations and move through significant surface residues without blockage. Seeding depth can only be maintained by careful tracking of the soil surface by the seed opener.

Maintaining surface residues is the main long-term benefit from no-tillage, especially for reducing erosion and temperature fluctuations and increasing soil fauna and infiltration. Residues are an equally important ingredient in short-term biological performance of seedling emergence and vigour. No-tillage does not offer the option to 'till out' last season's mistakes of vehicle ruts, animal paths, washed gullies, hardpans, etc. It is critically important to avoid creating field surfaces that are not mechanically manageable the following cropping season.

No-tillage seeding machines not only must physically handle residues consistently without blockage but must also have the ability to micro-manage those residues close to the slot and to utilize them for the benefit of the sown seeds and plants (Baker and Choudhary, 1988). Conversely, the inability of any opener to do these things significantly increases the risks from no-tillage, since the residues themselves are an important ingredient in creating a favourable habitat for seeds and seedlings. A positive utilization of crop residues in no-tillage is considerably different from tillage farming in that residues are seen as beneficial rather than a hindrance to machine performance. Since tilled soils, almost by definition, have minimal surface residues, they do not benefit in comparison with good utilization of residues by no-tillage openers, but they may compare well with no-tillage where residues are not utilized.

Similarly, the ability to uniformly track the untilled soil surface for uniform seeding by no-tillage drills will greatly determine the biological risks associated with poor seedling stands and vigour. These aspects are discussed in greater detail in Chapter 8, but in summary it should be acknowledged that there is a need for no-tillage openers to follow the surface better than their tillage counterparts, or the risk of poor crop stands will increase.

No-tillage drills encounter much higher forces and wear of components than their tillage counterparts. Since some of the critical functions, such as residue handling and slot formation, are often dependent on the mechanical wear remaining within narrow limits, maintenance of no-tillage machines is more important than for conventional drills. To put it another way, the absence of adequate maintenance on no-tillage drills may increase the risk of malfunction disproportionately.

None of the physical functions described above, however, has any relevance to risk unless its successful implementation has an identifiable biological function with regard to the sown seeds and emerging plants. Somewhat surprisingly, many of the early 'desirable functions' listed for no-tillage openers (e.g. Karonka, 1973) failed to define any biological objectives at all. Failure to recognize these biological-engineering linkages alone probably increased the level of risk of early no-tillage and accounted for much of the 'hit-and-miss' reputation the technique acquired in its early days.

Ritchie *et al.* (2000) summarized the biological risks associated with six critical functions that no-tillage drill openers must perform. Their modified chart is shown in Fig. 3.3. Each criterion was assigned a risk rating of 1 to 10 (1 being low-risk and 10 being high-risk) according to published scientific data and engineering principles.

Several commonly used drill openers were ranked using the criteria of Fig. 3.3 and are shown in Table 3.1. The risk-assessment of the disc version of winged openers closely matches actual field surveys of users in New Zealand, which have consistently found a 90–95% success rate over several years and hundreds of thousands of hectares of field drilling (Baker *et al.*, 2001). But the most commonly used opener throughout the world (vertical double disc) ranks poorly. This helps explain the many no-tillage failures associated with this opener.

Chemical risks

Chemical risks have many of the same implications as physical risks. They are linked to the resultant biological risks that arise from them. Two stand-alone chemical risks are the effectiveness of weed control by herbicide application and the risk of toxicity or 'seed burn' from inappropriate placement of fertilizer in the seed slot relative to the seed placement.

Weed control

Weed control with herbicides must be as effective as that with mechanical means or the risk of impaired crop performance will increase. The principal variables determining herbicide effectiveness are as follows.

APPLICATION OF ACTIVE INGREDIENT. The ability of operators to properly interpret the labels and literature supplied with various herbicides and pesticides has much to do with the success of applications. In addition, operators need to be able to recognize weed species and to be able to reliably calibrate their spraying machines. All of these operator choices are more risky than corresponding tillage operations. Nor are spraying mistakes as forgiving as tillage mistakes, which can often be 'repaired' the next day.

SELECTION OF APPROPRIATE CHEMICAL. The selection of tillage tools can follow a trial-and-error routine where: (i) the non-performance of one implement becomes obvious within a short time; (ii) the consequences are seldom of great magnitude; and (iii) rectification using an alternative implement is accomplished quickly. Few, if any, of these flexibilities are available when choosing appropriate chemicals for a given weed or pest situation. Occasionally a mistaken choice can be rectified by the application of another chemical, but the options are fewer than with tillage and the risks are therefore greater.

WEATHER. Some chemicals require several hours without rain to be fully effective, while others are virtually 'rain-fast'. Since most chemicals involve a significant outlay of cash and, unlike tillage tools, are not reusable, the risk from untimely rain and wind is greater than with tillage.

WATER QUALITY. Some foliage-applied herbicides, especially those that are inactivated upon contact with soil, such as glyphosate, have their efficacy altered by impurities in the mixing water. Of particular concern is water derived from storage dams or underground bores that is contaminated with particles or iron and carbonates. Some chemical effectiveness is quite variable with water acidity levels. Similarly, impurities on the leaves of target foliage, such as mud and dust from stock or vehicle traffic or recently applied lime, may inactivate some herbicides.

VIGOUR OF WEEDS. The vigour of the target weeds at application time is important. Some herbicides (e.g. glyphosate) work best when sprayed on to healthy, actively growing plants. Others (e.g. paraquat) work best

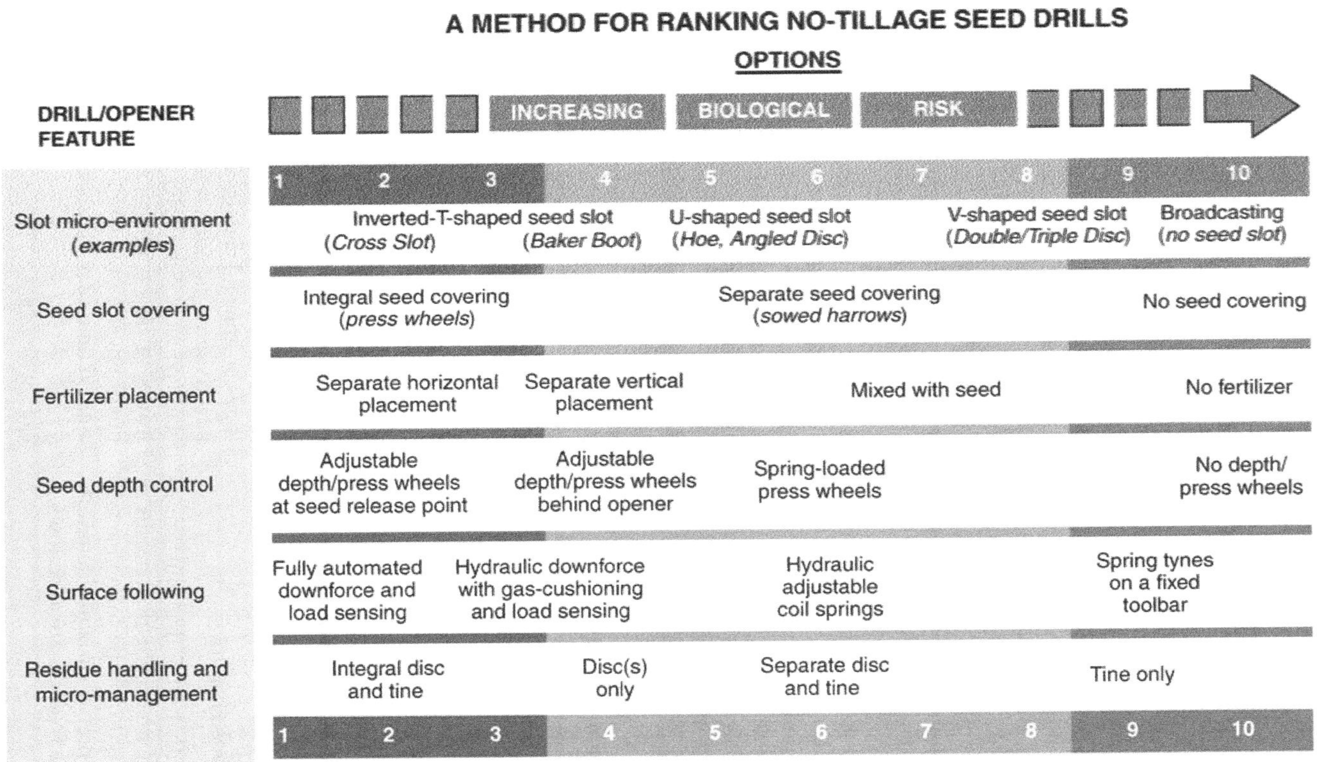

How to use this chart: Assess each of the six drill/opener features listed down the left-hand side of the chart. Assign a score from the colour bars that relates to the option the drill or opener exhibits. A high total score indicates a higher level of risk (Example: A hypothetical drill may have inverted-T openers (3), require towed harrows (6), have no fertilizer capability (10), no press/depth wheels (10), hydraulically adjustable coil springs (6) and separate disc and tine (6) = Total 41/60).

Fig. 3.3. A biological risk-assessment chart of drill opener designs (after Ritchie *et al.*, 2000).

Table 3.1. Examples of how some common no-tillage openers rank in terms of biological risk.

	Disc version of winged opener	Vertical angled disc	Slanted angled disc	Shank and sweep openers	Vertical double disc	Simple winged tine[a]
Slot micro-environment	1	4	4	3	7	2
Slot covering	1	3	2	2	7	4
Fertilizer placement	1	3	3	2	7	7
Seed depth control	2	1	1	9	3	8
Surface following	1	4	4	9	5	9
Residue handling	1	3	3	7	3	10
Total out of max. 60	7	18	17	32	32	40
Chance of impaired biological performance[b]	11%	30%	28%	53%	53%	67%

[a]Simple winged tine openers are designed to be used predominantly in smooth pasture. Comparing these openers for all no-tillage (including arable) penalizes them unfairly but they are nevertheless included here to illustrate how Fig. 3.3 exposes the limitations of such openers.

[b]The figures represent the chances of obtaining an impaired biological performance from using any of these openers. For example, the table suggests that use of the disc version of winged openers will result in an 11% chance of a poor crop, whereas use of shank and sweep openers will result in a 53% chance of a poor crop unless there is little residue present and the fields are smooth and flat.

Put another way, the table suggests that in heavy residues on less-than-smooth ground there would be about five times as much chance of getting an impaired crop using shank and sweep type openers as compared with the disc version of winged openers.

when the target plants are already stressed. Knowledge of these requirements is essential if effective weed control is to take place.

OPERATOR ERROR. During tillage, driving errors by an operator are seen immediately but they are seldom sufficiently serious to show up in the subsequent crop as an area of impaired yield. With once-over spraying, errors do not show up immediately. Paraquat is the most rapid to take effect but even then it is days before mistakes become visible. Most other herbicides take at least a week to show any visible effect, by which time the crop may have been sown, making remedial action virtually impossible without adversely affecting the sown crop.

Toxicity of fertilizers

There are two risks from inappropriate fertilizer placement at sowing. If fertilizer is broadcast on to the ground surface rather than placed in the soil at the time of drilling, there is a serious risk of impaired crop performance and yield as a result of limited plant availability (see Chapter 9). On the other hand, when fertilizer is sown with the seed there is a danger of the fertilizer damaging or 'burning' the seed under no-tillage unless the two are effectively separated in the soil. The latter risk increases with increased soil dryness. Separation is more difficult to achieve in no-tillage than in tilled soils, but it has been shown to be quite possible with the correct equipment without increased risk.

Economic risk

All forms of risk during no-tillage are finally measured as economic risk. But economic risk should not be centred on cost savings alone. Indeed, focusing on cost savings may increase rather than decrease both real and imagined economic risks. This is for two reasons:

1. Where farmers already own tillage equipment, they see the acquisition of no-tillage equipment or even the use of contractors (custom drillers) – no matter how cheap – as duplication of an existing cost.
2. Purchasing inferior no-tillage equipment for cost savings may well result in lowered crop yields, even if only temporarily. Such a result may indeed be less cost-effective than either tillage or no-tillage undertaken with more expensive (and probably superior) equipment that maintains or even improves crop yields.

We shall examine both scenarios below.

The costs of tillage versus no-tillage

The costs of several alternatives for adopting no-tillage under a double cropping system (two crops per year, e.g. wheat followed by a winter forage crop for animal consumption) in New Zealand were analysed and compared with the costs of tillage (C.J. Baker, 2001, unpublished data).

These were:

1. Engaging a tillage contractor (custom driller) versus engaging a no-tillage contractor.
2. Purchasing new tillage equipment versus purchasing new no-tillage equipment.
3. Retaining ownership of used tillage equipment versus purchasing used no-tillage equipment.
4. Retaining ownership of used tillage equipment versus purchasing new no-tillage equipment.
5. Retaining ownership of used tillage equipment versus engaging a no-tillage contractor.

Fixed costs were included, such as interest on the investment, depreciation, insurance and housing, and expressed as a per-hour cost of annual machine use. Drills and planters are used for a shorter period each year to plant the same area under no-tillage than under a tillage regime. Thus the per-hour costs increase even though the per-hectare and per-year costs decrease. The analysis also assumed that a single large tractor and driver would be required for no-tillage compared with two or more smaller tractors and drivers for tillage.

For simplicity, the study assumed that the no-tillage drill being compared was of an advanced design, which ensured that crop yields would remain unchanged regardless of which option was chosen. Such an assumption is reasonable when applied to advanced no-tillage drills (which cost more anyway) but is unrealistic for inferior drills (see below).

The cost analysis did not account for taxation issues, subsidies or other purchase incentives of any nature. These could otherwise be expected to favour no-tillage since many countries have incentives to encourage the practice because of its conservation value. Thus the results could be considered conservative in terms of the benefits recorded for no-tillage.

A more detailed account of the economic analysis is given in Chapter 18.

Operating costs strongly favoured no-tillage. In all of the above options (1) to (5), the costs favoured no-tillage by between US$16 and US$40/ha/year.

The greatest advantage (US$40/ha/year) was shown by option (2) – purchasing new tillage equipment versus purchasing new no-tillage equipment. This was mainly because of reduced running costs of the no-tillage equipment since the total capital outlays in each case were very similar.

The least advantage (US$16/ha/year) was shown by option (4) – retaining ownership of used tillage equipment versus purchasing new no-tillage equipment. Clearly the advantage would increase for this option when and if a decision was eventually taken to sell the existing tillage equipment, provided that a market for such equipment still existed. But realistically, the costs of purchasing no-tillage equipment would probably remain additional to the costs of retaining ownership of existing tillage equipment for a period.

Farmers often see retention of their existing tillage equipment as 'insurance' while they gain the knowledge and skills necessary to master the new no-tillage technique to a stage where they can abandon tillage altogether. Other farmers claim that by going 'cold turkey' (i.e. selling the tillage equipment at the same time as they purchase the no-tillage equipment) the learning process is achieved faster and more effectively. This study took the conservative approach.

The advantage for no-tillage from option 1 – engaging a no-tillage versus tillage contractor – was US$36/ha/year. The advantage for no-tillage from option 3 – retaining ownership of used tillage equipment versus purchasing used no-tillage equipment – was US$30/ha/year and for option 5 – retaining ownership of used tillage equipment versus engaging a no-tillage contractor – was US$34/ha/year. Cost advantages for no-tillage would be expected to increase when sale of the existing tillage equipment became feasible.

Machine impacts on crop yields and economic risk

The effect of any one no-tillage drill design on crop yield and risk (and therefore economic returns) will be more important than its initial cost, when compared with either tillage or cheaper no-tillage alternatives. This belief has caused the research and development of improved no-tillage machines and systems as a means to reduce the risks associated with the practice, almost regardless of cost. The following analyses of machine capability versus expected crop yields and the resulting economics clarifies this belief.

The per-hectare charges that no-tillage contractors (custom drillers) make for their services are a good barometer of the relative costs associated with different no-tillage machines and systems. If we take New Zealand contractors as an example, we find that those with advanced (expensive) no-tillage drills in 2004 charged between US$72 and US$96/ha for their services, whereas those with lesser (cheaper) drills charged between US$36 and US$60/ha.

Differences between the ranges of charges are attributable mainly to differences in the initial costs of the two classes of machines and the different sizes of tractors needed to operate them. Differences within both ranges of costs reflect differences in the costs of competing options (such as tillage) together with differences in work rates and maintenance costs brought about by different field sizes, shapes, topographies and soil types (including abrasiveness).

Taking the midpoint of each scale, the premium a farmer therefore paid in New Zealand in 2004 for access to a more advanced drill was about US$36/ha. Actual contractor charges in other countries will differ from these figures but the relativity between the costs associated with advanced machines and lesser machines is likely to be similar.

So a key question is: How much does an advanced no-tillage drill have to increase crop yields in order to justify the US$36/ha premium paid for the better technology under 2004 price conditions?

Wheat sold in New Zealand in 2004 for approximately US$170/t. The average yield of spring-sown wheat in New Zealand in 2004 was 5.7 t/ha and the average autumn-sown wheat yield was 7.4 t/ha (N. Pyke, Foundation for Arable Research, 2004, personal communication). Gross returns for average spring- and autumn-sown wheat crops in 2004 were therefore US$969/ha and US$1258/ha, respectively.

To recover an additional US$36/ha in the costs of no-tillage drilling would require an increase in yield of 0.21 t/ha (or 210 kg/ha). This represented a 3.7% increase in yield of a spring-sown wheat crop or a 2.9% increase in an autumn-sown wheat crop.

Such yield increases have been common. For example, the US Department of Agriculture obtained an average of 13% wheat yield increase in seven separate experiments over a 3-year period in Washington State by switching to a more advanced no-tillage drill compared with the best 'other' no-tillage drill that was then available (Saxton and Baker, 1990). Similarly, the New South Wales Department of Agriculture

(Australia) recorded an 11-year average of 27% yield advantage from soybean sown annually after oats using the same advanced no-tillage openers, compared with tillage (Grabski *et al.*, 1995).

Commercial field experience over a 9-year period in New Zealand, the USA and Australia suggests that such research-plot measurements have been a realistic reflection of field expectations. Wheat and other crop yields approaching twice the national averages have become common from no-tillage practised at its most advanced level.

Conclusions

It can be said that, when comparing the economic risks of tillage and no-tillage, more management and more sophisticated machinery are needed to undertake no-tillage correctly and successfully. But, if the appropriate management and machinery are used and the reasons for these choices are understood, there will be no more and often less economic risk with no-tillage than with tillage. All of the various forms of risk come together in the multiple-year rotations required of modern farming in an integrated management system. Figure 3.4 illustrates the results of a comprehensive assessment of financial risk made during 6 consecutive years of experiments by Young *et al.* (1994) in Washington State, USA.

These experiments compared the combined results of conservation tillage, which included several consecutive years of no-tillage, versus conventional tillage, the effects of maximum, moderate and minimum weed control and crop rotations, all under a high level of agronomic management. Considering all treatments and 6 years of variable weather factors, conservation tillage had the smallest economic risk due to conserved moisture, good yields and low inputs. They concluded that the winter wheat–spring barley–spring peas rotation at maximum or moderate weed management

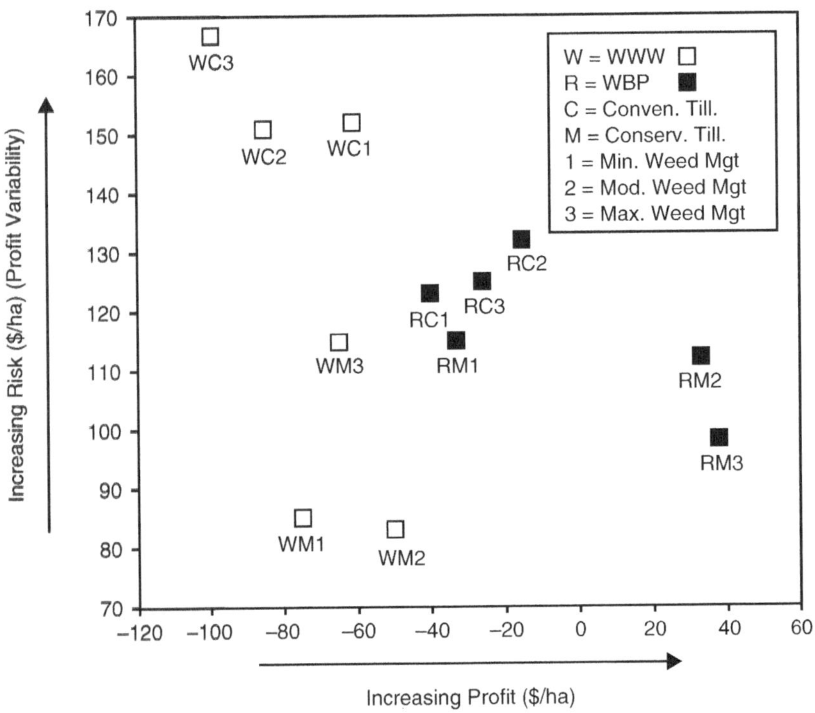

Fig. 3.4. Profit and risk analyses for 12 cropping systems in the Palouse area, Washington, 1986–1991 (from Young *et al.*, 1994). WWW, wheat, wheat, wheat rotation; WBP, wheat, barley, peas rotation.

levels (RM3 or RM2) dominated all other systems in profitability (profit of $30–40/ha) and had the lowest economic risk or 'profit variability'.

Summary of the Nature of Risk in No-tillage

1. The perception that no-tillage involves greater risk than tillage is one of the greatest impediments to its more widespread adoption.
2. The combination of all the components of risk manifests them as economic risk.
3. The components of risks in no-tillage are biological, physical and chemical.
4. Biological risks relate to pests, toxins, nutrient stress, seed vigour, seedling vigour, disease and impaired crop yield.
5. Physical risks relate to weather, slot micro-environment and machine performance and reliability.
6. Chemical risks relate to the supply and availability of plant nutrients, seed 'burn' from fertilizers and the effectiveness of application of chemical herbicides and pesticides.
7. The function and design of no-tillage seed drills can have an influence on pests, toxins, nutrient stress, diseases, fertilizer 'burn', slot micro-environment, machine performance and durability and the supply and availability of plant nutrients.
8. Performed correctly with appropriate equipment, no-tillage has no more, and often less, total risk than tillage, even in the short term.
9. Performed incorrectly with inappropriate equipment, no-tillage has greater associated risk than tillage.
10. It is often 'false economy' to cut costs in no-tillage, particularly in machine effectiveness, as the savings in cost may be much less than the reductions in crop yield that are likely to result.

4 Seeding Openers and Slot Shape

C. John Baker

Very few no-tillage openers were originally designed for untilled soils. Most are adaptations of conventional openers for tilled soils.

A seeding opener is the soil-engaging machine component that creates a 'slot', 'furrow' or 'opening' in the soil into which seed and perhaps fertilizer and insecticide are placed. Different shapes of soil slots may be created by conventional and no-tillage openers. The most important feature is the cross-sectional shape, as if you had cut across the opener path after its passage with a knife and were looking at the vertical exposed face.

Openers are the only components of a no-tillage drill or planter that actually break the soil surface. In no-tillage seeding, they are required to perform all of the functions necessary to physically prepare a seedbed as well as sow the seed and perhaps fertilizer. In contrast, in conventional tillage a succession of separate tillage tools are used to prepare the seedbed, and the seed drill then only has the relatively simple task of implanting the seed and perhaps fertilizer into a pre-prepared medium.

A large amount of scientific evidence shows that the most important aspect of the mechanics of different no-tillage opener designs is the shape of the slots they create in the soil and their interaction with seed placement and seedling emergence and growth. Generally, there are three basic slot shapes created by no-tillage openers and two other ways of sowing seed that do not involve creating a continuous soil slot at all: (i) V-shaped slots; (ii) U-shaped slots; (iii) inverted-T-shaped slots; (iv) punch planting (making discrete holes in the ground and sowing one or more seeds per hole); and (v) surface broadcasting (seeds randomly scattered). Only one slot shape, the inverted-T slot, is used in no-tillage that has not been an adaptation of a slot shape already used for tilled soils.

Figure 4.1 is a diagrammatic representation of slot shapes i–iii as created in a silt loam soil at three different moisture contents (Dixon, 1972). The mechanics of each of these seeding methods and the resulting characteristics will be further discussed in detail in the following sections.

Several authors (e.g. Morrison *et al.*, 1988; Bligh, 1991) have compiled lists and diagrams of openers and in some cases compared observations of field performance. But few detailed scientific studies have been made in which all but the important variables being studied have been controlled or accurately monitored. Such studies (which also included some new and innovative designs) are reported below.

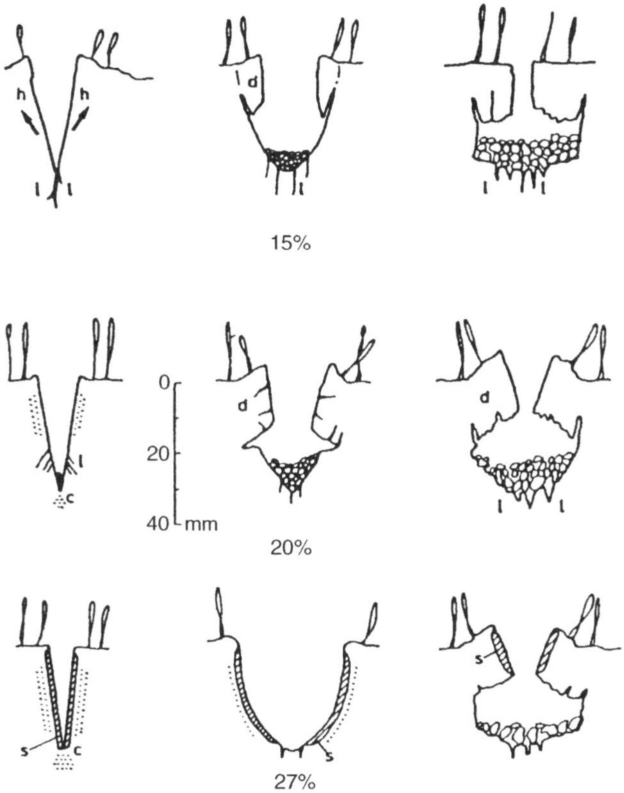

Fig. 4.1. Typical profiles of vertical V- (left), U- (centre) and inverted-T- (right) shaped no-tillage seed slots in a silt loam soil at 15%, 20% and 27% moisture contents (from Dixon, 1972).

Vertical Slots

V-shaped slots

In untilled soils, V-shaped slots are almost invariably created by two discs that touch (either at their bases or behind this position) and are angled outwards towards their tops. The two discs are not always of equal diameter. The included angle (the angle of the V) is usually about 10°, but this is not critical. Seed is delivered into the gap between the two discs, preferably rearwards of the centre 'pinch point', so as to prevent the seed from being crushed as the discs come together.

When arranged so that both discs are at the same angle to the vertical, the slot has a vertical V shape and is created by each of the angled discs pushing roughly equal amounts of soil sideways. The front edges of the two discs at the ground-surface level are apart from one another, which can cause a problem if residues enter the gap. To avoid this they are usually configured in one of the following three forms.

Double disc: offset (Fig. 4.2)

In this form one of the two angled discs (there is no third leading disc) is positioned forward of the other so as to present a single leading cutting edge and deflect residue. The second disc still forms the other side of a vertical V but its leading edge is nestled behind that of the first disc, thus avoiding residue blockage and reducing the magnitude of downforce required for penetration.

Double disc: unequal size (Fig. 4.3)

By placing the smaller of the two discs alongside its larger neighbour, the leading edge of the larger disc becomes the leading edge of the whole assembly in much the

Fig. 4.2. Typical offset double disc no-tillage openers that create vertical V-shaped slots.

Fig. 4.3. Typical unequal-sized double disc no-tillage openers that create vertical V-shaped slots.

same way as for the offset design. Often, the smaller disc is also offset.

Triple disc (Fig. 4.4)

In this form a third vertical disc is placed ahead of, or between, the two angled discs. This additional disc cuts the residue sufficiently for the two following discs to deflect it sideways. The third disc, however, adds to the amount of downforce required for penetration.

All forms of double disc and triple disc openers create vertical V-shaped slots since the actual slot shape is created by the two angled discs, regardless of their sizes or offsets. The third (leading) disc in the triple disc configuration mainly cuts the residue and influences the slot in a minor way. The triple disc design with the leading disc operating slightly below the bases of the two angled discs reduces some of the detrimental effects of 'hairpinning' (see Chapter 7, 'Drilling into Wet Soils') and root penetration problems common to both double and triple disc configurations. Similarly, by using a wavy-edged leading disc (sometimes referred to as a 'turbo disc'), a degree of soil loosening will usually be achieved ahead of the two angled discs and this helps offset the compacting tendencies of the following double discs.

The action of vertical double disc openers in the soil is to wedge the soil sideways and downwards in a V formation. They do not normally heave or raise the soil upwards. In some very sticky soils that cling to the outsides of the discs, some of that soil will be torn away and carried upwards, leaving a disrupted slot (Fig. 4.5).

Figure 4.6 shows the zones of compaction created by a vertical triple disc opener operating in a normal manner in a silt-loam soil (Mitchell, 1983).

From a dry soil perspective, the most distinguishing feature of the slot is the neatness of the vertical V-shaped cut, unless the soil is friable, in which case this neat cut may collapse. But even friable soils progressively become more structured and less friable (as organic matter levels and microbial action increase under no-tillage). Thus, with time, most vertical V-shaped slots become more clearly defined and less likely to collapse of their own accord after passage of the opener.

Fig. 4.4. A typical triple disc no-tillage opener that forms a V-shaped slot (from Baker, 1976b).

Fig. 4.5. A slot created by a vertical double disc opener in wet sticky soil in which the soil has stuck to the outside of the disc and been pulled up from the slot zone.

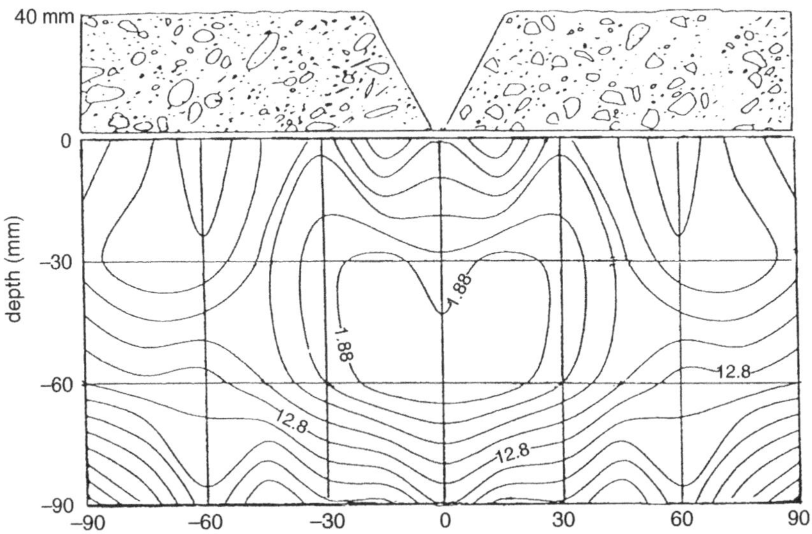

Fig. 4.6. The pattern of soil strength around a vertical V-shaped no-tillage slot as created by a triple disc opener in a damp silt loam soil (from Baker *et al.*, 1996).

Because of its wedging action, there is often little or no covering material available to cover seeds placed into the bottom of the V slot. This is even more of a problem when the opener is used in a moist, non-friable soil. Figure 5.1 illustrates such a situation. The plastic nature of the moist soil prevents the formation of loose soil crumbs, which might otherwise fall back over the seed as covering material (see Chapter 5).

The usual recourse is to follow vertical double disc openers with some configuration of V-shaped press wheels arranged so that they squeeze the soil in the opposite direction to the discs after the seed has been deposited (Fig. 4.7). Unfortunately, this action is also one of compaction, albeit in the opposite direction to the original forces. In an untilled soil, the wedging action of vertical double disc openers does little, if anything, to create a favourable environment for seeds.

The greatest advantages of vertical double disc openers are: (i) their construction is relatively simple and maintenance-free, although the latter attribute depends on the use of good bearings and seals; and (ii) their ability to pass through surface residues without blockage.

The most important disadvantages are: (i) the high penetration forces required; (ii) their poor performance in suboptimal soil conditions; (iii) their tendency to tuck (or 'hairpin') residue into the slot, which in dry soils interferes with seed–soil contact and in wet soils results in fatty acid fermentation that kills germinating seeds (Lynch, 1977); and (iv) the inability of individual openers to separate seed from fertilizer in the slot. Indeed, due to the shape of the slot, vertical double disc openers tend to concentrate the seed and fertilizer together at the base of the slot more than other openers (Baker and Saxton, 1988; Baker, 1993a, b).

Despite these shortcomings, vertical double disc openers have been included on more no-tillage drill designs than any other opener design to date. Unfortunately, however, because of their dependence on favourable soil conditions to achieve acceptable seeding results (or, more correctly, their intolerance of unfavourable conditions), they have also been responsible for much of the perception that risk increases with the practice of no-tillage.

It is important to emphasize the distinction between tilled and untilled soils and to illustrate the dangers inherent in deriving designs of no-tillage machines from those that had been successful in tilled soils. Tilled soils are naturally soft before seeding and the wedging action of vertical double disc openers is generally beneficial,

Fig. 4.7. Press wheels arranged in a V configuration for closing no-tillage slots created by vertical double disc openers (from Baker, 1981a, b).

especially when the soil is dry. It consolidates the soil alongside and beneath the seed, which results in increased capillary movement of water to the seed zone. Covering is seldom a problem in tilled soils, because the entire seedbed is comprised of loosened soil. Thus, in many ways, V-shaped openers are an advantage in tilled soils, whereas they have serious shortcomings in untilled soils.

Other mechanical forms of vertical V-shaped openers for tilled soils simply do not work in untilled soils because they will not penetrate in the less friable conditions. These include sliding shoe-type openers and V-ring roller openers (Baker, 1969b). Further consideration of these designs is not justified since they simply cannot effectively seed no-tilled soils.

Slanted V-shaped slots

To reduce the compaction tendencies of vertical V-shaped slots, some designers have slanted double or triple disc openers at an angle to the vertical, and sometimes also angled to the direction of travel. When they are slanted vertically, the uppermost disc pushes the soil partially upwards, thus reducing the compaction that otherwise results from the soil being displaced only sideways by vertical double disc openers. The lowermost disc on slanted double disc openers, however, is then forced to displace soil in a more downward direction, adding to its compaction tendency. Since roots mainly travel in a downward direction, it is debatable whether or not the slanting of double or triple disc openers overcomes the disadvantages inherent from their tendency to compact the slot in the root zone. On the other hand, slanting of V-shaped slots undoubtedly makes them easier to cover, since a near-vertical press wheel is required to shift soil more in a downward direction than sideways.

Two slanting double discs can be combined in such a way that the front pair of discs (which are angled vertically in one direction) sow fertilizer and the rear pair of discs (which are angled vertically in the opposite direction) sow seed at a shallower depth. Not only does this effectively separate seed and fertilizer in the vertical plane, but additionally the zone that would normally be compacted below the seed by the lowermost disc of the rear opener is pre-loosened by the uppermost disc of the front opener, thus partly negating the undesirable compaction effect of the seeding opener. Figure 4.8 shows a pair of slanted double disc openers.

Single discs that are angled in relation to the direction of travel (and sometimes also slanted vertically) are discussed below.

U-shaped slots

There is a wide range of opener designs that form U-shaped slots (Baker, 1981a, b): (i) angled disc-type openers; (ii) hoe openers; (iii) power till openers; and (iv) furrowers.

The slots made by all of these designs are distinguishable from V-shaped slots by the slot bases being broad rather than pointed like a V. The slot-making action of

Fig. 4.8. A pair of slanted discs at opposing angles. The front discs place fertilizer and the rear discs place seed at a shallower depth. (From Baker *et al.*, 1996.)

each of these openers is quite different, even though they all result in a similarly shaped slot, but none of the openers has the downward wedging action of double disc openers. Thus there is less soil compaction associated with all U-shaped slots than with V-shaped slots.

Angled disc-type openers mostly scrape soil away from the centre of the slot; hoe- and furrow-type openers burst the soil upwards and outwards; power till openers chop the soil with a set of rotating blades; and furrow-type openers scoop the soil out from the slot zone. Further, all of the designs produce some loose soil on the surface near the slot, which can be used to cover the slot again, although in all cases this usually requires a separate operation to drag this soil back over the slot (see Chapter 5) and its effectiveness is soil-moisture-dependent.

Angled disc-type openers

The action of angled discs is mostly (although not entirely) one of scuffing. Vertical angled discs are angled slightly to the direction of travel (normally about 5–10 degrees). Seed is delivered to a boot located at or below ground level, close to the rear (lee) side of the discs where it is largely protected from blockage by residue because of the angle of the disc. There are two forms of angled vertical disc opener.

ANGLED FLAT DISCS (FIG. 4.9). This type uses a vertical flat disc (i.e. it has no undercutting action) angled to the direction of travel. The disc and supporting bearings need to have considerable inherent strength since the side forces are quite large, especially when operating at some speed and/or in plastic soils that resist sideways movement. Because the discs continually have a sideways force, they are often configured in pairs with each pair of discs at opposite angles so that the side forces of the entire machine cancel (see Fig. 4.9).

Where the discs are not arranged in pairs, difficulty is sometimes experienced in turning corners in one direction with the drill, while turning in the other direction poses no problem. This is another example in which the requirements of no-tillage are different from tillage, since the soil forces in tilled soils are sufficiently low to not cause problems when cornering with angled disc-type openers.

Fig. 4.9. A pair of angled flat disc no-tillage openers (from Baker *et al.*, 1996).

Relatively steep side-slope drilling causes machine 'tailing', in which the whole machine pulls at an angle to the direction of travel because of gravity pulling the drill sideways. This poses a problem for drills arranged with half of the openers angled in each direction. That part of the drill in which the openers are caused to travel with no angle creates very small, ineffective seed slots, while the other openers double their angle and create extra wide slots that are difficult to cover.

ANGLED CONCAVE DISCS. This type uses a slightly concave, near-vertical disc set at an angle to the direction of travel (Fig. 4.10). The strength derived from the curvature of the disc allows thinner steel to be used in its construction, assisting in soil penetration. The axle of angled dished discs can be either horizontal or slightly tilted from the horizontal in either direction.

If the axle is tilted downwards on the convex (back) side of the disc, the action of the disc will be to undercut the soil like a disc plough. The benefits of this action are that the displaced soil is not thrown to one side where it is otherwise often difficult to retrieve again for covering purposes, as it is lifted, hinged and inverted. The disadvantages are that, in soils that are held together by plant roots (e.g. pasture), a soil flap is produced, which falls back over the seed. Since the seed is placed under the 'hinged' end of this flap, this can restrict seedling emergence. Figure 4.11 shows an angled dished disc that has had a small scraper added to attempt to slice this flap off.

If the axle is tilted upwards on the convex (back) side of the disc, it has the effect of confining the disc action to one of scuffing only, with little or no undercutting. Because of the disadvantages of undercutting, this has become the most commonly preferred option with concave disc openers for no-tillage, along with arranging them with the disc axle horizontal.

TILTED AND ANGLED FLAT DISCS. Some designers have tilted as well as angled the flat discs on their openers (Fig. 4.11). This has mainly been to reduce the throwing action of angled discs so that there is less soil disturbance and also to provide more of a mulch cover than where the discs stand vertically upright. Tilting the discs may also help

Fig. 4.10. An angled dished disc no-tillage opener (from Baker *et al.*, 1996).

Fig. 4.11. An angled dished disc no-tillage opener with both vertical and horizontal angle. This opener also features a scraper to cut and remove the turf slice. (From Baker *et al.*, 1996.)

penetration and reduce the hillside operation problem discussed above. But it does nothing to reduce the tendency of such openers to hairpin residues into the slot, which interferes with seed germination and/or seedling emergence. Nor do such openers solve the problem of fertilizer placement, since no more opportunity exists to separate fertilizer from seed than with any other configuration of angled disc.

The actions of all angled discs (flat or concave, upright or tilted) are very much dependent on their operating speed. Because all variations depend on at least angulation to the direction of travel (if not also angulation to the vertical) for much of their slot-creating actions, the speed with which they approach the soil has a marked effect on the amount of soil throw and therefore the width and shape of the resulting slot. At higher speeds, the slots tend to be wider and shallower than at slower speeds and the loose soil available for covering tends to be thrown further to one side, where it is more difficult to retrieve. In common with discs that travel straight ahead, the penetration of angled discs is also reduced with increasing speed, but this can be countered by simply increasing the downforce to achieve penetration.

The two biggest advantages of all angled discs are their ability to handle surface residues without blockage and their avoidance of compaction or smearing of the slot at the base and on at least one side wall. They are also relatively cheap, simple and maintenance-free.

The biggest disadvantages of angled disc openers are: (i) they tuck or hairpin residues into the slot in a similar manner to double disc openers; (ii) they make U-shaped slots, which, especially if wide at the top, dry easily despite the presence of loose soil; (iii) they are often difficult to set for correct operation; (iv) they may angle and operate poorly on hillsides; (v) they are not able to separate seed from fertilizer in the slot; (vi) they are affected by the speed of travel; and (vii) they wear rapidly.

Hoe- or shank-type openers

The term hoe or shank describes any shaped tine or near-vertical leg that is designed to penetrate the soil. Seed is delivered either down the inside of the hollow tine itself or down a tube attached to its back.

The shapes of hoe or shank openers range from winged (Fig. 4.26, p. 54), which are often also designed to separate seed and fertilizer simultaneously in the slot, through blunt bursting openers (Fig. 4.12) to sharp undercut points, which are designed to make a relatively narrow slot and penetrate the soil easily (Fig. 4.13). Sometimes a pair of narrow shanks is arranged with horizontal offset to separately place seed

Fig. 4.12. A blunt hoe-type no-tillage opener (from Baker *et al.*, 1996).

Fig. 4.13. A sharp hoe-type no-tillage opener (from Baker, 1976b).

and fertilizer (Fig. 4.14). One of the problems with hoe-type openers is that they wear rapidly; thus, the original shape seldom lasts long. Because of this they may take on several new shapes during their lifetime, making it difficult to generalize on the basis of slot shape.

Generally, all hoes scrape out a roughly U-shaped slot by bursting the soil upwards from beneath. In moist conditions they tend to smear the base and sometimes the side walls of the slot, but this only affects seedling root systems if the soil is allowed to dry and thus become an internal crust (see Chapter 5).

The bursting action produces considerable loose soil alongside the slot, which may be helpful when covering but can also leave severe ridging between rows. Because of this latter problem most shank-type openers are operated at low speeds (maximum 6–9 kph, 4–6 mph).

The nature and extent of the loose soil alongside the slot is also dependent on soil moisture content. Often, in damp plastic soils, no loose soil will be produced at all, while at other times a few hours of drying after drilling will produce crusty edges to the slots, which can then be brushed with a suitable harrow or dragged to at least partially fill the slot with loose soil. The most appropriate covering action after passage of hoe-type openers is therefore a matter of judgement at the time, which is one of their inherent disadvantages.

The biggest disadvantage of hoe or shank openers, however, is the fact that they can only handle modest levels of residues without blockage (also see Chapter 10), especially when arranged in narrow rows. The placement of a leading disc ahead of a hoe or shank opener, regardless of how or in what position it is placed relative to the hoe, cannot make a group of such openers arranged in narrow rows able to handle residues satisfactorily.

The most successful hoe or shank drill configurations for residue clearance have been to space the openers widely apart in multiple rows (ranks) in the direction of travel. This is based on the observation that, unless the residue is particularly heavy or damp or becomes wedged between adjacent openers, the inevitable accumulation of residue on each tine will usually fall off to one side, as a function of its own weight. If sufficient clearance is built into the spacing between adjacent tines, the falling off of clumps of residue will not block the machine – at least, not as often. These

Fig. 4.14. A pair of shank openers with horizontal offset. The front shank applies fertilizer while the rear shank applies seed offset to one side and sometimes shallower.

clumps of residue can cause problems for seedling emergence and later at harvesting, so it is questionable whether this action can be described as handling residue at all. Unfortunately, wide spacing demands undesirable dimensions from the whole drill, which compromises other functions such as the ability to follow the ground surface and seed delivery. Figure 4.15 shows a shank-type no-tillage drill with widely spaced openers.

Hoe or shank openers have several advantages: (i) they are relatively inexpensive; (ii) they can be made to 'double-shoot' seed and fertilizer relatively simply; (iii) they do not tuck (or hairpin) residues into the slot; in fact, they brush the residue aside, although this is a disadvantage for controlling the microclimate within the sown slot, as described in Chapter 5.

Their major disadvantages are: (i) a high wear rate; (ii) their poor residue handling ability; and (iii) their inability to separate seed from fertilizer in the slot (see Chapter 9).

Power till openers

Power till openers are an enigma in no-tillage. Because most people had become accustomed to tilling the soil before planting seeds, it seemed natural to till the soil in strips for no-tillage. Thus, power till openers consist of miniature rotary cultivators that are power-driven from a common source and literally till a series of narrow strips for the seed. While the tillage ensures that seeds will become well covered with loose soil, it has long been known that rotary tillage is one of the least desirable ways of tilling soil. Its main disadvantages, when applied either to general seedbed preparation or discretely in strips, is that it stimulates weed seed germination, is very destructive of soil structure and is power-demanding (Hughes, 1975; Hughes and Baker, 1977).

The actual placement of seed varies with design. With some, the seed is scattered into the pathway of the rotating blades and thus becomes thoroughly mixed with the soil, but depth of placement becomes random. With others, separate conventional openers for tilled soils (shoe, hoe or disc type) operate behind the rotating blades as if they were drilling into a fully tilled seedbed.

The advantages of power till openers are that the downforces required for penetration are little more than those commonly

Fig. 4.15. A no-tillage drill with widely spaced shank-type openers designed to 'clear' residues.

required for tilled soils. Power till openers substitute power applied through the tractor power take-off (PTO) shaft to the rotors for the downforces and draft forces more common to other non-rotating types of no-tillage openers. They create U-shaped slots, they do not tuck residue into the slot, they generally cover the seed well and, in cold climates, where there might otherwise be a disadvantage from the slow decomposition of surface residues, they chop up this residue and incorporate it into the soil.

On the other hand, because they physically dispose of the surface residues in this manner, power till openers do little to micro-manage the residues close to the seed, which is one of the most important functions that successful no-tillage openers should perform. Further, few of them separate the seed from the fertilizer in the slot, although, because of the amount of loose soil in the slot, there is more mixing of fertilizer with soil, which provides partial separation from the seed.

Power till openers are relatively complex mechanical devices when compared with other opener designs. They have a particular problem with wear, surface following and damage from stones and other obstructions.

Early designs were adaptations of conventional field rotary cultivators. The normal wide L-shaped blades, which were mounted on a common axle driven by the tractor PTO, were replaced with sets of narrow L blades corresponding to the desired row width and spacing. These created the discrete rows of tilled soil. The width of the tilled strips varied from about 20 mm to 200 mm, depending on the objectives. Figure 4.16 shows the effects from a narrow set of blades, while Fig. 4.17 shows wider tilled strips.

In early designs each set of blades was mounted on a common axle, so it was impossible for each tilled strip to maintain a constant depth while traversing the normal undulations of the ground. Even the use of independently articulated seed-depositing openers, which followed in the tilled strips, could not fully compensate for areas of soil that had missed being tilled altogether because the machine had traversed a small hollow, for example. Figure 4.16 shows a common-axle-type power till drill with independently mounted seed-depositing openers.

Fig. 4.16. Narrow tilled strips left by a power till 'no-tillage' opener (from Baker *et al.*, 1996).

Fig. 4.17. Wide tilled strips left by a strip-tillage machine (from Baker *et al.*, 1996).

Fig. 4.18. Power till no-tillage openers arranged in pairs (from Baker, 1981a, b).

Later designs attempted to mount each rotating set of blades independently so that they were capable of following the soil surface. This proved to be inordinately expensive, because, while each set of blades required its own flexible drive train, it also had to offer some protection from stone damage. Belt drives allowed slippage in these circumstances.

Some designs compromised by mounting blade sets in pairs. Figure 4.18 shows the head of a twin rotor model in which the rotors are able to articulate up and down. Other designs attempted to power each rotor individually through a chain driven by a wavy-edged ground-engaging disc ahead of the rotor in the hope that the disc would slip in the soil when a stone was encountered by the rotor. Figure 4.19 shows such a device.

Although power till openers have been an obvious design route for many engineers, with a number of models released for commercial production around the world, very few have been commercially successful due to the disadvantages mentioned above. Perhaps their greatest use is where other openers cannot function. An example of such a condition is the revegetation of high-altitude pastures where ambient temperatures remain sufficiently low to discourage complete decomposition of organic matter. The result is a build-up, over centuries, of a mat of undecomposed vegetation, which can be several centimetres thick (Fig. 4.20) and which simply resists the operation of any other no-tillage opener except designs that physically chop it up and mix it with soil. In these conditions, the objective is to drill improved pasture species by no-tillage to increase animal carrying capacity on otherwise low-producing fragile farmland.

Power till openers in general create more short-term mechanical aeration within the slot than any other type of opener, although the benefits of this are usually temporary in comparison with openers that encourage natural aeration by earthworms (see Chapter 7). They have a tendency to compact the base of the slot, but, unlike double disc openers, this does not seem to cause difficulties for seedling roots.

Furrowers

One opener, designed in England especially for pasture renovation, consisted of two vertical discs, spaced laterally several centimetres apart, which cut two vertical slits. The discs were followed by a miniature mouldboard-plough, which scooped out the soil between the slits at the same time as it created a small track in the base of the broad U-shaped slot, where seed was deposited (Haggar, 1977; Choudhary *et al.*, 1985). The function of scooping was to eliminate weed competition in the seed zone without spraying and to allow early seedling development to take place in a sunken zone physically protected from

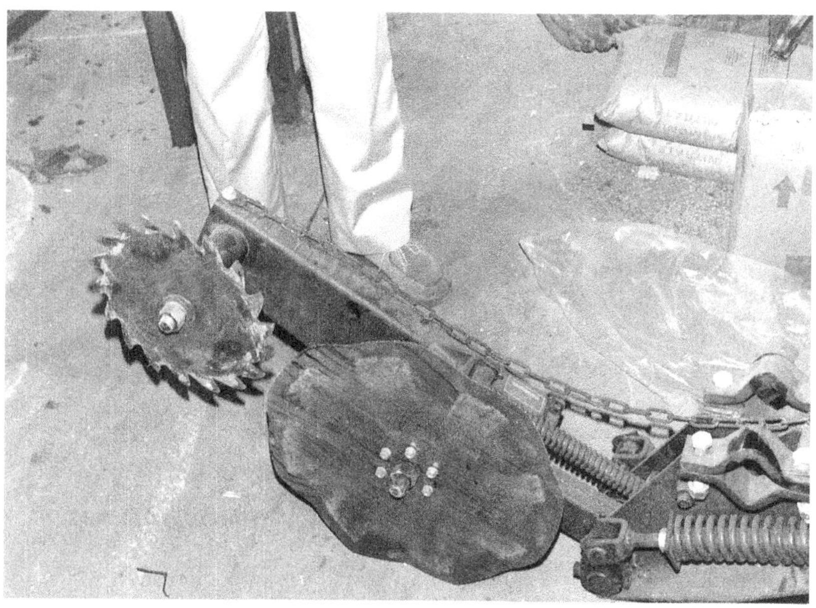

Fig. 4.19. A power till no-tillage opener driven by a ground-engaging wavy disc (from Baker *et al.*, 1996).

Fig. 4.20. Undecomposed sod in the Scottish Highlands (from Baker, 1981a, b).

Fig. 4.21. A furrowing no-tillage opener (from Choudhary *et al.*, 1985).

treading by cattle (Fig. 4.21). Seed cover in the damp English climate is not a high priority, but such openers are regarded as specialist tools designed solely for one intended purpose.

Vibrating openers

Several designers have attempted to reduce the downforces required to push the discs and other components of no-tillage openers into the ground by causing the openers to vibrate. Such a task has been particularly demanding when applied to a disc since the vibration mechanism needs to operate on the disc hub as it rotates, as well as moving up and down in response to natural undulations in the ground surface. Individual vibrating hydraulic motors have been used on individual openers, which increase the cost, complexity and power requirement considerably. Very slow operating speeds and difficulty in keeping all bolts and nuts tight because of the general vibrations generated throughout the machine have also been disadvantages.

In the end, it is the shape and action of the soil-engaging components that determine the biological success or otherwise of no-tillage openers more than the forces required for penetration of draught. Vibrating openers do nothing to improve biological reliability. Most designers have found it cheaper to add weight and/or use a larger tractor to overcome penetration and draught forces than to engage relatively complex vibrating devices.

Horizontal Slots

Inverted-T-shaped slots

All of the openers discussed so far have been adaptations of openers designed originally for tilled soils (with the exception of the specialized furrow and vibrating openers). The modifications to such openers, when employed for no-tillage, have mostly consisted of increased robustness, with only minor changes in function.

The inverted-T-shaped slot is the only known horizontal no-tillage slot shape and is one of very few slot shapes that have been developed specifically for no-tillage purposes, with few functions applicable to tilled soils.

The inverted-T principle was developed when researchers explored geometrical alternatives to the more common V and U slot shapes to overcome several of their inherent disadvantages (Baker, 1976a). The researchers reasoned that the most radically different shape would be to invert the wide-top narrow-base V shape and to create instead a narrow-top, wide-base slot. Practicality dictated that the simplest way to achieve this was to construct an opener consisting of a vertical shank with sub-surface wings that were horizontal in the lateral plane but inclined downwards towards the front in the longitudinal plane.

The other reasoning behind the winged concept was that the designers wanted to be able to fold the residue-covered soil back over the slot for moisture conservation and seedling protection. Since wings tended to undercut the surface layer of soil with a horizontal slicing action, this would allow the formation of horizontal shelves on either side of a vertical slit. In most conditions the wing action also created horizontal flaps of residue-covered soil with which to cover the shelves. It was a major objective of the inverted-T concept to create horizontal slots with a high degree of control and predictability.

Two winged opener concepts were developed, both of which created essentially the same inverted-T-shaped slots.

Simple winged opener

The first simple winged opener design consisted of a vertical shank attached to the bottom of a hollow tine (Baker, 1976a, b). Figure 4.22 shows the original winged opener design. The opener was hollow, to allow passage of seed, and open at the back. The shank curved out at its base on both sides to form a pair of symmetrical wings, which were downwardly inclined towards their fronts by 10° and projected laterally approximately 20 mm either side.

A leading vertical flat disc was used ahead of the shank to provide a neat vertical cut through pasture. The leading disc was not expected to give the opener an ability to clear lying surface residue (see Chapter 10), but to ensure passage of the opener through standing pasture (sod) with minimal tearing and disruption to the soil surface.

A commercial company in New Zealand successfully adapted the winged opener concept for pasture renovation purposes. This market opportunity was based on

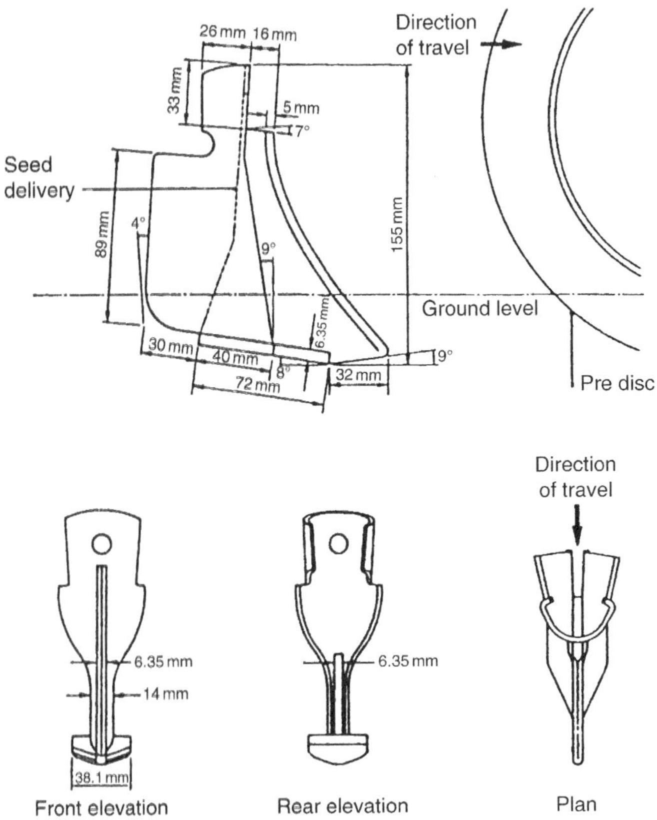

Fig. 4.22. The original inverted-T-shaped no-tillage opener (from Baker, 1976b).

knowledge that there is six times as much area of the world's surface under grazing land as there is under arable crop production (Kim, 1971; Brougham and Hodgson, 1992), although, of course, not all of the world's grasslands are accessible to tractors.

The design was simplified by fashioning the shank from plate steel and welding a vertical flat plate to its rear edge so as to entrap a wedge of soil ahead of this zone. Seed delivery was altered from the hollowed opener itself to a permanent tube positioned behind the vertical flat plate. The reasoning had been that, as the opener became worn and was eventually discarded, the modified design would allow the seed tube component to remain and only the minimum possible amount of replacement component would be discarded, thus reducing the cost. It was also reasoned that the soil wedge trapped ahead of the flat plate would reduce wear of the opener in that zone. Later, other designs also provided reversible and replaceable leading edges and tungsten overlays on the opener in an attempt to further reduce the effects of wear. Figure 4.23 illustrates a number of versions of the same modified opener, which eventually became known generically as the 'Baker boot', after the originator of the inverted-T principle.

Unfortunately, some of these benefits in the modified designs were achieved at the cost of retaining control over the exact shape of the slot. The thickness of the soil wedge that is retained by the vertical flat plate is a function of soil type, stickiness and moisture content. As a result, in sticky soils it is common for this soil wedge to become wider than the wings beneath it, with the result that the intended function for the wings to undercut the surface layer of soil is lost and the opener at times functions more as a wedge, creating a U-shaped slot.

Fig. 4.23. Several versions of the 'Baker boot' inverted-T-shaped no-tillage opener (from Baker *et al.*, 1996).

Although several manufacturers produced almost identical versions of the modified opener, not all of them provided a leading disc as originally envisaged, with the result that the slot edges were often torn and inconsistent, making controlled closure of the slot difficult, if not impossible. Since low cost was a primary objective with this simple opener, most designs attached the opener to very simple drills that had limited depth control (Fig. 4.24). One design provided a vertical pivot ahead of each opener to assist with cornering (Fig. 4.25).

Despite these shortcomings, the modified version of the simple inverted-T-shaped opener succeeded in its intended purpose of pasture renovation. Its principal advantage has been that the inverted-T-shaped slot, however poorly made, is demonstrably more tolerant of dry and wet soil conditions (see Chapters 6 and 7) than nearly all other opener designs, with the result that the success of the pasture renovation process improved noticeably.

The largest disadvantages of this opener have been that, by being a rigid shank, it has poor residue-handling qualities and speed of operation is limited. Where it is incorporated on simple drills, surface-following ability is limited.

Other designers have utilized the winged opener concept to separate the discharge of seed and fertilizer into two or more horizontal bands (double or triple shoot).

Fig. 4.24. A simple drill featuring 'Baker boot' inverted-T-shaped no-tillage openers (from Baker *et al.*, 1996).

Figure 4.26 shows a double-shoot winged opener.

Winged opener based on a central disc

Given the superior biological results obtained with the inverted-T-shaped slot concept from numerous experiments conducted in

Fig. 4.25. A simple drill featuring a self-steering version of the 'Baker boot' inverted-T-shaped no-tillage opener.

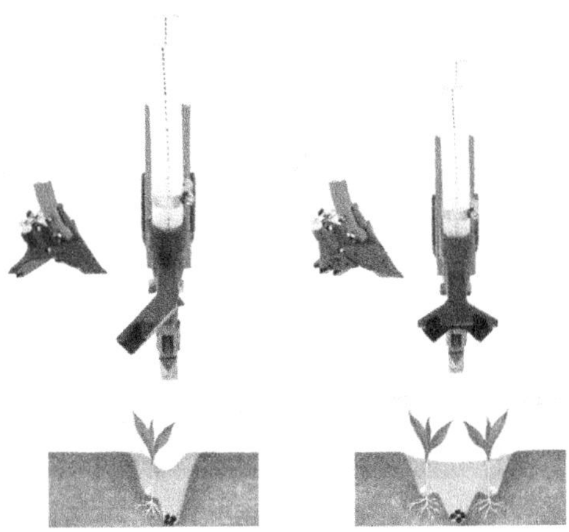

Fig. 4.26. Two versions of double-shoot winged openers.

New Zealand, Canada, Australia, Peru and the USA, it became imperative that the shortcomings of the simple opener should be overcome by designing a version that would suit arable agriculture as well as pasture-land. After all, it is the repeated tillage of arable land that has damaged the world's most productive soils. The potential of no-tillage to reverse this process is fundamental to the long-term sustainability of world food production.

A number of functional principles were considered essential if such an opener were to become fully capable of such an assignment:

1. The most important aspect was to maintain the inverted-T shape of the slot itself, even at high forward speeds and shallow seeding depths.
2. Ability to reposition loose residue on top of loose soil to cover the horizontal slot, as well as to fold back more structured, previously untilled material such as flaps of turf.
3. Effective separation of seed and fertilizer in the slot with a single opener, and to perform this function reliably over a wide range of soil type, moisture contents and forward speeds.
4. Handling without blockage of surface residue, even when configured in narrow (150 mm, 6 inch) rows, in difficult conditions ranging from dry or wet crop stubble to tangled, well-rooted sod, on soils ranging from soft and wet to hard and dry.

5. Self-closure of the slot without undue soil compaction for seedling emergence.
6. Capability to maintain a constant seeding depth by consistently following the ground surface.
7. Replacement parts to be inexpensive and easily removed and replaced in the field.

The resulting design, shown in Fig. 4.27, has working principles quite unlike other openers designed for either tilled or untilled soils (Baker *et al.*, 1979c). Essentially, the disc version of the winged opener arose from splitting the simpler winged opener both vertically and longitudinally and rubbing the insides of the leading edges of the two sides against a central disc. It is centred on a single flat vertical disc (smooth or notched) running straight ahead to cut the residue and the vertical portion of the soil slot. Two winged side blades are positioned so that the interior of their leading edges rubs on either side of the central disc. This patented principle effectively sheds residue from the side blades without blockage.

Fig. 4.27. A disc version of the winged opener for creating inverted-T slots.

The winged side blades cut horizontal slots on either side of the disc at seeding depth by partially lifting the soil. Seed and fertilizer flow down special channels between the side blades and the disc on either side, respectively, and are placed on the horizontal soil shelf. To achieve this, the side blades are held sufficiently clear of the disc at their rear edges to form a passageway for seed or fertilizer. Such a gap is narrow in comparison with other opener designs but movement of even large seeds is facilitated by the fact that one side of the passageway comprises the moving face of the revolving disc.

The fertilizer blade can be made slightly longer than the seed blade so that the fertilizer can be separated vertically from the seed as well as horizontally, i.e. diagonally, although in most circumstances horizontal separation has proved to be sufficient, if not preferable (Fick, 2000).

Two angled semi-pneumatic wheels follow the blades to reset the raised soil and residue, thereby positively closing the slot. They also regulate the depth of each opener independently for excellent soil surface tracking and thus precise seed depth control. Each opener is mounted on parallel arms necessary to maintain the shallow wing angle at seeding depth for tracking the soil surface.

Figure 4.28 is a diagrammatic representation of the horizontal seed and fertilizer separation (double shoot) with the disc version of a winged opener. Separating seed and fertilizer and sowing both with the same opener greatly simplify the design of no-tillage drills and reduce power demand. Fertilizer banding has become an essential function of successful no-tillage seeding for most crops (see Chapter 9). Few, if any, other no-tillage openers effectively and simultaneously achieve these important functions in a wide range of soils and at realistic forward speeds.

The opener is designed especially for no-tillage into heavy surface residues and grass sod where simultaneous sowing of seed and fertilizer is a priority. Because the incline on the wings is set at only 5° to the horizontal (compared with 10° for the

Fig. 4.28. Horizontal separation of seed and fertilizer by the disc version of a winged opener.

simple inverted-T version), it is capable of drilling at depths as shallow as 15 mm. It functions equally well, without modification, in heavy crop residues, pastures and sports turf (Ritchie, 1988), and can be used unmodified to sow the full range of field crops and pasture seeds, as well as for precision drilling of vegetables (Ritchie and Cox, 1981), maize and horticultural crops. It commonly retains 70–95%, of surface residues intact. Figure 4.29 shows 95% residue retention after passage of a disc-version inverted-T opener.

The main advantages of the disc version of the inverted-T-shaped opener are that it fulfils all of the design objectives listed above, without compromise. The same opener can be used unmodified for precision seeders, as well as grain drills and pasture renovation machines, in tilled and no-tillage farming.

Its disadvantages are that it has a slightly higher draught requirement, is relatively expensive to construct and requires a heavy drill frame design to ensure proper functioning. The relatively high cost can be weighed against its ability to maximize and even improve crop yields beyond those commonly experienced with other no-tillage openers and even tillage (Saxton and Baker, 1990). An apparent economic disadvantage when put in the fuller context becomes very cost-effective.

Fig. 4.29. Almost complete replacement of residues over the slot created by a disc version of an inverted-T-shaped opener (Class IV cover).

Punch Planting

Punch planters make discrete holes into which one or more seeds are placed before moving on to the next hole. Ancient farmers used pointed sticks to make the holes because there was insufficient energy to make continuous slots and utilize the convenience of continuous flow of seed and fertilizer into them.

Modern engineering has attempted to mechanize punch planting so that it can be performed with less human labour and with greater accuracy and speed. The devices created have mostly consisted of steel wheels with split spikes attached to their rims. The split spikes are hinged at their bases so that they can be forced to open in much the same way as a bird's beak. Figure 4.30 shows an example of a prototype mechanized punch planter.

In operation, the opening and closing functions are actuated by an internal cam and synchronized with a seed dispenser.

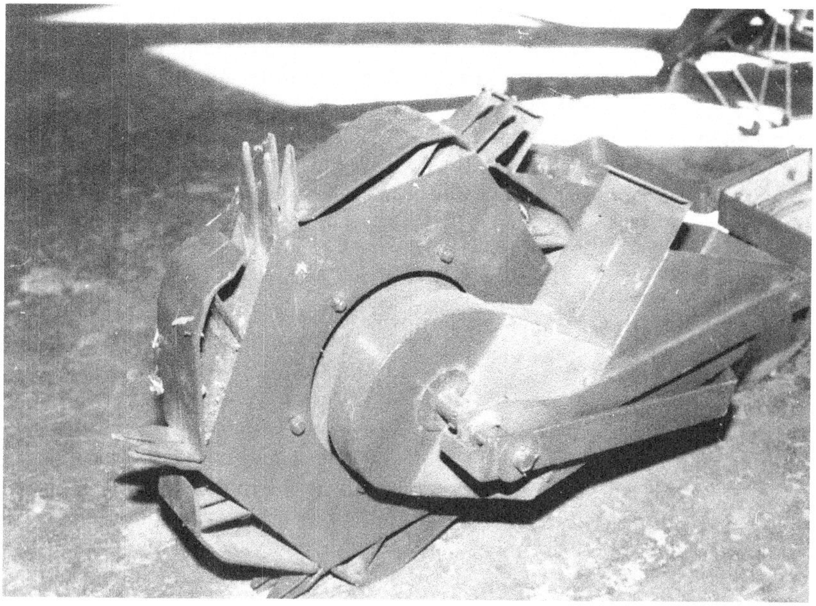

Fig. 4.30. A prototype mechanized punch planter (from Baker, 1981a, b).

After each spike has become fully embedded in the soil, a single seed or small group of seeds is directed from the dispenser tube, located in the centre of the wheel, through a hole in the rim of the wheel into the opened spike and deposited in the soil at a controlled depth and spacing from its neighbours.

Mechanized punch planters were seen as sensible solutions to mechanizing an ancient practice. Their relative mechanical complexity, however, has prevented their widespread adoption to date. The creation of V-shaped holes has all of the biological disadvantages of continuous V-shaped slots. This includes the tucking (hairpinning) of residues into the holes, difficulty in closing the holes and the wedging action of the spikes, which compacts soil under and alongside the seed zone.

Surface Broadcasting

There is little need to elaborate on the practice of surface broadcasting. Again, it is derived from an ancient practice brought about by the absence of energy sources for more mechanized solutions. Certainly, modern machinery is capable of mechanizing the broadcasting process with much increased speed and accuracy, but the absence of positive placement of seed beneath the soil significantly increases the biological risks from desiccation and bird, insect and rodent damage.

Broadcasting is not a recommended practice except in low-energy situations, and only then where local rainfall and humidity are so predictable and reliable that germination and rooting are assured. One thing in favour of no-tillage is that the retention of dead surface residues provides a protective canopy beneath which the humidity is likely to be higher than the surrounding ambient air (see Chapter 6). Research for many years has shown that effective seed and fertilizer placement beneath the soil produces crop yield advantages that surface broadcasting cannot duplicate.

One solution to broadcasting that reduces risk is 'auto-casting' in which seed is broadcast mechanically behind the pickup table of a combine harvester. The objective is to allow the straw and chaff to fall on top of the seed at the rear of the machine. This in turn ensures that there is a degree of cover over the seed, but success with this method is still very weather-dependent and there is no opportunity to strategically place fertilizers at the time of seeding. A dry period following harvest increases the risks of failure. Figure 4.31 shows an auto-casting system attached to the rear of a combine harvester table.

Summary of Seeding Openers and Slot Shape

The important functions of seeding openers are their abilities to:

1. Create a suitable seed/seedling micro-environment.
2. Avoid compacting and smearing of the slot walls.
3. Handle surface residues without blockage.
4. Micro-manage the surface residues so that they are positioned where they are of most advantage to the sown seeds/seedlings as well as the field in general.
5. Band seed and fertilizer simultaneously in the slot but separate them so as to avoid 'seed burn'.
6. Either avoid hairpinning altogether or avoid hairpinned residues affecting seed germination.
7. Self-close the slot.
8. Accurately control the depth of seeding.
9. Faithfully follow surface undulations that occur naturally in no-tillage.

The variety of slot shapes made by no-tillage seeding openers can be summarized as:

1. Vertical or horizontal.
2. Vertical slots are either V- or U-shaped.
3. Horizontal slots are usually inverted-T-shaped.
4. V- and U-shaped slots may also be slanted as well as vertical.
5. Compared with continuous slots, seeds can be sown in discrete holes (punch planting) or by surface broadcasting, mostly used where energy is limiting.
6. Most vertical and some slanted V- and U-shaped slots are adaptations of slots originally designed for tilled soils.
7. Most horizontal inverted-T-shaped slots were designed specifically for no-tillage seeding.
8. V-shaped slots are mainly created by double or triple disc openers.
9. U-shaped slots may be created by hoe, angled flat disc, angled dished disc, power till or furrow openers.
10. Inverted-T-shaped slots are created by winged openers.

Fig. 4.31. 'Auto-casting' of seed behind the pickup table of a combine harvester.

11. The practices of punch planting and broadcasting have ancient origins but have also been mechanized.
12. There are higher risks of poor plant establishment associated with surface broadcasting than where seed is sown beneath the soil by openers.

The action of openers on the soil varies by the opener design as:

1. Vertical double disc openers in the soil predominantly cut, wedge and compact.
2. Slanted double disc openers cut and heave on the uppermost side and compact on the lowermost side.
3. Punch planters wedge and compact.
4. Hoe openers mostly heave and burst, plus cut if preceded by a disc.
5. Power till openers cut, mix and pulverize.
6. Vertical angled flat disc openers cut, scuff and throw.
7. Angled dished disc openers and slanted angled flat disc openers cut, scuff, fold and/or throw.
8. Winged openers heave and fold, plus cut if associated with a disc.

The advantages and disadvantages of various openers by design are:

1. *Double and triple disc* openers are low-maintenance and have good residue handling. Their disadvantages are V-shaped slots, especially when configured vertically; unreliable seedling establishment; high penetration forces; compaction and smearing of soil; difficulty in covering; no separation of seed and fertilizer (unless doubled up); seed implantation into hairpinned residue.
2. *Punch planter* openers are low-energy and maintenance. Their disadvantages are mechanical complexity, slowness, hole compaction, difficulty in covering and no separation of seed and fertilizer.
3. *Hoe* openers are low-cost, no hairpinning of residue and reasonable penetration forces. Their disadvantages are poor residue handling, high wear rates, smearing in wet soils and no separation of seed and fertilizer unless doubled up.
4. *Power till* openers mix undecomposed organic matter with soil, do not hairpin residue, low penetration forces, burial of seed and dilution of fertilizer with soil. Their disadvantages are poor residue handling, residue destruction, tillage, slot-base compaction, difficulty in handling stones and sticky soils, cost, mechanical complexity, weed seed stimulation, high maintenance and no ability to separate seed and fertilizer.
5. *Vertical angled flat disc* openers have reasonable penetration forces, scuffing action, residue handling and no smearing or compaction. Their disadvantages are seeding into hairpinned residue, no separation of seed and fertilizer (unless doubled up) and affected by forward speed.
6. *Angled dished disc* openers and slanted angled flat disc openers have scuffing action, residue handling and no smearing or compaction. Their disadvantages are high penetration forces, seed implantation in hairpinned residue, no separation of seed and fertilizer (unless doubled up) and affected by forward speed.
7. *Simple winged* openers provide horizontal inverted-T-shaped slots that are easily closed, reliable seedling emergence, no compaction, reasonable penetration forces and do not hairpin residues. Their disadvantages are poor residue handling, high wear rates and no separation of seed and fertilizer.
8. *Disc versions of winged* openers (centred on a vertical disc) provide horizontal inverted-T-shaped slots, self-cover slots, reliable seedling emergence, horizontal or diagonal separation of seed and fertilizer, good residue handling and micromanagement, no seed implantation in hairpinned residue, capable of high forward speeds, low compaction, low weed seed stimulation, good depth control and low maintenance. Their disadvantages are high initial costs, high penetration forces, and high draught.

5 The Role of Slot Cover

C. John Baker

In no-tillage, nothing influences the reliability of seedling emergence more than the nature of the slot cover.

If you stand on the ground and look down on a seeded soil slot ('furrow' or 'groove'), after passage of a no-tillage drill or planter, you will see varying types of seed and slot coverage, which we have described in five 'classes' (Baker *et al.*, 1996):

1. Class I: visible seed (Fig. 5.1). Little or no loose soil covering the seed.
2. Class II: loose soil (Fig. 5.2). Loose soil and perhaps a small amount (less than 30%) of surface residue or mulch that has been induced back into the slot to cover the seed.
3. Class IIIa: intermittent mulch and soil (Fig. 5.3). There is a variable amount (30% or more) of residue or mulch on top of the loose soil covering the seed.

   Class IIIb: a mixture of residue and soil (Fig. 4.17). Thirty per cent or more of residues or mulch is mixed in with, rather than on top of, the loose soil covering the slot.
4. Class IV: complete mulch and soil (Figs 5.4 and 5.5). Soil and a covering of at least 70% of residue or mulch has been induced back over the slot in roughly the same layering positions as they were prior to drilling, i.e. with the mulch covering the soil, which in turn covers the seed.

The basis of these classifications was described by Baker (1976a, b, c) and Baker *et al.* (1996), who observed that, where an intermittent mulch/soil cover

Fig. 5.1. Visible seed in Class I no-tillage slot cover (from Baker *et al.*, 1996).

Fig. 5.2. An example of Class II no-tillage slot cover (from Baker *et al.*, 1996).

Fig. 5.3. An example of Class IIIa no-tillage slot cover (from Baker *et al.*, 1996).

(Class IIIa) occurred under dry conditions, seedlings were seen to emerge from under a flap of dead turf (mulch) or even a piece of random residue and soil, but had not emerged from where the seed cover was confined to loose soil alone or where there was no cover at all. This suggested that loose soil may not have been the ultimate seed cover as had been previously assumed.

Fig. 5.4. An example of Class IV no-tillage slot cover in heavy standing wheat stubble and scattered straw (from Baker *et al.*, 1996).

Fig. 5.5. An example of Class IV no-tillage slot cover in sparse close-growing weeds. Note the replacement and layering of whatever residue is available in its original position and the absence of soil inversion after passage of the drill. (From Baker *et al.*, 1996.)

In fact, some engineers and agronomists continue to mistakenly assume, even today, that the best cover for seeds is loose soil (Class II). This assumption comes from what has been provided in a tilled seedbed for centuries. Residues do not exist to any degree on well-tilled soils. Generally, they have been buried or burnt prior to tillage.

The only other resource available for covering in addition to clean, loose soil is perhaps a press-wheel effect to provide slightly compacted soil, but even the benefits of that are dubious. So loose soil has been regarded as the 'ultimate cover', at least in a tilled soil.

Based on the 'loose-soil-is-best' assumption, some engineers therefore postulated that all that was needed for no-tillage was to till the soil in a series of strips and sow seed into the tilled strips as you would in a generally tilled soil, but, in this case, leaving the rest of the seedbed untilled between the strips. This is one form of strip (or zone) tillage, which has been described previously in Chapter 4.

Unfortunately, this simplistic view has no scientific basis and it is now known that it destroys several of the very special resources close to the seed that most untilled soils have, such as a mulch covering, an unbroken macropore system within the seed zone and an equilibrium soil humidity near 100%.

The Role of Soil Humidity

The atmosphere in the macropores within an untilled residue-covered soil has an equilibrium humidity of very near 100% (Scotter, 1976) at almost all moisture levels down to 'permanent wilting point', which is when a soil is too dry for plants to survive. In fact, it is 99.8% even at wilting point (1500 kPa tension). In no-tilled seeding, the soil is only broken at the surface by strips (slots) where the drill or planter openers have travelled. The greatest loss of humidity from the soil to the atmosphere occurs at these broken strips (slots). The aim, therefore, of drilling into dry soils should be to create slots that do not encourage loss of humidity from these zones, since they are also the zones where the seeds are placed, which require moisture to initiate plant growth.

The classification of covers listed above is arranged in order of ascending humidity retention. A 'complete' (70% or greater) mulch/soil cover (Class IV) retains more humidity than an intermittent mulch and soil cover (30 to 70% residue – Class III), which is better than loose soil (less than 30% – Class II), which itself is better than no cover at all (Class I).

Choudhary (1979) and Choudhary and Baker (1981b) measured the daily loss of relative humidity (RH) from a range of different slot shapes under controlled dry conditions with constant temperature. They used the average daily RH loss for the first 3 days following seeding to compute an index value for the ability of a slot to retain humidity, moisture vapour potential captivity (MVPC).

$$\text{MVPC} = 1/(\text{average 3-day RH\% loss})$$

Table 5.1 lists results from two separate experiments in which Choudhary placed a small humidity probe in positions that would normally be occupied by the seeds within drilled slots in a dry soil. Undisturbed soil bins (weighing 0.5 t each) were placed within climate-controlled rooms at a

Table 5.1. Effect of no-tillage slot shape and cover on slot drying rates and MVPC.

	V-shaped slot (Class I cover)		U-shaped slot (Class II cover)		Inverted-T-shaped slot (Class IV cover)	
	Daily loss of RH%	MVPC	Daily loss of RH%	MVPC	Daily loss of RH%	MVPC
Experiment 1	4.23%	0.24	2.78%	0.36	2.34%	0.43
Experiment 2	3.13%	0.32	2.03%	0.49	1.02%	0.98
Mean	3.68%	0.28	2.41%	0.43	1.68%	0.71

MVPC, moisture vapour potential captivity = 1/(average 3-day RH% loss).

constant ambient temperature and constant RH of 60%.

Relative humidity is a measure of the amount of water vapour in the soil atmosphere at any one temperature. The source of supply of water vapour in the drilled slots is from the surrounding soil since its equilibrium relative humidity is always near 100%, but the rate of escape of water vapour to the atmosphere outside the soil (which is usually less than 100% RH unless it is raining or there is thick fog) is controlled by the diffusion-resistance to gases passing through the covering medium in or on the slot. For at least a few days after drilling, the soil temperatures (even in the slot) can be expected to remain at reasonably constant levels (Baker, 1976a). Therefore, measurements of relative humidity in the slots at these constant temperatures closely reflect the amount of water vapour (or the water vapour pressure) in the slot at the time.

The higher MVPC values (or lower daily losses of RH%) for Class IV covers indicate that such a slot had a higher potential to retain in-slot water vapour than Class II cover, for example, which itself had a higher water vapour retention and lower daily loss of RH% than Class I cover. The Class IV cover in these experiments was, in fact, 65% better than Class II and 154% better than Class I in retaining in-slot humidity. No Class III cover was included in this experiment.

The effects of moisture transfer from the slot micro-environments was also studied by varying the overlying air humidity at a constant temperature (Choudhary, 1979; Choudhary and Baker, 1980, 1981b). The humidity within the slots increased differently as the ambient RH was raised from 60% to 90%. Those slot shapes that increased most rapidly with a rise in ambient humidity will obviously decrease (dry) most quickly after sowing and be less favourable to seed germination and plant establishment. The most rapid change was in the open V-shaped slots (Class I cover), which increased at the rate of 8% RH per day, followed by the U-shaped slot (Class II cover), followed by the inverted-T-shaped slot (Class IV cover), which increased by only 1% RH per day.

For the inverted-T-shaped slot (Class IV cover), the rate of re-moistening was about the same as its rate of drying (i.e. approximately 1% RH per day), but for the V-shaped slot (Class I cover) the rate of re-moistening was about twice that of its drying. This confirmed that Class I cover had done little to isolate the slot micro-environment from changing ambient conditions, while Class IV cover had effectively isolated the slot from such climatic changes and retained a highly humid slot atmosphere throughout.

From a practical point of view, if seeds are sown into a favourable soil and the following week is dominated by hot dry winds, a slot that might have presented an ideal habitat for the seeds at the time of sowing can soon turn into a hostile environment unless the slot is protected from such climatic changes by adequate slot cover. Choudhary and Baker (1982) showed that no-tillage slots with Class IV cover allowed seed germination and seedling emergence from soils that were otherwise too dry to germinate seeds sown by either conventional tillage or with other no-tillage openers and slots.

A field experiment in Manawatu, New Zealand, before Class IV cover had been fully evaluated (Baker, 1976a, c) illustrated that loose soil (Classes II and III cover) is much better than no cover at all (Class I cover). In this experiment, a barley (*Hordeum vulgare*) crop was sown in late spring using hoe openers (U-shaped slot) in a silt loam soil with adequate moisture. One half of the sown rows was covered by pulling a bar harrow over the slots (Class IIIa cover) and the other half was left as the drill had created the slots (essentially uncovered, Class I). The period after drilling was hot, dry and windy. Eight days after sowing the Class IIIa covering had 205 plants/square metre, compared with the Class I cover, which had only 22 plants per square metre.

An experiment conducted at the same time and in the same soil showed that increased seed size did not compensate for poor covering. Where larger seeds might have been expected to have more vigour and therefore be able to compensate for emergence difficulties, the opposite seemed to have happened under no-tillage. In this experiment a small-seeded species, lucerne (*Medicago sativa*), and a large-seeded species, maize (*Zea mays*), were substituted for barley and no-tilled in exactly the same manner. After 10 days, the small-seeded lucerne had 118 plants per m^2 under Class IIIa cover and 87 under the Class I cover. After a similar length of time the maize had 4.6 and 0.3 plants per m^2, respectively, for the two classes of cover.

While Class IIIa cover still increased seedling emergence with both the larger and smaller seeds, the increase was less with lucerne than with either maize or barley. The smaller lucerne seeds apparently had a better chance of finding themselves covered with a small piece of soil or mulch, which produced a favourable micro-environment for them, even in a Class I situation, than did the larger barley seeds, which were better placed than the even larger maize seeds in this respect.

A few days after the measurements of this experiment, rain ensured that all seeds germinated in all three of the experiments and the differences between treatments disappeared. Thus, the effects of cover were only important when the soil was dry or drying, although, as described in Chapter 7, cover is also important in wet conditions for other reasons.

As further evidence of the importance of cover in both wet and dry soils, Table 5.2 summarizes the 'best' and 'worst' treatments of 30 experiments conducted in New Zealand between 1971 and 1985. Each experiment, amongst other things, compared the effects of different openers and classes of cover under different soil moisture conditions on seedling emergence of a range of crops (Baker, 1979, 1994).

There are several clear trends to be seen in the Table 5.2 data, and the experiments are grouped accordingly. The first is a tendency towards improving seedling emergence with Classes III and IV covers, where surface residues were present and the soils were either very dry (experiments 1–12) or very wet (experiments 25–30). As the moisture conditions became more optimal (experiments 13–18) and/or when surface residues were not present (experiments 19–24), the difference between the classes of cover generally became less or non-existent.

Perhaps just as important was the magnitude of some of the differences. Two- to 14-fold differences are rare in agricultural experimentation, suggesting that slot shape and cover have a major influence on the reliability and success of no-tillage practices, a fact not formerly recognized or reported. Even a ratio of 1.2 : 1 represents a 20% advantage for the 'best' treatment.

It is also notable that, where Classes I and II covers were included in the comparisons, they were almost invariably classed either as the 'worst' treatment or as 'no better than' the other treatments. They seldom outperformed any other treatment, the exceptions being in two very wet soils without residue, where seedling emergence was low with all of the openers compared. On the other hand, Classes III and IV cover were never bettered by any other treatment in the presence of surface residues in wet, optimum or dry soils.

The Table 5.2 data include only the 'best' and 'worst' treatments for simplicity. Comparisons of other intermediate treatments between these two extremes are not shown. Almost invariably, however, Class IV cover produced greater seedling emergence than Class III cover, which in turn outperformed Class II cover, especially in dry conditions. More detailed descriptions of these comparisons are given in Chapters 6 and 7.

Methods of Covering Seed Slots

There are several principles involved in covering slots after the passage of no-tillage openers, and these are often combined with pressing to obtain soil–seed contact. These methods are:

1. Squeezing – attempting to move soil sideways into the slot by a wedging action to cover and to obtain soil–seed contact.

Table 5.2. Effects of slot cover on seedling emergence in 30 experiments.

	Year	Soil[a]	Crop	Soil moisture and residue status[b]	Best and worst treatments and classes of cover (best) : (worst)[c]	Ratio of seedling emergence counts (best) : (worst)
1	1979	S/L	Wheat	V. dry	(R) inv. T/C (IV) : t.d. V/C (I)	14 : 1
2	1971	S/L	Maize	Dry	(R) hoe U/C (III) : hoe U (I)	14 : 1
3	1971	S/L	Barley	V. dry	(R) hoe U/C (III) : hoe U (I)	9.5 : 1
4	1972	S/L	Barley	V. dry	(R) inv. T/C (IV) : hoe U/C (II)	6 : 1
5	1979	FS/L	Wheat	V. dry	(R) inv. T/C (IV) : t.d. V/C (I)	5.5 : 1
6	1976	FS/L	Wheat	Dry	(R) inv. T/C (IV) : t.d. V/C (I)	3 : 1
7	1971	S/L	Kale	Dry	(R) hoe U/C (III) : hoe U (I)	2 : 1
8	1979	S/L	Wheat	V. dry	(R) inv. T/C (IV) : t.d. V/C (I)	1.7 : 1
9	1979	FS/L	Wheat	Adeq.	(R) inv. T/C (IV) : t.d. V/C (I)	1.6 : 1
10	1979	S/L	Lucerne	V. dry	(R) hoe U/C (III) : hoe U (I)	1.4 : 1
11	1979	S/L	Wheat	V. dry	(R) inv. T/C (IV) : t.d. V/C (I)	1.3 : 1
12	1979	S/L	Wheat	Dry	(R) inv. T/C (IV) : t.d. V/C (I)	1.2 : 1
13	1978	S/L	Wheat	Adeq.	(R) inv. T/C (IV) : t.d. V/C (I)	no diff.
14	1978	S/L	Lupin	Adeq.	(R) inv. T/C (IV) : t.d. V/C (I)	no diff.
15	1979	S/L	Wheat	Adeq.	(R) inv. T/C (IV) : t.d. VN (I)	no diff.
16	1979	S/L	Wheat	Dry	(R) inv. T/C (IV) : t.d. V/C (I)	no diff.
17	1979	S/L	Wheat	Adeq.	(R) inv. T/C (IV) : t.d. V/C (I)	no diff.
18	1979	S/L	Wheat	Adeq.	(R) inv. T/C (IV) : t.d. V/C (I)	no diff.
19	1985	S/L	Barley	Adeq.	(NR) inv. T/C (IV) : t.d. V/C (I)	no diff.
20	1985	S/L	Barley	Adeq.	(NR) inv. T/C (IV) : t.d. V/C (I)	no diff.
21	1985	S/L	Barley	V. wet	(NR) p.t. U/C (III) : p.p. U/C (I)	4.2 : 1
22	1985	S/L	Barley	V. wet	(NR) inv. T/C (IV) : t.d. V/C (I)	1.7 : 1
23	1985	S/L	Barley	V. wet	(NR) t.d. V/C (I) : inv. T/C (IV)	1.6 : 1
24	1985	S/L	Barley	V. wet	(NR) t.d. V/C (I) : inv. T/C (IV)	1.2 : 1
25	1985	S/L	Barley	V. wet	(R) inv. T/C (IV) : t.d. V/C (I)	4.4 : 1
26	1985	S/L	Barley	V. wet	(R) inv. T/C (IV) : t.d. V/C (I)	2.9 : 1
27	1985	S/L	Barley	V. wet	(R) inv. T/C (IV) : t.d. V/C (I)	2.7 : 1
28	1985	S/L	Barley	V. wet	(R) inv. T/C (IV) : t.d. V/C (I)	2.5 : 1
29	1985	S/L	Barley	V. wet	(R) inv. T/C (IV) : t.d. V/C (I)	1.5 : 1
30	1985	S/L	Barley	V. wet	(R) inv. T/C (IV) : t.d. V/C (I)	1.4 : 1

[a]Soil types: S/L = silt loam; FS/L = fine sandy loam.
[b]Soil moisture and residue status: V. dry = Very dry; Adeq. = Adequate: V. wet = Very wet.
[c](R) = surface residues present; (NR) = no surface residues present; (I), (II), (III) and (IV) = the classes of cover in each experiment. Drilling and covering treatments: t.d. V = triple disc opener, vertical V-shaped slot, not covered; t.d. V/C = triple disc opener, vertical V-shaped slot, covered; hoe U = hoe opener, U-shaped slot, not covered; hoe U/C = hoe opener, U-shaped slot, covered; inv. T = winged opener, inverted-T-shaped slot, not covered; inv. T/C = winged opener, inverted-T-shaped slot, covered; p.t. U = power till opener, U-shaped slot, not covered; p.t. U/C = power till opener, U-shaped slot, covered; p.p. U = simulated punch planter, U-shaped holes, not covered; p.p. U/C = simulated punch planter, U-shaped holes, covered.
Sources: Experiments 1, 5, 8, 9, 11, 12, 15, 16, 17 and 18 (Choudhary, 1979); Experiments 2, 3, 4 and 10 (Baker, 1976a); Experiment 6 (Baker, 1976b); Experiment 7 (Baker, 1971), Experiments 13 and 14 (Mai, 1978); Experiments 19, 20, 21, 22, 23, 24, 25, 26, 27, 28, 29 and 30 (Chaudhry, 1985).
Note: In all experiments where the slots were covered, the covering material was the best available as provided by the shape of the slot and opener action.

2. Rolling – pressing vertically on the soil alongside the slot with a roller of some description.
3. Pressing – selectively pressing on or in the slot zone itself, including non-vertical rolling or pressing mainly to obtain seed–soil contact, but can also include an element of covering.
4. Scuffing – scraping up loose surface material from the slot zone and directing it to fall back into the slot, solely for covering.
5. Deflecting – discretely deflecting soil from a particular part of the slot, solely for covering.
6. Tilling – loosening the ground behind the opener, usually so that it can be more easily manipulated by one of the other devices previously listed.
7. Folding – folding soil and/or residue back from whence it came, solely for covering.

Often two or more of these actions are combined in one covering/pressing device or system.

To a casual observer, there might not seem to be much difference between the various actions described above. However, a description of the advantages and disadvantages of each principle will illustrate why cover and, to a lesser extent, pressing are such an important factor in reducing the risks associated with no-tillage.

Squeezing

Squeezing is the principle applied by many manufacturers of vertical double disc openers (see Chapter 4). It usually involves pressing down with a V-shaped wheel alongside the slot after its formation in such a manner that the mass of soil is pushed bodily sideways without actually loosening it. The aim is to squeeze the slot closed by moving the soil back from whence it came. Figure 4.7 illustrates squeezing wheels behind double disc openers. The advantages are that such wheels are simple, require little adjustment and are not inclined to block with residue.

The disadvantages are that there is almost as much downforce required on the pressing wheels as was needed on the opener to create the slot in the first place, adding to the weight requirements of the drill; the pressing action further compacts the soil next to the seed; its ability to close the slot is highly dependent on soil plasticity and moisture content; any useful effect may be undone quickly if the soil dries and shrinks after pressing. Slots made in soils that do not squeeze easily might not be adequately closed, although with soils of this nature there is little else that can be done to remedy the situation. With soils in which the slot can be squeezed back together, there is a risk of so tightly trapping the seeds with compacted soil that emergence of seedling shoots is restricted.

Rolling

General rolling of a field after drilling is often undertaken in an attempt to produce some of the squeezing action described above in a random manner, without directing the action to any specific zone. It works best where slot formation results in considerable hinged upheaval of the soil such as with hoe openers and some simple inverted-T-shaped openers. The vertical forces from the roller tend to squash any raised ridges of soil downwards and, to a limited extent, sideways. Since most of the raised portions of soil will be alongside the slots, a degree of covering often results, although, as with squeezing, the final result is highly dependent on soil moisture content and plasticity.

Both flat and ringed ('Cambridge') rollers are used. The problem with ringed rollers is that the points of the rings apply more pressure than the shoulders. If the point of a ring happens to coincide with the centre of a sown row it may help to bury the seed too deeply or at least it may seal the exit zone so tightly as to restrict seedling emergence. For these reasons flat rollers are preferred to 'Cambridge' rollers.

The main advantage of rollers is that they are generally readily available implements and easy to use, and their downforces are derived from their own weight rather than the drill. They also leave a relatively flat finish to the field, which might be important at harvesting.

The disadvantages are that covering must be done as a separate operation and that much of the loose soil and debris is not adequately moved sideways into the slot zone but is instead 'trampled' down where it lies, in which case it might not contribute to covering at all. This latter disadvantage is more of a problem with hoe openers than with simple inverted-T-shaped openers, because the latter hinge up a flap of soil rather than bursting it out bodily sideways in the manner of hoe openers.

Pressing

Pressing is really rolling in a discrete zone and perhaps at a discrete angle in or on top of the slot. The slot can be pressed either after it has been covered by some other means (e.g. scuffing) or prior to the covering action. The object of pressing alone is to effect the covering action and it is particularly useful with slanted double disc openers. Pressing in association with another covering device improves soil–seed contact, but there is little scientific evidence to show that this results in an improvement in seedling emergence under no-tillage except perhaps by improving the consistency of seeding depth (Choudhary, 1979; Choudhary and Baker, 1981a).

Pressing before covering, on the other hand, has been shown to be of major benefit with some openers such as hoe and vertical double discs. Few manufacturers, however, have seen fit to provide press devices that act on the seed before covering of the slot. Figure 5.6 illustrates a ribbed press wheel designed to press in the base of the slot while simultaneously rolling on the undisturbed soil alongside. Figure 5.7 shows a packing device designed to firm the seed into the base of the slot at the same time that covering takes place.

The advantages of pressing are that it usually involves a wheel (or pair of wheels) that can double as a depth-control device. This double function, however, is not easy to achieve if the press wheel operates in the base of a slot, since the wheel then registers on a soil surface that has already been created by the opener and thus may have little reference to the true surface of the soil. On the other hand, pressing before covering does more to counteract the disadvantages

Fig. 5.6. A press wheel with central rib, which is designed to press in the base of the slot at the same time as it locates on the soil surface (from Baker *et al.*, 1996).

Fig. 5.7. A shank-type opener with packing device to firm seeds into the base of the slot at the same time as covering takes place.

of U- and vertical V-shaped slots than any other known method (Choudhary, 1979; Choudhary and Baker, 1981a). The effect seems to be to press the seeds into the undisturbed soil at the base of the slot so that their emerging roots do not need to negotiate the slot wall in order access soil water.

The disadvantages are that pressing alone is not always a covering action at all. It is usually done after or before covering is achieved by some other means, so two separate mechanisms are necessary. Also, because pressing after covering is easier to achieve and the press wheels are able to roll on the undisturbed soil alongside the slot and thereby achieve depth control at the same time, this has become the preferred option. It does not, however, achieve as much biologically as pressing before covering (see also Chapter 6).

Scuffing

Scuffing is probably the easiest and most effective general slot covering option that can be performed by a separate machine after drilling, regardless of the type of drill opener used. It usually involves a heavy, wide, flexible harrow of some nature, which is pulled across the ground, preferably parallel to the drill slots. The harrow scrapes up the general loose soil spilled from the slots and other debris, and pushes this material back over the slots in a random manner. Its action depends on the untilled ground between the rows being able to support the weight of the device so that it does not cut into the soil and thereby accumulate excess soil and debris. Some of the heavy harrows used in no-tillage are therefore not applicable to tilled soils.

Various harrows have been used, ranging from chain harrows with the points facing upwards to avoid gouging seed out of the slots, truck tyres that have been split longitudinally with the cut surfaces facing downwards, oyster nets, heavy chains and short lengths of railway iron chained together. Figure 5.8 shows a bar harrow made of railway iron operating in a friable soil after a drill with hoe openers (Baker, 1970). Figure 5.9 is a plan of such a harrow, suitable for a 2.4 m wide drill.

The advantages of harrows are that they are virtually foolproof to operate, simple and inexpensive. For many slot shapes created in damp soils, harrowing is best delayed a few hours to allow some dry crumbs to develop, which can then be scraped up as friable covering material. A separate harrow is ideal for such situations.

The disadvantages are that if no crumb is formed when drilling, for example, with vertical double disc openers operating in a damp soil, even harrows will be ineffective to provide cover. Their use constitutes another operation, although, if a time delay is not appropriate, they can be attached behind the drill; and with severe residue they can become blocked.

A variation of scuffing and rolling is provided by spiral-caged rollers, as shown in Fig. 5.10. These devices combine the pressing effect of a roller with the scuffing effect of a harrow, since the spiral nature of the rolling ribs ensures that some sideways scuffing takes place as the roller rotates. They are easy and convenient to use but do not move as much debris and soil as a true harrow.

Deflecting

With some hoe openers, small deflecting devices are incorporated on the rear of the

Fig. 5.8. A simple bar harrow for covering no-tillage slots (from Baker *et al.*, 1996).

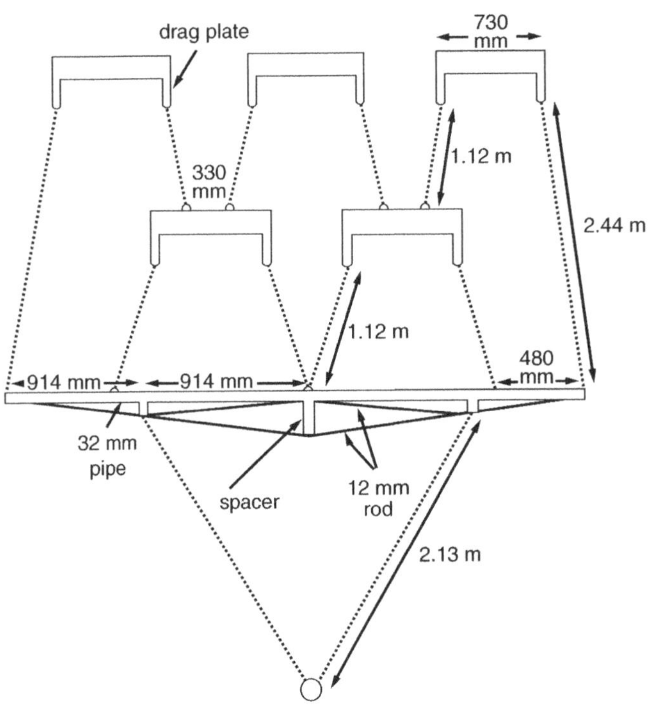

Fig. 5.9. Plan for a simple bar covering harrow (from Baker, 1970).

opener so as to scrape a small slice of soil from the slot wall and allow it to fall on to the seed and/or fertilizer. One of the purposes of doing this has been to attempt to get a soil covering over a deposit of fertilizer in the base of the slot before seed is deposited on top of the soil, thus separating them vertically within the slot (Hyde *et al.*, 1979, 1987).

Fig. 5.10. A spiral-caged roller for covering no-tillage slots.

Unfortunately, the function of any fixed device, such as an internal scraper of this nature, is highly dependent on the position of the scraper relative to the slot walls. Since the slot walls themselves are never in exactly the same place in two different soils, or even in the same soil at different moisture contents or operating speeds, either the scrapers have to be manually adjusted for each new soil condition or the functional ability of the device will vary quite widely with the conditions. While successful deflectors facilitate vertical separation of seed and fertilizer in the slot, stationary scrapers often collect residue and cause blockages.

Tilling

Because of the difficulty of moving soil that has been squeezed sideways back in the opposite direction, some openers attempt to loosen the soil alongside the slot with the aid of spiked wheels or discs. Often, spiked discs are arranged alongside angled press wheels so that the loosening and reverse-squeezing actions are combined into one, such as those shown in Fig. 5.11.

The advantages are that the soil is more easily moved, and, because it is in a loosened state, the risk of further compaction, particularly over the seedling emergence zone, is reduced. The disadvantage is that any disturbance of this nature partly destroys the integrity of the residue and soil layering, and at best results in a random mixture of soil and residue as the covering medium.

Folding

Folding of material back over a slot presupposes that a horizontal slot has been created in a manner that hinged the original covering material up in the first place. Alternatively, the slot may have been created so that the original covering material has been displaced bodily sideways without inversion and mixing, in a manner that allows it to be retrieved and replaced as if it had not been moved in the first place.

Realistically, this applies only to inverted-T-shaped horizontal slots, slanted

Fig. 5.11. A pair of combined spiked discs and angled press wheels for covering no-tillage slots (from Baker *et al.*, 1996).

double disc openers and perhaps those angled dished disc openers that have a positive tilt angle. Even with inverted-T openers, the folding feature is more a function of how the slot is created than the action of the covering device. For example, the uplifted flaps of most inverted-T-shaped slots, when created in pasture, can be folded down again either by a scuffing harrow or by press wheels. Press wheels are more tolerant of different soil and pasture conditions, and are more predictable than scuffing harrows, but they need to be angled to combine the folding and pressing functions.

In non-pasture soils such as arable soils with loose or lying residue, the folding function can only be realistically performed by press wheels. It is even possible to refine the folding function sufficiently to allow stratified soil layers, e.g. a thin dry dust mulch that overlies more moist soil, to be replaced more or less in the same order that they were in before passage of the opener. Figure 4.27 and 4.29 show a pair of folding wheels, which also function as depth-gauging wheels, on a disc version of a winged opener.

The advantages of folding are that the covering function is predictable and reliable and usually does not require adjustment of opener components to cope with different soil or residue conditions. It can also result in complete mulch and soil cover (Class IV), so long as there was a mulch covering the soil in the first place.

The disadvantages are that excess pressure from press wheels on a damp pasture flap might close the slot so tightly as to make it difficult for seedlings to emerge. Since this is a function of the downforce applied to the openers, it is easily adjusted in the normal course of setting up a no-tillage drill.

Summary of the Role of Slot Cover

1. There are four distinguishable classes of slot covers, ranging from no cover (Class I), loose soil (Class II), soil and a small amount of mulch or residue (Class III), to complete (greater than 70%) soil and mulch (Class IV).

2. In Class III, the small amount of mulch or residue in the covering medium may be either in intermittent clumps (Class IIIa) or a thoroughly mixed combination of residue and soil (Class IIIb).

3. Class I–IV covers are ranked in ascending order of their abilities to retain slot water vapour.

4. The benefits of covering in terms of seedling emergence are ranked in ascending order of Classes I–IV.

5. Principles of covering slots and/or obtaining soil–seed contact involve squeezing, rolling, pressing, scuffing, deflecting and/or folding soil and/or mulch.

6. Some covering methods involve separate operations and machines that are used after drilling, in which case the weather and soil plasticity after seeding become important.

7. Other covering methods involve simultaneous functions by the openers themselves, in which case the nature and speed of slot formation become important.

8. Vertical double disc and triple disc furrow openers and punch planters usually produce Class I or II cover.

9. Slanted double and single disc openers and winged openers are capable of producing Class IV cover.

10. Hoe, angled vertical flat disc and angled vertical dished disc openers tend to produce Class II or IIIa cover, depending on the speed of travel.

11. Power till openers tend to produce Class IIIb cover, regardless of speed.

12. Angled dished disc openers sometimes produce Class IV cover at slow speeds.

13. The disc versions of winged openers are designed to produce Class IV cover regardless of speed, soil moisture conditions or residue conditions.

6 Drilling into Dry Soils

C. John Baker

A dry untilled soil has more potential to germinate seeds and allow seedlings to emerge than a dry tilled soil; but very few no-tillage openers are capable of harnessing that potential.

Most of the world's agriculture involves growing plants in soils that become dry at some point in their growing cycles. If farmers could predict exactly when the soil was going to become dry, they would plan accordingly. In many climates an approximate idea of the onset of rain allows farmers to match the planting of crops to expected weather patterns. These matchups, however, are seldom accurate to better than a few weeks, if that.

When sowing seeds into untilled soils, a matter of a few days either way may make the difference between successful crop establishment or failure. This is not to say that untilled soils are less forgiving than tilled soils; indeed, most have the potential to be more forgiving. The problem is that most people have not yet learned how to harness that tolerance to their advantage.

With little guarantee that it will rain on a particular day after drilling, farmers are unlikely to attempt to drill seed into an already dry soil. On the other hand, if a farmer drills seed into a soil that appears to have adequate moisture but then finds the next week dominated by hot dry winds, what had been an optimum environment for seeds may soon become a hostile environment.

None the less, so long as there is sufficient weight for penetration of the drill openers and sufficient energy to pull the machine through the soil, it is possible to operate a no-tillage drill in a dry soil. This contrasts with wet soils (see Chapter 7), where operation of machinery is often simply not possible.

How Soils Lose Moisture

To understand the tolerance of untilled soils to dry weather, it is necessary to distinguish between an untilled soil that is covered with a mulch and an untilled soil that has a bare surface. It is also important to compare the ways in which tilled and untilled soils transport water to the surface for evaporation.

A tilled soil will lose moisture more rapidly than an untilled soil, at least initially. But because of the increased porosity of tilled soils, the loss of moisture from the upper zones will not be quickly replenished from deeper zones. The capillary rise of water is poor through the large voids and pores that result from tillage.

 No-tillage Seeding and Conservation Agriculture, 2nd edn. (eds C.J. Baker and K.E. Saxton)

Because of this, a dry layer may be formed at the top of tilled soils. In some climates a dry dust mulch layer is deliberately formed by repeatedly tilling the surface layer of soil until it becomes a super-dry dust with very low moisture and thermal conductivities. The rationale behind such a practice is that, in the absence of any other form of surface mulch, there is a net saving in moisture loss by sacrificing a small amount of water to form a 'dust mulch' in the interest of conserving the greater amount of water lying beneath it.

An untilled soil, on the other hand, will usually have a well-developed capillary system from the surface to some significant depth, which acts as a continuous 'wick', transporting water upwards during periods of drying at the surface. This internal transport system will become more effective with time as soil structure improves. Thus, while the initial loss of moisture will be slower from the surface of a bare untilled soil than from a tilled soil because the surface is smoother and therefore does not create as much air turbulence or allow air to enter as easily, it may continue supplying water to the surface for evaporation for a much longer time than a tilled soil that is covered with a dust mulch. This, then, is where the presence of an organic residue mulch and the action of the drill openers that operate in an untilled soil become important.

The Role of Vapour-phase Soil Water

All soils contain both liquid-phase water and vapour-phase water in the form of humidity. The equilibrium relative humidity of the pore spaces between the particles of undisturbed soil is virtually 100% at all liquid moisture levels down to permanent wilting point (Scotter, 1976). The permanent wilting point (PWP) is the point where the soil is considered too dry to sustain most plant life. The status of liquid soil water is often expressed as the tension by which water films are held by the soil particles. At PWP this tension is –15 bar. The important point is that plants wilt and die at PWP and will not recover if watered again. However, it is important to remember that, even at that moisture content, the soil macropores contain 99.8% relative humidity.

Like hair on the skin of an animal, an organic mulch traps a layer of still air close to the soil surface, which slows down the exchange of water vapour between the soil and the atmosphere. Most importantly, the humidity within that mulch layer will remain much higher than the atmosphere above it, unless, of course, it is raining or the atmosphere is at a high humidity anyway.

On a hot, dry day, for example, if one were to take a rapid-response humidity probe and carefully slide the probe under a single large leaf lying on bare untilled soil without moving the leaf, there would be a noticeable rise in the humidity reading as the probe moved under the leaf and then a drop when it was removed. The same thing would happen under a piece of plastic or paper. This demonstrates that a localized high-humidity zone is possible under a mulch at the soil surface. This mulch zone can be quite small in area and unaffected by another un-mulched zone nearby that has a much lower humidity. This is a very important phenomenon and is one of the major differences between no-tillage openers.

Every farmer in the world can recognize whether or not a tilled soil has sufficient liquid-phase water for germination. The judgement is usually made on the basis of the colour of the soil – darker-coloured soil is wetter – or the temperature of the soil – colder soil is wetter.

Soil humidity is rarely accounted for in a tilled soil. Nor should it be. Unless the soil humidity is at least 90%, germination will mainly occur through uptake (imbibition) of liquid-phase water from the soil by the seed (Martin and Thrailkill, 1993; Wuest, 2002). The humidity in the surface layers of a tilled soil is likely to approach 90% only on a very humid day or immediately after rain. As will be explained below, the humidity in the drilled slot of an untilled soil is even more important than in the general soil matrix (Choudhary, 1979; Choudhary and Baker, 1981a, b).

Response / Opener	Triple (double) Disc V	Hoe U	Winged Opener ⊥
Germination	✓	X	✓
Subsurface seedling survival	XX	✓	✓✓
Seedling emergence	✓XX	✓X	✓✓✓
Loss of humidity from slot	High	Medium	Low
Covering the seed	✓X	✓✓	✓
Pressing on cover	X	X	X
Pressing on seed before covering	✓	✓✓	X

Fig. 6.1. Summary of the responses of contrasting no-tillage slot shapes to dry soil conditions. (✓ = good, X = poor.)

Figure 6.1 illustrates what generally occurs when seeds are drilled into dry untilled soils with vertical double disc openers (V-shaped slot, Class I cover); hoe openers (U-shaped slot, Class II or III cover); and winged openers (inverted-T-shaped slot, Class IV cover). The following explanations are relevant for each line on Fig. 6.1.

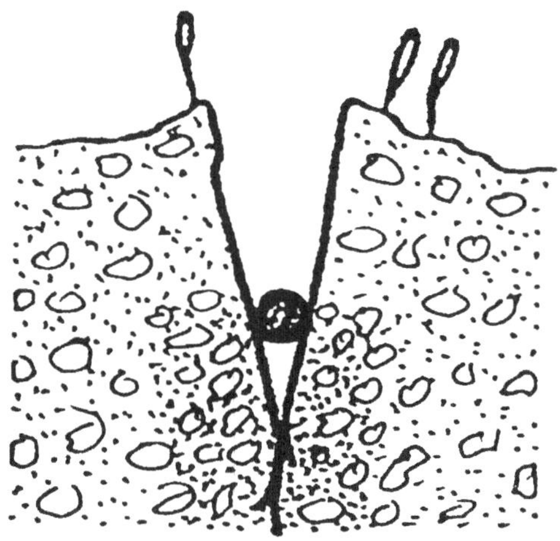

Fig. 6.2. The position a seed takes in a vertical V-shaped no-tillage slot.

Germination

Germination can occur from uptake of either liquid-phase water or vapour-phase water (humidity), or both. For liquid-phase water uptake to occur the seed must have physical contact with water-bearing soil by adequate soil–seed contact.

When seed is wedged in the base of a V-shaped slot (vertical or slanted) in a dry soil, the transfer of water from the soil to the seed is generally adequate, even though the contact zones with each wall of the slot may be relatively small (Fig. 6.2). The smooth, and often compacted, slot walls are a ready source of liquid-phase water, which is otherwise scarce in a dry soil. Thus germination within a V-shaped slot in a dry soil (Class I cover) can be 'good'.

With U-shaped slots, there is usually more loose soil within the slot, which also has a broader base for the seed to lie upon (Fig. 6.3). These two factors cause poor transfer of scarce liquid-phase moisture to the seed. Even when loose soil covers the slot and seed, there is little liquid-phase moisture in this covering medium because of its loose nature. It remains dry and acts in

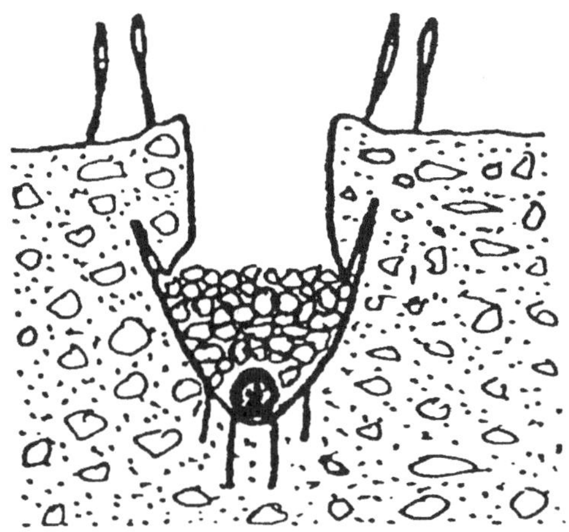

Fig. 6.3. The position a seed takes in a U-shaped no-tillage slot.

a similar manner to a dust mulch, as described above. Thus germination within a U-shaped slot in a dry soil (Class II or III cover) is often 'poor'.

With inverted-T-shaped slots, the supply of liquid-phase water to the seed is little different from that with U-shaped slots (Fig. 6.4). The Class IV cover, however, results in the seed being surrounded by vapour-phase water of 90–100% humidity (see Chapter 4). The seeds take a little longer to germinate than where liquid-phase water is available, but eventually a high germination count results. Thus germination within an inverted-T-shaped slot in a dry soil (Class IV cover) is usually 'good'.

Subsurface Survival

The most overlooked and under-studied stage of development of no-tilled seedlings is the time between germination and when the juvenile plants finally emerge from the soil. All of this period is spent beneath the soil. To remain alive the seedlings derive nutrients from their seed reserves and moisture through the embryonic roots, which appear at the time of germination.

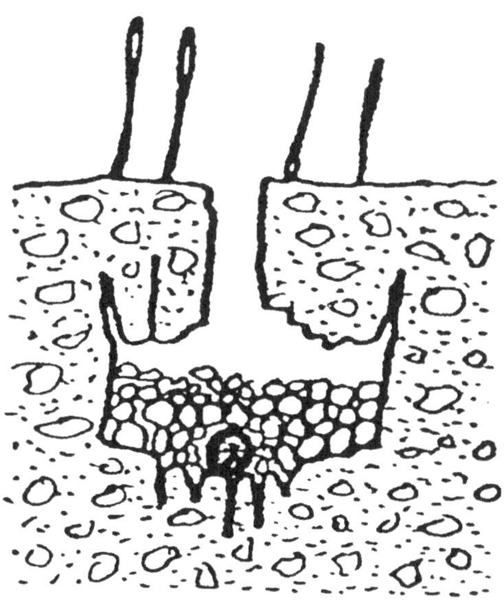

Fig. 6.4. The seed position in an inverted-T-shaped no-tillage slot.

These pre-emerged plants will not have developed the ability to photosynthesize food and energy from the sun's rays. There is only a limited need for them to draw water from the dry soil while they are beneath the surface, because it is mainly the sun that stimulates transpiration from plants. The subsurface seedlings, however, do respire (breathe), consuming moisture, and there may be subsurface water loss where the soil humidity, and therefore water vapour pressure, is lower than the corresponding water vapour pressure within the embryonic plants, which results in a diffusion loss through the cell walls.

Together with respiration, the end result is a tendency for subsurface seedlings to desiccate (dry out) unless they have an available source of soil water. With vertical V-shaped slots (Class I cover), many of the new seedlings become desiccated and die. Often they see sunlight very soon after germination because of the absence of covering material in the slot. But, even with Class II cover (loose soil), they may still die. The reason often is that the embryonic roots have to negotiate and penetrate the compacted slot walls before they can access liquid-phase water from the surrounding soil.

Since the slot walls are nearly vertical and there is little resistance against which the roots can base penetration forces, other than the weight of the seed, the roots tend to have difficulty penetrating the slot walls and instead spread sideways along the slot. The result is that seedlings after germination receive a poor water supply. Seedlings cannot stand the strong desiccation demand from a soil humidity that usually, at best, remains in the 60–80% range in vertical V-shaped slots. Therefore, many subsurface seedlings die before emergence in a vertical V-shaped slot in a dry soil.

It is useful to contrast this situation with a fully tilled dry soil. In a tilled dry soil, seeds are placed in a loose and friable medium. First, this medium probably does not transport enough liquid-phase water to the seed to bring about germination. But, even for those seeds that do germinate, there is no compacted slot wall for embryonic roots to penetrate. So subsurface seedling deaths in tilled soils are rare, similar to U-shaped no-tillage slots.

With U-shaped slots (Class II or III cover), although germination is often poor, the roots of those seedlings that do germinate have less trouble penetrating the uncompacted and broader base of the slots. If the slot can be covered to Class II or Class III standard, i.e. at least loose soil or a mixture of soil and residue, the likelihood of desiccation of subsurface seedlings is also reduced. Humidity is likely to remain in the 70–90% range. The result in U-shaped slots in a dry soil is that a reasonable percentage of the subsurface seedlings survive, although there may not be many that germinate until rain (or even dew) arrives, which means that seedling emergence may be spread over a long time.

Figures 6.5 and 6.6 show four wheat plants that were extracted from dry no-tillage plots in Australia. In Fig. 6.5, the plants are oriented so that the slot is running in the same direction as the wire fence (i.e. across the field of vision). The two plants on the left were sown with a vertical double disc opener (V-shaped slot) and the two on the right were sown with a wide, hoe-type opener (U-shaped slot). Root development along each of the rows is approximately equal for all four plants (i.e. for both slots).

In Fig. 6.6, all four plants have been rotated 90° and are now oriented with the drill rows running towards the camera. Clearly the roots of the plants on the left (vertical V-shaped slot) have hardly moved sideways out of the slot at all, but have stayed essentially within the slot walls. The roots of the plants on the right (wide U-shaped slot), on the other hand, have spread about as much sideways as they had lengthwise (Fig. 6.5). This illustrates the difficulty that young (and even, in this case, mature) roots have in penetrating the slot walls of some vertical V-shaped slots, compared with U-shaped slots.

With inverted-T-shaped slots (Class IV cover), humidity usually remains in the 90–100% range because of the residue-covered slot. While this will result in high (if sometimes slow) counts of germination, its most important function is that it removes most of the desiccation or transpiration stress from the subsurface seedlings, with the result that their survival rate is also high.

Embryonic root exploration out of the slot zone is no more restricted with inverted-T-shaped slots than with U-shaped slots.

Fig. 6.5. Wheat plants from a no-tilled crop in New South Wales, Australia; slot direction is parallel to the fence (from Baker *et al.*, 1996).

Fig. 6.6. Wheat plants from a no-tilled crop in New South Wales, Australia; slot direction is towards the camera (from Baker *et al.*, 1996).

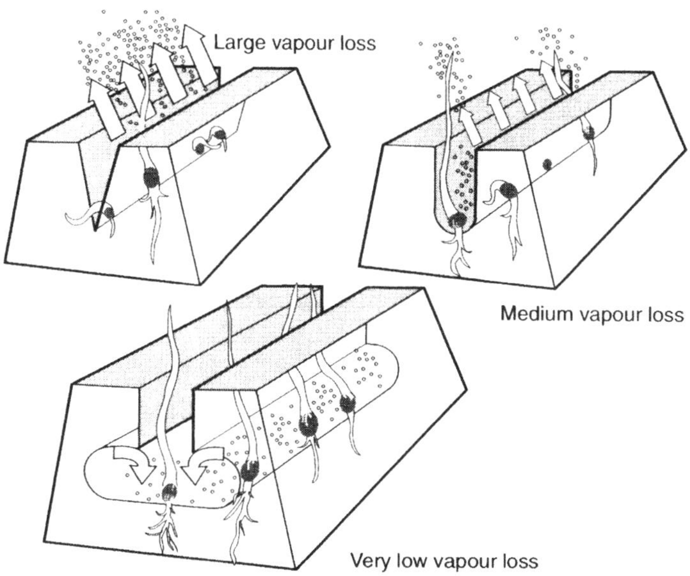

Fig. 6.7. The relative rates of loss of soil humidity from V-, U- and inverted-T-shaped no-tillage slots (from Carter, 1994).

The combined result is that, with inverted-T-shaped slots in a dry soil, most of the subsurface seedlings survive, leading to rapid and consistent seedling emergence.

Figure 6.7 illustrates the relative rates of humidity loss from the three contrasting slot shapes (Choudhary and Baker, 1994).

Scientists in New Zealand tried covering vertical V-shaped slots with strips of plastic to artificially trap water vapour in the otherwise open slots and create artificial Class IV cover (Choudhary, 1979). The humidity increased, but fungal growth soon also became evident in the slots, probably indicating that air circulation had been reduced. Therefore, nature had the perfect covering medium in the form of organic mulch and residue. Mulch breathes, as well as trapping humidity. Plastic does not breathe, even if it traps humidity, and it is quite impractical to cover every slot drilled with plastic strips.

It is little wonder, therefore, that deciduous trees flower, set seed and drop their seeds to the ground before they drop their leaves. Nature's intention seems to have been to cover the seeds with mulch.

Seedling Emergence

The more **X**s in the total for a slot in Fig. 6.1, the less effective that slot is at promoting seedling emergence from a dry soil. Conversely, the more ✓s in the total, the better the slot.

In summary, the order of ranking with regard to dry soils is:

1. Inverted-T-shaped slots – Class IV cover – excellent germination, excellent survival and thus excellent emergence.
2. U-shaped slots – Class II or III cover – poor germination, adequate survival and thus substandard emergence.
3. Vertical V-shaped slots – Class I or II cover – excellent germination, poor survival and thus poor emergence.

Table 6.1 (Choudhary, 1979) lists typical patterns of wheat (*Triticum aestivum*) seed and seedling responses to the three slot shapes in dry soils. These results illustrate the separate mechanisms of failure of vertical V- and U-shaped slots, i.e. subsurface seedling mortality and germination failure, respectively.

With vertical V-shaped slots, seedling emergence was poor (27%), although germination had been reasonably good. Only 9% of the seeds failed to germinate, the same as for the inverted-T-shaped slot. On the other hand, a high percentage (64%) of these germinated seedlings remained un-emerged beneath the soil in the vertical V-shaped slots, and most of them died.

With U-shaped slots, although a higher percentage (51%) emerged than with V-shaped slots, 23% of the seeds had not germinated in the first place. For those that did germinate, subsurface seedling survival was reasonably good. Only 26% of the seedlings remained un-emerged beneath the soil, similar to the inverted-T-shaped slots (27%).

The distinguishing feature of the inverted-T-shaped slots was that 64% of the seeds germinated and emerged. In addition, 27% germinated and remained alive beneath the soil, awaiting rain. Only 9% did not germinate in the first place.

Figure 6.8 shows typical seedling emergence patterns of wheat, no-tilled into a dry soil under controlled dry conditions (Baker, 1976b). Clearly the seeds sown in the inverted-T-shaped slots emerged in much greater numbers (78%) than from U- (28%) or vertical V-shaped slots (26%). There was a few days' delay before the seeds in the inverted-T-shaped slot started to emerge, possibly because they were taking up vapour-phase water rather than the liquid-phase water that the other two slots were supplying; but thereafter the emergence rate was very rapid compared with the other two slot shapes.

Table 6.1. Wheat seed and seedling responses to no-tillage openers and slot shapes in a dry soil.

	Double disc opener Vertical V-shaped slot Class I cover	Hoe opener U-shaped slot Class II cover	Winged opener Inverted-T-shaped slot Class IV cover
Seedling emergence	27%	51%	64%
Germinated seeds that had failed to emerge	64%	26%	27%
Un-germinated seeds	9%	23%	9%
Total seed pool	100%	100%	100%

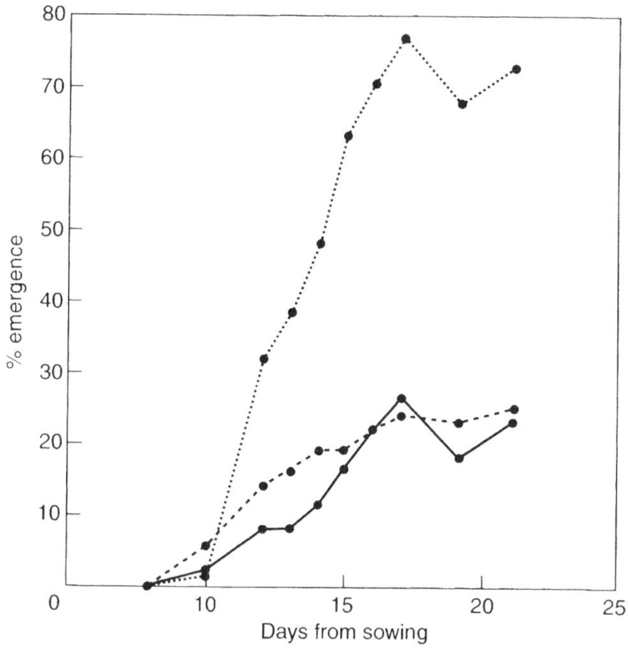

Fig. 6.8. Wheat seedling emergence patterns from V- (——), U- (- - - -) and inverted-T-shaped (·····) no-tillage slots in a dry soil (Baker, 1976b).

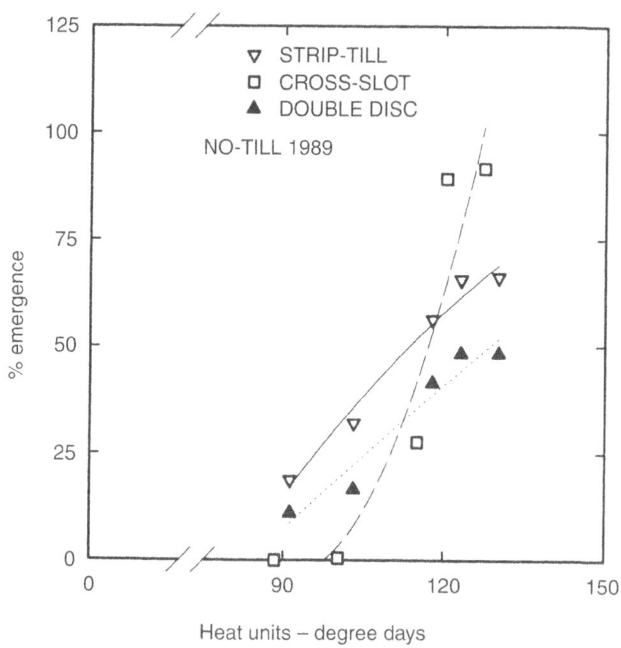

Fig. 6.9. Pea seedling emergence patterns from V-, U- and inverted-T-shaped no-tillage slots in a dry soil (from Wilkins *et al.*, 1992).

This phenomenon is also illustrated in Fig. 6.9, which shows field seedling emergence patterns of peas in a dry soil in Oregon, USA (Wilkins *et al.*, 1992). Vertical V-, U- and inverted-T-shaped slots were used, which were represented by 'double disc', 'strip-till' and 'cross-slot' openers, respectively.

Emergence from the U-shaped slots was spread over a 2–3-day period and reached a maximum of 65%, 5% better than V-shaped slots, which otherwise spread their emergence pattern over the same length of time. Seedlings in the inverted-T-shaped slots did not start to emerge until 1–2 days after the other two slots, but then almost all of the plants came up in a single day and attained a total of 90% emergence. The evenness and consistency of emergence shown by the inverted-T-shaped slot has important consequences for eventual crop maturity and yield; and, of course, 90% emergence contributes to greater yields than 50–65% emergence.

A further experiment by Choudhary (1979), shown in Table 6.2, illustrates the effectiveness of the three slot shapes in a dry soil compared with the same soil when rewetted. The most noticeable effect was that both the vertical V- and U-shaped slots responded positively when the moisture status of the soil was raised. Their seedling emergence counts increased by fourfold and twofold, respectively. The inverted-T-shaped slots increased by only 9% because their dry soil counts were reasonably high in the first place.

As in Table 6.1, vertical V-shaped slots had a high count (72%) of un-emerged seedlings in the dry soil, which decreased only slightly (to 58%) in more moist conditions, indicating that many seedlings had already died. U-shaped slots had a relatively high count (47%) of un-germinated seeds in the dry soil, which was later eliminated altogether (to 0%) when the soil moisture level was raised, indicating that all the un-germinated seeds had remained viable. This illustrates again that the causes of failure in a dry soil for vertical V- and U-shaped slots are quite different from one another. In the case of vertical V-shaped slots, it is failure of seedlings to survive beneath the soil, while, in U-shaped slots, it is failure of seeds to germinate in the first place. With inverted-T-shaped slots, most of the seeds had germinated even in the dry soil and about the same number as for U-shaped slots remained un-germinated beneath the soil.

The question arises as to what happens to the subsurface seedlings that have not emerged from a dry soil in field situations. The fate of such seedlings depends on two things: (i) how soon after drilling rain occurs; and (ii) how effectively the slot maintains the subsurface seedlings in a viable state awaiting that rain. The high humidity of inverted-T-shaped slots will maintain seedlings in a viable state for much longer than U-shaped slots, which are themselves better in this respect than vertical V-shaped slots. In the laboratory, germinated wheat seedlings have remained viable beneath a dry soil with Class IV cover for 3 weeks. In one field situation, however, on a very light soil of volcanic ash origin, ryegrass (*Lolium perenne*) seedlings survived beneath the surface of Class IV cover (inverted-T-shaped slot) for 8 weeks before rain finally fell, at which time they emerged, apparently none the worse for having spent that amount of time beneath the soil (S.J. Barr, 1990, unpublished data).

Provided that rain or irrigation occurs before the subsurface seedlings have died

Table 6.2. Wheat seed and seedling responses to no-tillage openers in a dry soil and soil of adequate moisture.

	Double disc opener Vertical V-shaped slot Class I cover		Hoe opener U-shaped slot Class II cover		Winged opener Inverted-T-shaped slot Class IV cover	
	Moist	Dry	Moist	Dry	Moist	Dry
Seedling emergence	42%	10%	70%	31%	68%	59%
Germinated seeds that had failed to emerge	58%	72%	30%	22%	32%	23%
Un-germinated seeds	0%	18%	0%	47%	0%	18%
Total seed pool	100%	100%	100%	100%	100%	100%

from desiccation, it might be possible to get a positive response to watering after drilling with both vertical V- and U-shaped slots. By irrigating 22 days after a dry soil had been drilled under no-tillage, Baker (1976a) obtained an increase in emergence counts from 21% to 75% with V-shaped slots, and from 38% to 92% with U-shaped slots. With inverted-T-shaped slots, the increase was much more modest, from 78% to 86%, again because seedling emergence had already been high when the soil was in a dry state prior to irrigation.

The Effects of Pressing

One of the most common practices in tilled seedbeds is to press on the rows after covering. The practice seeks to improve seed–soil contact and attract water to the seed by capillary action. Undoubtedly it improves seed–soil contact but its function in attracting water to the seed is dubious. Cross (1959) demonstrated that, in a dry soil, consolidation under the seed was more important than consolidation above the seed, and there has always been doubt about the real benefits of pressing on tilled soils anyway.

It seems that pressing after covering in an untilled soil is of even less benefit. Choudhary (1979) and Choudhary and Baker (1981b) conducted experiments that compared pressing on the soil after covering with covering alone and pressing on the seed before covering. They found no benefit at all for pressing on the covered slots in a dry soil. Most importantly, they found substantial benefits from pressing on the seeds in the slot before covering, but only in vertical V- and U-shaped slots. With inverted-T-shaped slots, seedling emergence was already high in the absence of pressing, so there was little improvement from any subsequent pressing action.

In U-shaped slots, pressing the seed into the base of the slot ensures that the seed has good contact with the water-bearing soil. Since there is usually insufficient water vapour in U-shaped slots to germinate the seed and seed–soil contact is otherwise poor for liquid water uptake, pushing the seed into the undisturbed soil ensures that at least liquid water uptake is available in much the same way as for V-shaped slots, as illustrated in Fig. 6.10.

In vertical V-shaped slots, pressing the seed into the base of the slot has a different effect. Embedding the seed directly into the undisturbed soil ensures that the radicle (first root) emerges directly into soil, from which it will derive its all-important water uptake (Fig. 6.11), thus bypassing the stress period when embryonic roots otherwise attempt to penetrate the slot wall. Thus, pressing on the seeds prior to covering of

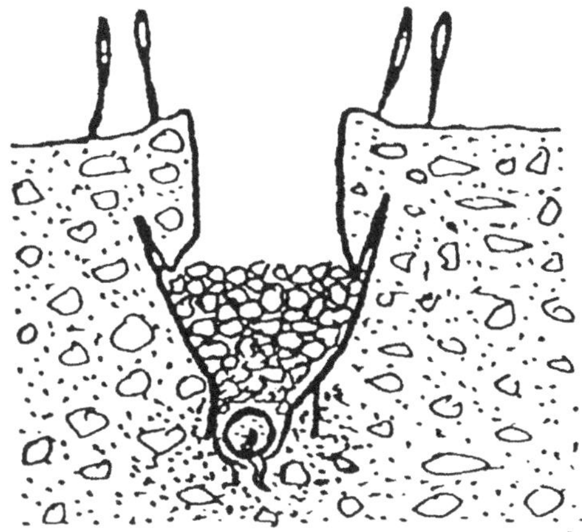

Fig. 6.10. The position of seeds after pressing in the base of a U-shaped no-tillage slot.

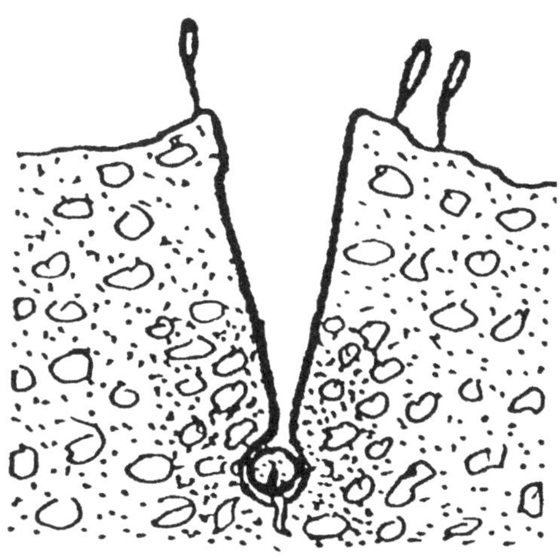

Fig. 6.11. The position of seeds after pressing in the base of a V-shaped no-tillage slot.

both U- and vertical V-shaped slots has significant benefit in terms of improving seedling emergence from a dry soil.

Field Experience

In New Zealand a field experiment sought to drill with three contrasting no-tillage opener types each second Monday for 6 summer months regardless of soil or weather conditions in order to gauge how often limiting conditions occurred in that region (Choudhary and Baker, 1982). By chance, on one occasion the soil moisture level was close to the permanent wilting point. On this occasion, inverted-T-shaped slots obtained 50% emergence of wheat, whereas U- and V-shaped slots in the same soil produced virtually no seedling emergence. It is doubtful if any seeds would have emerged from a tilled soil at or near PWP either.

It is little wonder, therefore, that repeat surveys of operators of drills with openers that created inverted-T-shaped slots in New Zealand, covering some 40,000 hectares per year in both spring and autumn sowing (Baker *et al.*, 2001), revealed a 99% success rating for the drilling process and technology.

Summary of Drilling into Dry Soils

1. The descending ranking of biological performance of slot shapes in dry soils is inverted-T-, followed by U-, then vertical V-shaped slots.
2. The descending ranking of effectiveness of slot cover in dry soils is Class IV to Class I.
3. Inverted-T-shaped slots trap water vapour within the slot, which germinates seeds as well as sustaining subsurface seedlings.
4. The predominant cause of failure of vertical V-shaped slots is subsurface desiccation of seedlings, not germination failure.
5. The predominant cause of failure of U-shaped slots is germination failure.
6. Pressing on the soil after covering the seed has negligible effect with any slot shape.
7. Pressing on the seeds in V- and U-shaped slots before covering improves their performance noticeably.
8. Surface residues are an important resource for promoting seedling emergence from dry soils, provided the openers utilize them correctly in the covering medium to trap humidity. Inverted-T- and slanted V- (but not vertical V-) shaped slots are most effective.
9. It is possible to obtain more effective seedling emergence from a dry soil using no-tillage rather than tillage, provided the correct technique and equipment are used.
10. With inverted-T-shaped slots, it is possible to obtain seedling emergence from untilled soils that are too dry to sustain effective crop growth.

7 Drilling into Wet Soils

C. John Baker

The biological ranking of no-tillage opener performance for wet soils is almost identical to that for dry soils, but for different reasons.

Unlike dry soils, it is usually impossible to physically drill into soils that are already very wet because of limitations in drill performance, limited traction or excessive compaction. Thus, in considering wet soil effects, it is important to distinguish between two different situations:

1. Drilling into soils that are sufficiently wet to make them sticky and/or plastic in nature and yet are still able to be drilled.
2. Drilling into soils that were not excessively wet at the time of drilling but that become very wet soon after drilling.

Drilling Wet Soils

The most pressing problem to drill an already wet soil without plugging (situation 1 above) from an operational point of view relates to the physical ability of openers. There are few common principles that distinguish one opener from another in this regard. In general, all openers with rotating components have limitations in wet soils, especially in wet soils that are also sticky. The use of subsurface scrapers on some disc openers will extend their tolerance of wet soils.

Where an opener employs press or gauge wheels of the semi-pneumatic ('zero-pressure') type, the operational limit of the whole opener in wet and/or sticky soils is the limit to which these tyres can continue to operate without plugging. Semi-pneumatic tyres are particularly good at shedding mud (see Chapter 10), so it is illogical to expect an opener to handle wet soils any better than its tyres.

Putting to one side the ability of different openers to operate without plugging, there are important biological effects that also arise as a result of the physical action of different openers in wet soils. The most important biological factor is the amount of compaction, smearing and crusting created by different openers. Smearing is very localized compaction within the slot (perhaps only 1–2 mm thick) and crusting is usually a smear that has dried hard.

Dixon (1972) illustrated the effects of vertical double disc openers (V-shaped slot), simple hoe openers (U-shaped slot) and simple winged openers (inverted-T-shaped slot) at different soil moisture contents, one of which was quite wet (27%) (Fig. 4.1). Several others have also studied the tendencies of different openers to compact the base and side walls of the slot

(Dixon, 1972; Baker and Mai, 1982b; Mitchell, 1983). From these studies and countless field observations, the compaction, smearing and crusting tendencies of different openers can be summarized as follows.

Vertical double (or triple) disc openers (V-shaped slots)

These have the strongest compaction tendencies of all no-tillage openers. Compaction occurs at both the base and side walls of the slot. They also have a strong smearing tendency, which is accentuated by the open slot. Because the smears are open to the elements, they often dry after passage of the opener and soon become internal crusts, which restrict root penetration.

In sticky wet soils, soil clings to the outside of the discs, which lift soil and seed from within the slots and deposit them alongside, thus negating the true V shape of the slots. Figure 4.5 shows a slot made by a vertical double disc opener in a sticky Australian soil. The slot has been severely disrupted by soil sticking to the disc.

Vertical double or triple disc openers have a strong tendency to tuck (or hairpin) residue into the slot, as described in more detail later. The slot cover is typically Class I.

Slanted double (or triple) disc openers (slanted V-shaped slots)

These are somewhat less likely to compact the seed zone but only if the seeding opener is preceded by another double or triple disc fertilizer opener slanted in the opposite direction. Because of the slant, the upper side of the slot wall created by the first opener actually heaves the soil upwards and loosens it somewhat. Although the second slanted opener actually compacts the soil beneath it more than if it had been operating in a vertical position, the pre-loosening of this soil by the first opener, which normally operates somewhat deeper than the second opener, negates most of the harmful effects.

Where a slanted double or triple disc opener is not preceded by a similar opener slanted in the opposite direction, the compaction beneath the opener will be greater than if the opener had been operating vertically. Compaction above the opener will be relieved, but loosening will have little effect on root penetration of seedlings, although it will improve the moisture-retention properties of the slot, which in turn will reduce the risk of the internal surfaces of the slot drying to form crusts.

Slanted double or triple disc openers otherwise have all of the same problems associated with their vertical counterparts, including hairpinning of residue into the slot zone and a tendency for sticky soils to cling to the outside of the disc and disrupt the integrity of the slot shape. The slot cover varies from Class II to Class IV.

Vertical angled flat (or dished) disc openers (U-shaped slots)

These have little or no compaction tendencies and little or no tendency to smear or lift soil in sticky conditions. Covering of the slots may be difficult, however, in continued wet weather, for the same reasons later outlined for hoe-type openers. Angled disc openers also tend to tuck (or hairpin) residue into the slot (see below). The slot cover is typically Class I or II.

Hoe-type openers (U-shaped slots)

These usually result in little compaction, unless they are of a design that has a large flat base, in which case they may compact the base of the slot, but not the side walls. In wet soils they almost invariably create smears on the base and side walls of the slot. These become important if the slot remains uncovered after drilling and the smears are allowed to dry to form crusts.

Covering is a particular problem. Hoe openers rely on the covering device collecting up the spilled soil alongside the slot and brushing it back over the slot as covering material. In a wet soil, such covering material is unlikely to become crumbly, so the slot is difficult to cover at all, encouraging eventual crust formation.

If covering needs to be a separate operation, its effectiveness depends on allowing sufficient drying for crumb to form in the debris alongside the slot, but not so much drying as to allow any smears within the slot to become crusts. Thus, although hoe openers can be used successfully in wet soils, they require a high level of skill to overcome their shortcomings. Hoe openers can experience problems in sticky soils if soil accumulates on the sides of the opener and changes its shape and dimensions. The slot cover is typically Class I.

Power till openers (U-shaped slots)

These mostly compact the base of the slot and may smear that zone as well. This smearing and compaction, however, are seldom severe and, because the soil is not often spilled completely out of the slot, the smears are usually not at risk of becoming crusts unless a very severe drying period follows drilling.

Power till openers mechanically aerate the soil more than any other opener type, which can be beneficial in wet soils with low residue levels and only small populations of earthworms. On the other hand, some power till openers may become totally inoperable in sticky wet soils due to 'plugging' between the cutting blades. The slot cover is typically Class IIIb.

Winged openers (inverted-T-shaped slots)

These smear the base of the slot about as much as most hoe openers but result in minimal compaction. Like power till openers, winged openers have an advantage in that they either close the slot themselves or make closure by a separate device easy and not dependent on moisture or weather. Thus, smears do not become crusts and therefore do not restrict root growth.

Winged openers handle sticky soils reasonably well. The disc version of the opener uses subsurface scrapers to overcome the tendency of sticky soils to cling to the disc. Figure 7.1 shows the benefits of scrapers used on a winged opener in the same sticky Australian soil as depicted in Fig. 4.5. The integrity of the slot and the residue cover have remained intact. The slot cover is typically Class IV.

Figures 7.2 and 7.3 show sections of soil in the side walls of two no-tillage slots photographed with an electron microscope (Mai, 1978). The lighter grey areas in the uncompacted soil in Fig. 7.2 are natural voids and macropores. In addition, much organic matter in the form of roots and buried residue is visible. In contrast, the compacted soil in Fig. 7.3 has almost no macropores and little visible organic matter. Instead, it contains only a few cracks in which soil oxygen can circulate. It is obvious why earthworms prefer soil surrounding inverted-T-shaped slots to that which surrounds V-shaped slots.

Soil type is also important in wet-soil seeding. If a small handful of soil can be 'ribboned' by rubbing it between the thumb and forefinger, it will probably become smeared by those openers that have smearing tendencies. In general, sandy soils and well-structured loamy soils with reasonably high levels of organic matter seldom take on smears or become permanently compacted by passage of no-tillage openers. Many clay soils take on a smear readily when wet. Montmorillonitic clays may become sticky instead. Silty soils lie in between clays and sands.

Many of the sticky montmorillonite clays produce good crops because of their incredible water-holding capacity. They also have a strong tendency to shrink when drying. This produces internal cracking, forming quite deep fissures in the soil. During the early stages of drying and cracking, the soil mass breaks itself into smaller particles by shrinkage, almost as if it had been tilled. Such soils are said to be self-mulching. They produce a dilemma for tillage practices. Because they are so sticky when wet, they are difficult to work in that

Fig. 7.1. Class IV slot cover remaining intact after passage of a winged opener, equipped with scrapers (inverted-T-shaped slot), through a damp sticky soil (compare with Fig. 4.5).

Fig. 7.2. Electron-microscopic section of soil from the wall of an inverted-T-shaped slot (from Baker and Mai, 1982b).

state with tillage equipment. But waiting until they dry and are easier to work risks sacrificing valuable soil water during the drying and tillage periods.

No-tillage offers a realistic option for such soils, since it allows sowing directly into the untilled soil with minimal disturbance. This is best done when only a small

Fig. 7.3. Electron-microscopic section of soil from the wall of a V-shaped slot (from Baker and Mai, 1982b).

amount of surface drying has occurred. Avoiding inversion of the deeper, more moist layers during drilling then becomes an important function of the no-tillage openers, both because such inversion brings up wet soil that sticks to everything and because it results in unnecessary loss of soil moisture. This contrasts with continuous tillage, in which the resistance of soils to compaction and smearing declines with time and continuous working. Vehicle traffic exacerbates the situation, leading to a cumulative decline in the usefulness of such soils when they are worked in a wet state. Since the practice of no-tillage gradually increases SOM levels and structure over time, many soils are likely to become less liable to smear or compact with time and therefore better able to be drilled when wet.

Drilled Dry Soils that Become Wet

Drilling dry or moist soils that have yet to become wet will not create substantive smearing or compaction problems with any design of opener. Thus, the differences between openers reflect the abilities of the various slot shapes to create micro-environments that will remain beneficial to seeds, seedlings and growing plants even after the soils have subsequently become wet. The most important criterion is their effect on the oxygen status of the soil, since roots breathe, and saturation by water will otherwise drown both seedlings and beneficial soil fauna.

Wet soils, especially when they have not been tilled, have a complex relationship with seeds. For example, if the soil has not been tilled for some time and has a reasonable population of earthworms, the earthworms will have an important effect on oxygen diffusion in the seed zone and water drainage. Their burrowing activity provides channels for air entry and water exit.

Earthworms also need feeding. They respond rapidly to the presence or absence of food supplies. There are several species of earthworm and each species prefers to occupy a certain depth range of soil. Those that feed on surface residues (e.g. *Lumbricus rubellus* Hoff and *Allolobophora caliginosa* Sav) live near the surface and are the first to react to excess water on the soil surface. They also react to the presence or absence of residues, which comprise their food supply, even to the extent that their burrowing and casting will reflect the presence of surface residues only a few centimetres apart.

In experiments with no-tillage openers in soils that were to become wet, Chaudhry (1985) tested the effects of the presence or absence of surface residues. 'Residue' plots had long, rank ryegrass (*Lolium perenne*) growing on them, which was sprayed. 'Non-residue' plots had this grass removed at ground level just before drilling. Within 24 h of mowing, the earthworm populations in the 'non-residue' plots had halved, presumably as a response to the removal of their principal food source.

It has also been observed that earthworms appear to have a preference for the disturbed slot zone in a soil after drilling, as opposed to the undisturbed soil alongside, but only if this slot zone is covered with a ready source of food (residue) and only if it is not compacted. Presumably they find the loosened soil easier to burrow through and the covering of residue provides an improved environment and a convenient food source.

Table 7.1 shows the effects on seedling emergence of barley (*Hordeum vulgare*) in a wet soil by the three common slot shapes with and without surface residues (Chaudhry, 1985; Chaudhry and Baker, 1988). The table also shows the numbers of earthworms recovered from 120 mm diameter × 100 mm long soil cores centred on the drilled rows. The index of earthworm activity, measured as the percentage of the area of ground covered by earthworm casts, showed similar trends to the numbers of earthworms counted in the soil cores. To create very wet conditions after drilling in this experiment, the soil was irrigated with 20 mm of simulated rainfall per day over a 4 h period, for 20 days (total, 400 mm in 20 days). In a field situation, such an intensity of repeat rainfall would be expected to produce supersaturated conditions and surface puddling in a short time span. In the free-draining bins used in this experiment, supersaturation did not occur but the soil none the less remained above 'field capacity' most of the time.

There were three strong trends in the data of Table 7.1. First, the greatest seedling emergence was promoted by the surface broadcast treatment (87%) and inverted-T-shaped slots created by winged openers (76%) (no statistical difference). Next were U-shaped slots created by hoe (65%) and power till (63%) openers. The vertical V-shaped slots created by double disc openers and the U-shaped holes created by a simulated punch planter performed poorly (24% and 17% seedling emergence, respectively).

Secondly, the number of earthworms found in cores of soil centring on the drilled rows mirrored very closely the seedling emergence counts. Most earthworms were found in the vicinity of the slots created by the winged (25), hoe (22) and power till (23) openers, together with surface broadcasting (22) and perhaps the punch planter (18), but the vertical double disc opener (9) performed poorly.

Thirdly, the presence or absence of residues had a very positive effect on both seedling emergence and earthworm numbers with the inverted-T- and some of the U-shaped slots and holes, but not with V-shaped slots or with surface broadcasting. Residues improved seedling emergence with the inverted-T-shaped slots from 48% to 76% and earthworm numbers from 13 to 25. The effect on U-shaped slots was not quite so marked, but residues none the less improved seedling emergence from 40% to 65% and earthworm numbers from 13 to 22 with the hoe opener.

In contrast, residues actually depressed seedling emergence with the vertical double disc openers (from 25% to 17%) and punch planter (from 17% to 14%), but had no effect with surface broadcasting or the power till openers. The latter phenomenon is not surprising since the power till opener chopped up the surface residues (and probably a number of earthworms) and incorporated them into the soil. With surface broadcasting, the seeds were left lying on top of the ground, making them less likely to be affected by earthworm activity taking place beneath the surface. Further, because moisture was not limiting, it is not surprising that residues on the soil surface had no direct effect on emergence with broadcasting.

Table 7.1. Effects of no-tillage openers on barley seedling emergence and earthworm numbers in a wet soil after drilling.

	Double disc opener vertical V-shaped slot Class I cover		Hoe opener U-shaped slot Class I cover		Winged opener Inverted-T-shaped slot Class IV cover		Power till opener U-shaped slot Class IIIb cover		Punch planter opener U-shaped holes Class I cover		Surface broadcast No slot Class I cover	
	R	NR	R	NR	R	NR	R	NR	R	NR	R	NR
% seedling emergence with earthworms	17	25	65	40	76	48	63	62	17	15	84	87
Earthworm number (per core)	9	8	22	13	25	13	23	14	18	10	22	14
% seedling emergence without earthworms	15	19	24	23	20	22	43	41	14	16	89	89

R, plots covered with surface residues, both before and after drilling; NR, plots with no surface residue covering, either before or after drilling.

These results suggest that all three observed trends are linked in a wet soil. Indeed they are. The third line of Table 7.1 illustrates emergence when earthworms were eliminated from the soil by poisoning in an otherwise identical experiment.

Without earthworms, seedling emergence was weakened with all drilling treatments. Most residue advantages with inverted-T- and U-shaped slots disappeared in the absence of earthworms, indicating a strong linkage between the three factors when they were present. This also demonstrates one of the longer-term benefits of no-tillage, that of building up earthworm numbers and organic matter, which work to the advantage of this farming system, provided that appropriate equipment is used to maintain and capitalize on those benefits.

The data of Table 7.1 also illustrates that mechanical aeration can to some extent substitute for the absence of natural aeration caused by earthworms and other soil fauna. The chemical treatment to kill earthworms also kills some of the other channel-forming soil fauna. Although the use of power till openers may only be of short-term benefit when drilling into soils that subsequently become wet, this was the only opener to promote more than 24% seedling emergence in the 'sterilized' soil. Even then, the 43% emergence obtained with this opener in residue and the 41% without residue cannot be regarded as satisfactory and do not compare with the 76% obtained with the winged opener in the presence of both earthworms and residues.

Surface broadcasting promoted the highest seedling emergence counts in the absence of earthworms (89% both with and without residue), presumably because seeds on the surface were unaffected by earthworm activity beneath it. But this treatment can hardly be considered a recommended field practice unless one can guarantee 400 mm of rainfall for the first 20 days after sowing. It was used in this experiment solely to compare the seed's need for oxygen and water.

Figure 7.4 illustrates similar responses to those just presented for inverted-T-shaped slots, hoe (U-shaped slots) and vertical double disc (V-shaped slots) openers. The most noticeable effects are that the seedling emergence trends follow the trends of earthworm numbers with all openers and that residues increased both emergence and earthworm numbers with the inverted-T and hoe openers but not with vertical double disc openers.

To further understand the interactions between opener types, the moisture status

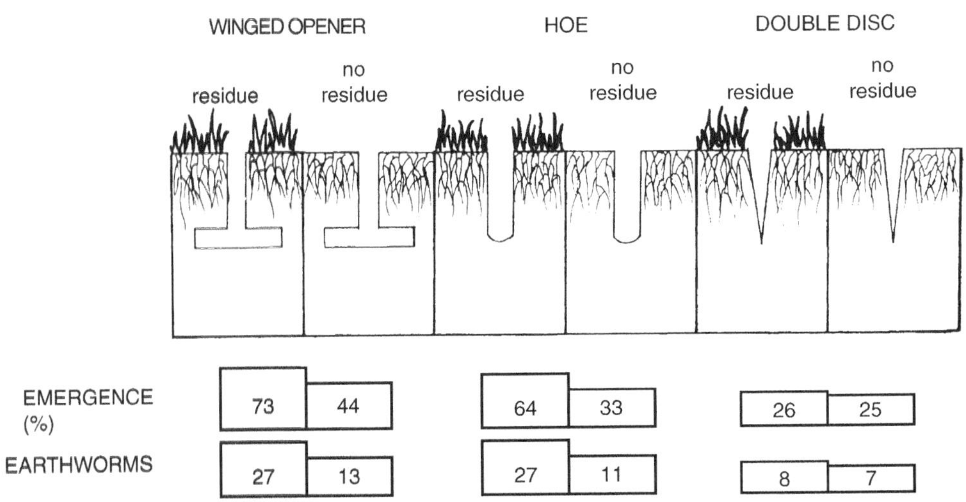

Fig. 7.4. Responses of seedling emergence and earthworm numbers to three contrasting, no-tillage slot shapes and surface residues in a wet soil (from Baker *et al.*, 1996).

of the soil and the level of residues present, Chaudhry (1985) conducted an experiment in which these factors were varied independently. The results are shown in Table 7.2.

The data show that most openers performed reasonably well in favourable soil moisture conditions, regardless of the level of residue (range, 65–90% seedling emergence). When the conditions became wet, however, the shortcomings of the vertical double disc opener (V-shaped slot) became progressively more apparent as the length of the residue increased. In the wet soil, emergence from the V-shaped slot dropped from 38% with no residue to 35% with short residue and 30% with long residue. The winged and hoe openers, in contrast, performed best when long residue covered the wet soil, which was attributable to the increase in earthworm activity in response to the long residue. As the residue length was reduced with these two openers, their advantages over the vertical double disc opener were reduced or eliminated.

Although the hoe opener responded positively to long residue, it is difficult to actually make a hoe-type opener function in long residue in the field. It is one thing to do this on a plot scale for experiments but, in the field, hoe openers soon block because of their raking action. In practical terms, therefore, of the two openers that performed well in wet soils with long residue, only the winged opener (inverted-T-shaped slot), which is able to handle residues in its disc form, can be regarded as a practical option.

Opener performance

The performance of various seeding openers in soil (that is, wetted after seeding) can be summarized as follows.

Power till openers (U-shaped slots)

These openers, in the absence of earthworms, will provide some compensatory mechanical aeration. The presence of earthworms, however, will not necessarily result in any improvement to seedling emergence because the gains that mechanical aeration brings to an earthworm-populated soil are offset by physical burial of the food source for any surface-feeding earthworms. There will also be some actual destruction of earthworms in the slot zone, but because the width of tillage by such openers is normally very narrow, it is likely that the slot zone will be rapidly recolonized by earthworms from the undisturbed soil alongside.

Punch planting (V- or U-shaped holes)

This is not likely to produce good results, with or without earthworms, although further work needs to be conducted with such openers. The poor performance of the punch planter in these experiments was somewhat surprising since the method used to make holes did not result in any compaction. In practice, punch planters almost invariably produce V-shaped holes, which could be expected to behave in much the

Table 7.2. Effects of openers, residue levels and soil moisture status on barley seedling emergence from a soil containing earthworms.

	Seedling emergence %								
	Vertical double disc opener V-shaped slot Class I cover			Hoe opener U-shaped slot Class I and IIIa cover			Winged opener Inverted-T-shaped slot Class IV cover		
	LR	SR	NR	LR	SR	NR	LR	SR	NR
Adequate moisture	65	84	82	86	70	76	90	76	82
Wet soil	30	35	38	68	36	42	75	43	47

LR, long residue; SR, short residue; NR, no residue.

same way as continuous V-shaped slots. In this case, however, a small coring device was used to remove cores of soil without compaction.

Vertical double disc openers (V-shaped slots)

These can be expected to perform poorly in wet soils for two reasons. First, compaction and smearing, together with crust formation, result in earthworms avoiding the slot area. Thus, not only does the opener disadvantage the seeds directly, it discourages natural processes (earthworms) from repairing the damage.

To examine the tolerance of earthworms to smearing, Chaudhry (1985) placed a number of earthworms on the surface of a damp, smooth soil contained in two high-sided pots (to prevent escape of the earthworms). Before placing the earthworms on the soil, he lightly smeared the surface of one of the plots with his finger. Overnight, all of the earthworms on the un-smeared soil had burrowed into the soil while only half had achieved the same result in the smeared soil, indicating the difficulty earthworms have in burrowing through smears.

Chaudhry (1985) also tested the tolerance of earthworms to compaction and found much the same result as for smears. Because wet soils are softer than dry soils, the action of vertical double disc openers acting through surface residues on wet soils is more one of pressing than cutting. This accentuates their compaction tendency. Slots that are both smeared and compacted are largely avoided by earthworms and do not benefit from their burrowing or nutrient cycling (Baker *et al.*, 1987, 1988).

Secondly, double disc openers tuck (or hairpin) residues into the slot. In wet soils, Lynch (1977, 1978) and Lynch *et al.* (1980) showed that the decomposition of this residue produces fatty acids, in particular acetic acid, which tend to kill seeds and germinating seedlings. They looked at ways of countering this problem, ranging from applying lime with the seed to neutralize the acid to separating the seed from the residue.

Apparently, separation of the two by only a small distance will largely avoid the problem since acetic acid is very quickly broken down in the soil by bacteria. The residue tucking problem is reflected in the negative response to the presence of residue by the vertical double disc opener and the fact that this negative response increased as the length of residue (and size of hairpins) increased.

Although slanted double disc and angled disc openers were not included in the above experiment, it is known that both of these openers also tuck residue into the seed zone, in much the same manner as vertical double disc openers. They can therefore be expected to experience acetic acid fermentation and its detrimental effects on seeds, but should experience fewer problems associated with smearing or compaction.

Winged openers (inverted-T-shaped slots)

These return most of the residue to a position over (not inside) the slot. This encourages earthworms to colonize the slot zone because when the residue is removed, the earthworm numbers decline noticeably. The central disc of the disc version of the winged opener will hairpin residues, in common with every other disc-type opener. But the winged side blades of this opener place the seed to one side of the central slit and therefore remove the seed from contact with the hairpinned residue. This is probably the only disc-type opener that effectively prevents seeds from lodging within hairpins and for this reason benefits from the presence of residues even in wet conditions. When long residue was positioned over the slot, the inverted-T slot produced more seedling emergence than any other design.

Hoe openers (U-shaped slots)

These behave in a similar manner to winged openers except that instead of placing the residue over the slot, they tend to push it to either side. As a consequence, although hoe openers will produce a positive response to the presence of residue (in terms of seedling emergence and earthworm numbers), that response is not likely to be as strong or as positive as for winged openers.

The seedling emergence responses of the various openers and surface broadcasting have also been reflected in root and shoot weights of the seedlings, as shown in Figs 7.5 and 7.6 (with and without earthworms, respectively).

Without earthworms, there were few differences between openers. Only the mechanical aeration of power till openers had any positive effect. With earthworms, however, the seedling growth closely paralleled the trends of seedling emergence and earthworm numbers.

Figure 7.7 shows typical oxygen diffusion rates within the soil containing earthworms associated with winged and double disc openers (Chaudhry, 1985; Baker *et al.*, 1987, 1988). Oxygen diffusion rate is measured by passing a current through platinum electrodes placed in a grid pattern around the sown slots and measuring the rate of consumption and replacement of oxygen in the vicinity of the electrodes (see Chapter 19).

Figure 7.7 shows that the winged opener had no negative effect on the oxygen status of the soil. The oxygen status surrounding the hoe, power till and punch planter openers (not shown) was very similar to that of the winged opener. In fact, all of these openers had similar patterns to that of the undisturbed soil, indicating that none of them had any detrimental effect on the oxygen diffusion rate of the soil. But, in all cases, the presence of residues moved the high-oxygen zones closer to the seeds, probably as a result of increased earthworm activity.

In contrast, the double disc opener had a marked negative effect on the oxygen status of the soil, regardless of the presence or absence of residues. Essentially, this opener, because of its wedging action, squeezes the high-oxygen zones away from the immediate vicinity of the seeds altogether and replaces them with compacted zones of low or, at best, medium oxygen diffusion.

Also of note is that the effects of wetness on the soil, both with and without earthworms, seems not to be related to how the soil becomes wet. For example,

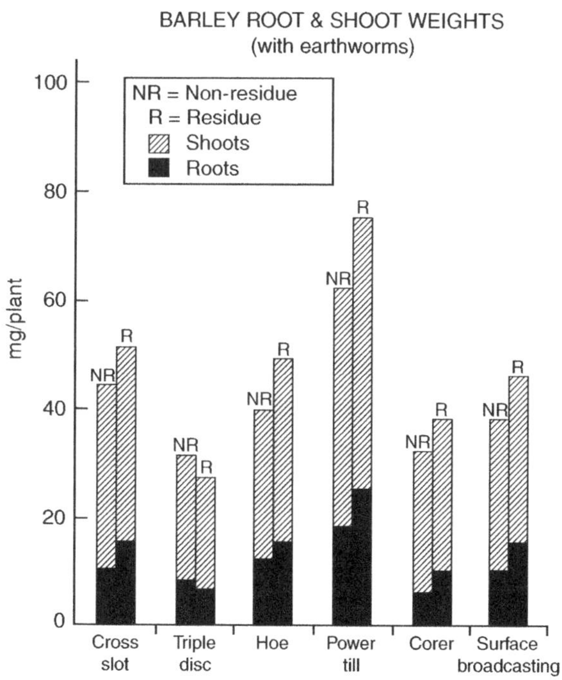

Fig. 7.5. Root and shoot weights of no-tilled barley seedlings in response to opener types and residue in the presence of earthworms (from Baker *et al.*, 1988).

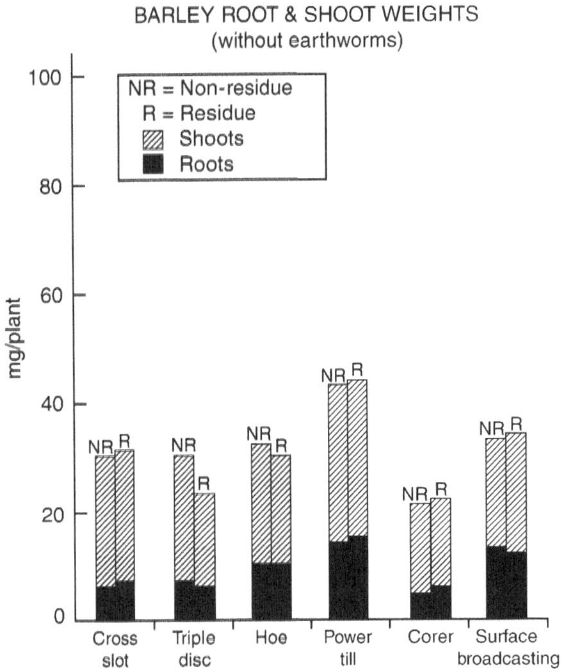

Fig. 7.6. Root and shoot weights of no-tilled barley seedlings in response to opener types and residue in the absence of earthworms (from Baker *et al.*, 1988).

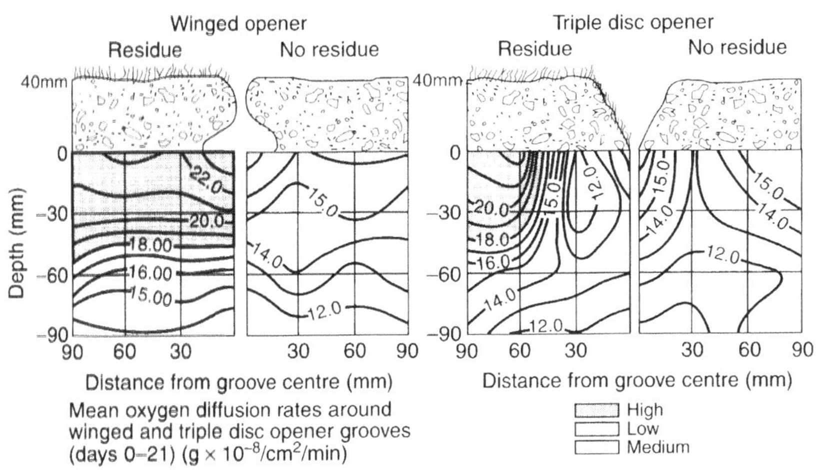

Fig. 7.7. Oxygen diffusion rate profiles around winged and double disc no-tillage openers operating in a wet silt-loam soil, in the presence and absence of surface residues (from Baker *et al.*,1988).

Chaudry (1985) had earlier conducted two experiments with earthworms and residue, identical in all respects except that one used simulated rainfall to wet the soil after drilling and the other used a rising water table. He was particularly interested in whether or not persistent rainfall had some sealing effect on the internal faces or the cover, or, alternatively, washed the seed out. He found no differences in barley seedling performance between wetting the soil from above or below, but both experiments confirmed the differences between openers and residue.

Later, Giles (1994) quantified the rate of accumulation of earthworm biomass in the top 100 mm of soil as a function of different levels of barley straw on the surface of the ground in New Zealand. He found an almost linear relationship, in which the total biomass of two surface-feeding species (*L. rubellus* Hoff and *A. caliginosa* Sav) had accumulated to 9 t/ha under 11 t/ha of straw and 5.1 t of earthworms under 6.4 t/ha of straw. During that period the recoverable biomass of the straw had decreased from 11 t/ha to 3.2 t/ha and 6.4 t/ha to 1.2 t/ha, respectively. For the first 6 months, the heavier rate of residue remained wetter than the lighter rate, which might help account for the faster decomposition of the former. At the termination of the experiment, a part of the residues appeared to have decomposed while another part had simply been buried by earthworm casts.

It should be appreciated that these levels of cereal straw were deliberately set very high to test the ability of earthworms to cope with 'overload' conditions under no-tillage. In general terms, such straw levels equate with grain yields of about the same magnitude.

Finally, experiments relating to wet soils would not be complete without also measuring the infiltration of water into the slot zones in the field. Figure 7.8 shows the results of a field experiment that compared the infiltration rates of a range of openers in a residue-covered silt-loam soil containing earthworms (Baker *et al.*, 1987). The results

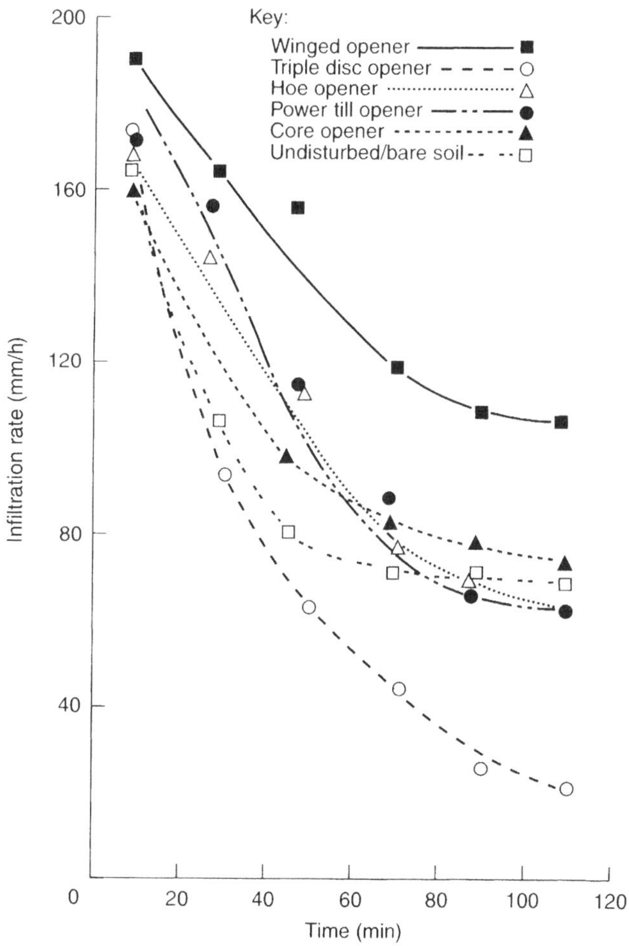

Fig. 7.8. Infiltration rates of no-tillage slots in a silt-loam soil (from Baker *et al.*, 1987).

reflect earthworm and seedling emergence trends. The winged opener (inverted-T-shaped slot) produced the most rapid infiltration (110 mm/h after 2 h), which is not surprising since it had promoted the greatest earthworm activity and seedling emergence. Next was a group of openers including hoe, power till (U-shaped slots) and punch planter (U-shaped holes), together with the undisturbed soil, all of which averaged 70 mm/h after 2 h. The poorest infiltration was with the double disc opener (V-shaped slots), with only 20 mm/h infiltration after 2 h. Water remained puddled in the V-shaped slots for hours after the experiment.

Summary of Drilling into Wet Soils

1. The ranking for the three basic slot shapes from poorest to best (V, U and inverted-T) in wet soils containing earthworms and residues is exactly the same as for dry soils, but for somewhat different reasons.
2. Seeds need ready access to oxygen in a wet soil, and different openers create different oxygen environments around the seeds in wet soils.
3. Double disc openers have an adverse effect on the oxygen diffusion rate of the soil surrounding the seed slot.
4. Inverted-T, hoe and power till openers, together with punch planters, have either a neutral or positive effect on oxygen diffusion around the slot.
5. Both earthworms and surface residues give clear-cut advantages if managed correctly. Both will increase with time under no-tillage and have an increasingly positive effect on aeration, drainage and infiltration.
6. Winged and hoe openers encourage earthworm activity in the slot zone.
7. Surface residues encourage earthworm activity, with the amount of activity being proportional to the amount of residue.
8. The ability of the inverted-T-shaped slot (winged opener) to retain residue over the slot is as important in wet soils as it is in dry soils because it encourages earthworm activity within and around the sown slot.
9. Double, triple and angled disc openers, together with punch planters, tend to tuck (hairpin) residue into the seed zone, where it has a negative effect on germination and seedling vigour. This is especially true of long, stringy and damp residue.
10. Winged, hoe, power till and furrow openers effectively separate decaying residue from direct contact with seeds.
11. In the absence of earthworms, mechanical aeration of the slot by power till openers may have a short-term benefit.
12. Surface broadcasting can perform well if regular daily rainfall is available for 3 weeks after sowing, but obviously this cannot be regarded as a practical option.
13. V-shaped slots and punch planter holes tend to be compacted and/or smeared. Class I cover (or lack of cover) allows these smears to dry to form crusts.
14. Smears and/or crusts discourage earthworm activity in the slot zone.
15. U-shaped slots created by hoe, power till and furrow openers may be smeared but only minimally compacted. If Class II cover or better is possible, the smears should not dry to become crusts.
16. U-shaped slots created by angled disc openers will not be smeared or compacted.
17. Inverted-T-shaped slots created by winged openers may be smeared but not compacted. Class IV cover will prevent drying of smears.
18. Excellent water infiltration is possible with inverted-T-shaped slots but infiltration is likely to be poor with V-shaped slots created by double or triple disc openers. But infiltration between the rows can be expected to be greater with no-tillage than with traditional tillage anyway, particularly with increased earthworm populations and organic matter.
19. Excellent seedling emergence can be obtained by inverted-T-shaped slots in wet soils, and satisfactory emergence can be obtained by most of the openers that create U-shaped slots.
20. Poor seedling emergence will result from V-shaped slots or holes in wet soils.

8 Seed Depth, Placement and Metering

C. John Baker and Keith E. Saxton

Accurate seed placement is more important in no-tillage than in tillage.

When an opener on a no-tillage drill or planter deposits seed, and perhaps fertilizer, into the soil, its ability to control the final placement and environment of each depends on a number of sometimes contradictory functions. The required combined capability of the drill or planter and soil opener includes:

1. Continuously following the soil surface of each row and maintaining precise seeding depth.
2. Dispensing seed under the soil, on the move, in a consistent band relative to the opener itself.
3. Covering the seed (and perhaps fertilizer) or at least making provision for effective covering after the opener has passed.
4. Separating the seed from the fertilizer if the two are being placed at the same time and optimizing the positions of each relative to one another so as to maximize biological responses.
5. Metering and dispensing seed at the desired spacing and in the desired pattern along the row.
6. Transferring seed from the metering mechanisms to the openers without disrupting the intended spacing or pattern.

Functions 1–3 are important for proper seed placement and function. Function 4 is important for fertilizer placement, as described in Chapter 9. Functions 5 and 6 (and, to some extent, 1) are dependent on the design of the whole drill or planter, especially the drag-arm configuration and downforce mechanism, as well as the openers.

Placing seed and fertilizer in the soil is a function of opener design. For optimum performance, openers need to have the ability to:

- Ignore or control soil disturbance beneath the ground surface (or lack of it when soils are wet).
- Ignore soil stickiness.
- Cope with stones and other obstructions beneath the surface.
- Avoid depositing seeds in hairpinned residue.
- Prevent seed bounce.
- Cover the slot to a consistent depth.

Covering might be a separate operation performed by a separate machine (e.g. harrows), in which case the openers should create the slots in such a way that the covering operation will result in a consistent depth of cover (see Chapter 5).

Seed metering is a function of the seed metering mechanism of the drill or planter and is not peculiar to no-tillage. In general, a precision planter is distinguished from a drill by the fact that a planter dispenses single seeds with the intention that the seeds are placed a predetermined distance apart. A drill, on the other hand, dispenses seeds in bulk so that a given number (or weight) of seeds is deposited in a given length of row (or area) in an approximately uniform distribution with no attempt at individual seed spacing.

Transferring seed from the metering mechanism to the opener might seem a mundane function, but, with precision metering especially, this transfer must maintain the continuity of metered seed timing for accurate spacing in the row. Agronomists argue about the effects of variations in seed spacing on crop yield, especially when this is traded off against the natural variation between plants and their abilities to compensate for imperfect spacing. But most experts agree that there is little agronomic disadvantage from having seeds spaced at precise intervals along the row. Recent evidence for maize suggests that uniform seeding depth and emergence are likely to be more important than plant spacing.

Seeding Depth and Seedling Emergence

Almost everyone agrees that seeding depth should be as consistent as possible. But surprisingly there have been few studies quantifying the target depths for seeds sown under no-tillage (as distinct from tillage) or the crop performance effects of variations around that target depth. Obviously, the importance of this factor will vary with the compensatory growth potential of any given crop or species.

To quantify the effects on seedling emergence of imperfect drilling depth under no-tillage, Hadfield (1993) measured the variations in germination and emergence of wheat (*Triticum aestivum*) and lupin (*Lupinus angustifolius*) drilled in inverted-T-shaped no-tillage slots at various depths. The results are shown in Table 8.1.

Hadfield concluded that the particular variety of wheat he used (cv. Otane) was less sensitive to depth of sowing than lupin in the 20 mm to 50 mm depth range, but both were seriously affected by depths greater than 50 mm. Overall, seedling emergence with this variety of wheat decreased by 4% for each 10 mm increase in drilling depth between 20 mm and 70 mm. But other varieties of wheat have been observed to have quite different tolerances of depth. In comparison, in these experiments lupin emergence declined by 17% for each 10 mm increase in depth between 20 mm and 70 mm. In both cases, the reduction in seedling emergence was not caused by failure of seeds to germinate but by subsurface mortality of seedlings that had already germinated. This confirmed earlier observations by Heywood (1977).

Campbell (1981, 1985) also studied drilling depths of a small-seeded pasture legume, red clover (*Trifolium pratense*), sown in inverted T-shaped no-tillage slots. He concluded that seedling emergence of

Table 8.1. Effects of drilling depth on seedling emergence of no-tilled wheat and lupin.

	Seedling emergence* (plants per square metre in parenthesis)	
Nominal drilling depth	Wheat	Lupin
20 mm	79% (209) a	93% (66) a
30 mm	80% (210) a	87% (62) b
50 mm	73% (192) a	60% (43) c
70 mm	61% (160) b	24% (17) d

Unlike letters in a column denote significant differences, $P < 0.05$.
*% seedling emergence = % of the estimated number of seeds sown from the known weights of seeds sown.

Table 8.2. Effects of drilling depth on seedling emergence of no-tilled red clover.

Nominal drilling depth	Seedling emergence* (%)
0 mm	53% b
13 mm	89% a
38 mm	56% b

Unlike letters in the column denote significant differences, $P < 0.05$.
*% seedling emergence = % of the estimated number of seeds sown from the known weights of seeds sown.

pasture legumes was particularly sensitive to drilling depths above and below his mid-treatment, 13 mm. The results are shown in Table 8.2.

Salmon (2005) examined the effects of seeding depths (from 0 to 50 mm) on the emergence of brassica seedlings when sown into a range of no-tillage soils in New Zealand using the disc version of winged no-tillage openers. He also sought interactions with seed treatments, which ranged from coated (Superstrike), insecticide-treated (Gaucho®), to bare (untreated) seed.

He concluded that, with this particular opener, which is known to create a favourable environment for both seeds and seedlings, depths of sowing from 10 to 25 mm had no significant effect on the rates or final counts of seedling emergence, but that zero depth and 50 mm depth reduced emergence markedly. There were no interactions between seeding depths and seed treatment.

Salmon was not able to test the effects of low seed vigour, other brassica species and/or other no-tillage opener types in these experiments. It is doubtful, however, if any of these factors would have improved the range of sowing depths found possible, which was considered to already be unusually broad in Salmon's experiments.

Maintaining Consistent Opener Depth

Maintaining a consistent depth of seeding is one of the most demanding tasks that any no-tillage machine must perform. This is for several reasons:

- The surfaces of untilled soils do not get smoothed in the same way that tilled soils do.
- Untilled soils are often harder than tilled soils and therefore have less cushioning effect, causing more bounce of the openers, especially at higher speeds.
- The harder soils require greater downforces to push the openers into the ground. Variations in ground resistance therefore result in larger variations in seeding depth than where soils are softer and smaller downforces are used.
- The hardness or strength of untilled soils usually varies across a field as a result of natural settling of the soils. Regular pulverization by tillage virtually eliminates these differences in soil strength.
- No-tilled soils are often covered with surface residues, which might interfere with the opener's ability to manipulate the soil beneath it and further accentuate the surface roughness.

We shall consider each of the above aspects separately.

Surface following

Control of opener depth is partly a function of the opener and partly a function of the supporting drill or planter frame. With no-tillage, there is little or no opportunity to smooth the soil surface prior to drilling. No-tillage openers must therefore have superior surface-following ability compared with their counterparts for tilled soils. The extent of vertical mechanical movement alone should increase from approximately ± 75 mm (total 150 mm) travel for tilled soils up to ± 250 mm (total 500 mm) travel for untilled soils.

Depth-gauging devices

One of the important contributions that openers make to controlling seeding depth is the presence or absence of depth-gauging devices (wheels, skids or bands), which 'track' the soil surface. Penetration forces are generally higher for untilled soils than for tilled soils. Further, the soil strength of tilled soils is usually quite uniform across the entire field as a result of the tillage process, while soil strengths of untilled soils vary quite widely on a metre-by-metre basis.

The result is that, if an opener relies solely on the penetration downforce reaching equilibrium with the soil's resistance to penetration in order to maintain a consistent seeding depth, as is common in tilled soils, seeding depths in untilled soil will vary just as widely as the soil strength. Consequently, any opener designed to operate at a consistent depth in an untilled soil will need at least some form of depth-gauging device. With such an attachment, a downforce can be applied in excess of that required to just attain target depth for that particular metre of soil. The additional force is carried by the gauging device without materially altering the depth of seeding.

Clearly, depth-gauging devices for untilled soils need to have the capacity to absorb quite large variations in applied force to operate satisfactorily in the inherent variability of such soils. Fortunately, untilled soils also have an inherently high ability to withstand surface loading and avoid furrowing.

There are differences in the accuracy of depth-gauging devices according to how close to the point of seed release the gauging device is located. Obviously, being closer to this position results in more effective depth control. The effectiveness of the device may suffer if it is located too far from the seed deposition zone since, for example, it may register on a small hump when the seed is being released into a small hollow.

There are often mechanical limitations to where the gauging device can be located on an opener in relation to where the seed is finally ejected into the soil. Probably the nearest any opener designs have come to gauging depth precisely at the seed exit points are those on which a specially shaped semi-pneumatic tyre operates alongside (touching) the base of a disc at the point where the seed is ejected. Figure 8.1 shows such an arrangement.

Where possible, it is desirable to combine the depth-gauging function of wheels with the additional function of slot covering and/or closure, so long as one function is not markedly compromised by the

Fig. 8.1. Depth-gauging wheels located alongside the point of deposit of seed in a no-tillage opener.

requirements of the other. The depth-gauging wheels on the disc version of winged openers are located close to, but slightly rearward of, the seed-ejection zone so that they can perform these dual functions without significant compromise to either (see Fig. 4.27). The wheels in Fig. 8.1 do not perform a slot-closure function.

Almost universally, the gauging devices most favoured by opener designers are wheels, although skids and depth bands are also used on less expensive opener designs. The problems with skids in no-tillage are that they gather and block with residue and the higher down forces result in high wear rates as they slide along the ground.

Depth bands are sometimes attached to the sides of discs to limit the depths of their penetration, but the depth of seeding cannot be conveniently adjusted for different crops without removing the band and replacing it with a band of different diameter. They also tend to accumulate soil in the corner between the band and the disc, effectively increasing the diameter of the band and decreasing the seeding depth.

Gauge wheels are not without their problems either. Because wheels can only be attached by their axles, designers have to trade off the disadvantages of attaching them behind the opener against the disadvantages of attaching them beside the opener, where they might interfere with residue clearance and are unlikely to be able to function in a slot-closure capacity as well.

Since most no-tillage openers for residue conditions involve a disc of some nature as the central component, the disadvantage of locating gauge wheels behind the opener can also take on a new and additional dimension because the distance from the seed zone then increases by at least the radius of the disc. Consequently, despite their advantages for controlling depth of seeding, many no-tillage opener designs do not use gauge wheels at all. With those that do, most are located either beside the opener or partly beside and partly behind it.

A further complication arises when gauge wheels are required to perform the additional function of covering the slot. Wheels that only function for covering are called 'press wheels', those that only gauge depth are 'gauge wheels' and those that perform both functions are 'gauge/press wheels'.

Few openers have gauge/press wheels. One reason is that, for accurate depth control, the wheel should operate alongside the seed deposit zone, while for effective pressing the wheel should follow behind the opener. Furthermore, the wheel must roll on undisturbed soil to maintain depth control, while for useful slot pressing the wheel should be on either the loose soil over the slot or in the slot itself (see Chapter 5). These somewhat contradictory requirements often lead either to two separate wheels or to one of the functions being compromised in the interests of cost and residue clearance. In general, if the wheels on openers are supported by springs, they will probably be there solely for the press wheel function rather than also as gauge wheels.

The wheel on the opener shown in Fig. 8.1 is solely a gauge wheel. A smaller separate press wheel can be seen operating at an angle behind the disc.

An example of combined press/gauge wheels is shown in Fig. 4.27, where two wheels are used on either side of a central disc and slightly rearwards of the seed zone. The wheels are sufficiently wide to register on the undisturbed soil alongside the opener (the gauge wheel function) but are also angled so that they fold the flaps of residue and soil back over the inverted-T-shaped slot and gently press on it (the press wheel function). Inverted-T-shaped slots do not require pressing on the seed in the slot, so there is no disadvantage from only pressing on the top of the covered slot (see Chapter 6). The depth-control function of this opener is slightly compromised because the wheels are not located exactly at the seed release point, but there are other systems employed with this opener (see below) that more than compensate for this shortcoming.

The value of semi-pneumatic tyres

It is appropriate here to pay tribute to semi-pneumatic tyres, which are used on most modern press wheels and gauge wheels.

This often undervalued invention is one of the most successful adjuncts to agricultural machinery. Until semi-pneumatic tyres were invented, all gauge/press wheels were either rigid wheels or, at best, solid rubber, plastic or fully inflated tyres.

Because press wheels on seed drills almost invariably operate at least partially in a disturbed soil zone, even in no-tillage, they are very inclined to accumulate mud in damp conditions. Flexure is the most effective means for a wheel to shed accumulated mud. Fully inflated tyres under normal pressures and rigid wheels do not flex sufficiently to shed mud. Some no-tillage situations may require enough downforce for a limited flexing by fully inflated tyres.

A method had to be found to combine flexure with maintaining the accuracy of the gauging radius of the wheel, i.e. it had to be able to flex but still retain a predictable loaded radius, regardless of the loading on it. This is where semi-pneumatic tyres excel. Although they are hollow (in a multitude of cross-sectional shapes), there is no air pressure within them. Indeed, most have a small bleed hole so that air cannot be permanently trapped inside. The distance between the outer wall and the inner wall (against the rim) is relatively small. In operation, where the footprint zone contacts the ground, the outer wall collapses temporarily and presses against the inner wall and thence the rim. As it leaves the ground, the resilience of the rubber causes the outer wall to return to its original position. In so doing, the outer wall continually flexes in and out, which dislodges mud. The operating radius remains predictable so long as there is sufficient force applied to collapse the outer wall against the inner wall and rim in the footprint zone.

Walking beams

Another adjunct to no-tillage openers is the use of 'walking beams' for mounting the gauge wheels, such that a pair of wheels can independently move vertically while continuing to share the down pressure. These are simple mechanical leverage systems, which are applicable where there are at least two gauge wheels. A single linkage, pivoted at its centre, joins the mounting brackets for the two wheels in a pivotal manner. The two wheels find their own positions by equalizing the footprint forces about the pivoting walking beam. The equalized positions of the two gauge wheels constantly change as each wheel in turn encounters changes in the soil surface. As one wheel moves upwards, the other wheel moves downwards.

The point of this arrangement is that as each wheel encounters a small rise or hollow the whole opener is forced to rise or fall by only half the height of the rise or depth of the hollow. Thus surface roughness is smoothed by a factor of a half, which is important for no-tillage in the absence of general smoothing by tillage.

Figure 8.2 shows a walking beam arrangement for a pair of gauge wheels.

Fig. 8.2. A walking beam arrangement for equalizing the loads carried by two independent gauge wheels.

Disc seed flick

The tendency of double disc openers to flick seeds out of the ground arises when seeds become clamped between the two discs at or near the pinch point where they touch. At speed, as the discs move apart again behind this point, the clamping action, followed by sudden release of the seeds, may propel them upwards and rearwards, expelling them from the slot.

The problem is overcome by dropping the seeds behind the pinch-point zone and/or by inserting covering plates in the zone between the two discs at their rearmost edges.

With all disc openers in sticky soils, at least one surface of the disc can become sticky. Seeds may either adhere to the disc and be lifted from the slot or soil may stick to the disc and carry seeds out with it.

With double disc openers, the seed is released against the inside surfaces of the discs that are not in contact with the soil. Thus, seeds seldom stick to the discs but soil sticking to the outside of the discs can seriously disrupt the integrity of slot formation and carry seeds, which have already been deposited, out of the slot (see Fig. 8.3).

With angled discs, the seed side of the disc is largely sheltered from soil contact, which helps to avoid seeds sticking directly to the disc.

The disc version of the winged opener has special subsurface scrapers designed to wipe sticking seeds off the disc below the ground (Thompson, 1993; Fig. 4.27).

Soil disturbance

With most disc openers, even when operating in non-sticky soils, a certain amount of soil disturbance occurs as the disc leaves the bottom of its rotation. This also occurs with hoe openers as the rigid shank moves forward in the soil. While seeds might not be flicked out of the soil by this soil movement, it may redistribute the seeds so that they occupy more random vertical positions within the soil than would otherwise be expected.

With some power till openers, the soil is deliberately disturbed and the seed is deposited into the rotor area while slot tilth is being formed, with the intention of thoroughly mixing the seed and soil. While this undoubtedly achieves its aim, the resulting variation in the depths of individual seeds does little for consistency of germination, emergence and maturity.

Residue hairpinning or tucking

The tendency of discs in any configuration to hairpin, or tuck, residue into the slot without actually cutting the residue often leaves the seeds embedded in or on this residue rather than in contact with clean soil. Many poor no-tillage plant stands have resulted from the hostile seed environment created by residue tucked directly into the seed slot. This occurs with both dry and wet residues, although the cause of the problem is different in the two cases.

With tough resilient residue, such as wet maize stover, the residue may quickly straighten out again after passage of the disc, in which case it may flick a portion of the seeds out of the slot. Figure 8.3 shows a soybean (*Glycine max*) seed that has been flicked completely out of a slot by a maize stalk after passage of a vertical double disc opener.

But, even if seeds are not flicked out, when they become embedded in dry hairpinned residue, they will not have effective seed–soil contact, this affects imbibition and germination. In wet soils, the fatty acids that are the products of decay of the residues cause seed and seedling mortality (see Chapters 6 and 7).

Opener bounce

Hoe-type and simple winged openers, which are under considerable downforce for penetration, often bounce in response to variations in soil strength, particularly at high speeds, disrupting the accuracy of seed ejection into the soil.

But disc-type openers are not immune either. Any opener is capable of leaving

Fig. 8.3. A soybean seed (lower centre) that has been flicked completely out of a no-tillage slot made by a double disc opener (from Baker, 1981a, b).

seeds on the surface after encountering stones in the soil. Hoe-type openers tend to push stones aside or flick them out of the ground, whereas disc-type openers tend to rise up and over stones and deposit seeds on top of the ground.

Seed bounce

As a result of high operating speeds and seeding into dry cloddy soils, large seeds often bounce upon contacting the soil. In severe cases, some seeds bounce right out of the slot.

The problem is accentuated with some air delivery systems when excessive delivery velocity of the air and seeds is used, which, combined with a high forward speed of the opener, may cause severe seed-bouncing problems.

Slot closure

Problems such as seed bounce can be largely overcome if the opener self-closes the slot immediately after it has been opened to receive the seed. Some winged openers, slanted double disc openers and power till openers are examples of openers with good self-closing abilities.

Drill and Planter Functions

Downforce mechanisms

The most common downforce mechanisms for conventional drills and planters are springs. But springs change their loading forces in a linear fashion with changing length (i.e. they change their forces by the same proportion as their lengths change). This might be acceptable for tilled seedbeds because: (i) the loads applied are relatively small and the springs are not significantly compressed; (ii) the variations in ground surface and therefore spring lengths are relatively small; and (iii) springs are relatively cheap and trouble-free.

For no-tillage seedbeds, however, the opposite is true: (i) spring loads are high; (ii) surface changes can be quite large; and (iii) no-tillage drills are generally more robust and expensive. Because spring loads are high, no-tillage drills tend to use either very heavy and unresponsive springs or smaller-section, longer springs compressed to short lengths. Because the changes in spring force are related to a spring's compressed length at the time, having a spring compressed to a short length to achieve opener penetration magnifies the force changes relative to length changes. Accordingly, some no-tillage drills and planters are designed with inordinately long springs (Fig. 8.4), or, alternatively, the springs are positioned near to the pivot points of the drag arms so that dimensional changes are minimized.

The force relationship with the length of springs applies equally well if the springs are arranged to be working in tension or in compression. Compression is more common, as it is difficult to overload a spring in compression compared with a spring in tension. For reasons of compactness, a few no-tillage drills and planters use springs acting in tension.

Either way, it is virtually impossible to maintain constant downforces with springs. A number of innovative designs have been used with the objective of reducing the shortcomings of springs. Some of these are illustrated in Figs 8.5 and 8.6. In Fig. 8.5, the mechanical springs have been replaced with rubber buffers acting very close to the pivot (fulcrum) of the drag arms to reduce the required travel of the springs for any one change in position. Rubber acts in an almost identical manner to spring steel with regard to the force it exerts in relation to changes to its compressed length. But problems from prolonged exposure of rubber to ultraviolet light and retention of 'memory' after long periods of compression have made this an unpopular choice.

In Fig. 8.6, the designers have attempted to better equalize the spring forces across the drill, to accommodate, for example, passing over a hump on one side of the drill, by dividing the bar that compresses the springs into shorter articulating lengths. The effect is similar to walking beams described above for press wheels.

Another way to overcome the disadvantages of springs for downforce application is to provide the gauge wheels with very large footprints and then apply excessive downforces to ensure that the spring force is sufficiently large to allow for lengthening of the springs for the deepest hollow likely to be encountered by the openers.

Fig. 8.4. Long compression springs on a no-tillage drill.

Fig. 8.5. No-tillage openers pressed into the soil with rubber buffers acting close to the fulcrum of the drag arms.

Fig. 8.6. A no-tillage drill with an 'equalizing' spring arrangement.

Figure 4.24 illustrates a design that has gone to the other extreme. In this case, the total vertical opener travel has been restricted by the use of spring tines that move largely horizontally (backwards) in response to increases in loading. The ground surface-following ability of such drills is poor, restricting their use to relatively smooth fields and/or seeds that are very depth-tolerant.

Regardless of the measures outlined above, springs are generally an unsatisfactory, though still the most common, way of applying downforces to no-tillage openers. Characteristically, their shortcomings can regularly be seen in the field as too shallow drilling through hollows and too deep drilling over humps, leading to poor seedling emergence in both situations. Figure 8.7 shows the travel of a no-tillage opener with superior surface-following ability. Unfortunately, not all no-tillage drills are capable of achieving this degree of surface following.

Compressed air

Fortunately, there are alternatives to springs. The two most useful to date have been the use of air and oil (hydraulic) pressure, acting through rams or cylinders (Morrison, 1988a, b). The air pressure option uses large volumes of air acting on large-diameter cylinders attached to the drag arms. Because it is difficult to compress air to sufficiently high pressures to allow small-diameter cylinders to be used, there are limits to the amount of downforce obtainable with compressed air.

On the other hand, air is free and large volumes can be compressed, with the result that changes in volume resulting from movement of openers up and down can be designed to have a minimal effect on the magnitude of the downforces. It should be appreciated that any gas under pressure has the same characteristics as mechanical springs. At any given temperature, a change in volume of the compressed gas will be linearly proportional to its pressure. With air, however, the volume can be made so large that pressure changes with movement of the openers can be minimized.

The biggest disadvantages of using air directly are the limited amount of pressure that can be practically obtained and the fact that the oxygen in air under high pressure can be explosive and that high-pressure air cylinders need to be independently lubricated, which is a problem in a semi-static system such as this. Lubrication is easiest where a continuous flow of compressed air

Fig. 8.7. An example of excellent surface following through a hollow by no-tillage openers (gas-over-oil downforce).

is used, such as with air tools. But in this case the compressed air is contained within a closed system, so lubrication is difficult.

Gas-over-oil systems

A more workable option has been to use oil in a hydraulic system in equilibrium with a compressed inert (non-explosive) gas (usually nitrogen) contained in one or more accumulators. This is referred to as a 'gas-over-oil', 'oil-over-gas' or 'nitrogen-cushioned hydraulic' system. The volume of gas in the accumulator(s), when the system is at its likely operating pressure(s), needs to be sufficiently large to reduce changes in pressure, arising from changes in opener position, to a minimum.

In reality, if the hydraulic cylinders on all openers are connected in common (parallel) to the hydraulic system, when one opener rises in response to a rise on the soil surface, another opener is likely to be falling in response to a hollow somewhere else across the drill or planter. Thus these two openers simply exchange oil between them without affecting the overall volume of oil or pressure of the system to any great extent.

Because of this, the need for large volumetric changes by the hydraulic system as a whole is much reduced. In contrast, mechanical springs can only work with individual openers unless a very complicated linkage is used to obtain some measure of combined action, as illustrated in Fig. 8.6.

Another advantage of the gas-over-oil system is that, if the individual hydraulic cylinders are of the double-acting type (i.e. they can be powered in both directions), these downforce cylinders can also be used to lift the openers for transport. This eliminates the need for a separate lifting assembly on the drill or planter.

The biggest advantage of either gas-over-oil or air cylinders is that they can be arranged so that the downforce on the openers remains virtually unchanged throughout the entire length of opener travel upwards and downwards because the force exerted by the cylinders remains constant throughout their entire stroke length. This in turn allows much greater vertical travel to be designed into the openers for surface following and depth control.

Figure 8.8 shows a no-tillage drill with a gas-over-oil downforce system sowing at the same depth on the top of an irrigation

Fig. 8.8. An illustration of the extraordinary surface-following ability of a gas-over-oil opener downforce system on a no-tillage drill.

border dyke as on the flat surface alongside, and even part-way up the slope. Tillage drills are never required to provide this much opener travel and many simple no-tillage drills do not achieve it either.

Automatic down force control (ADF)

A further refinement to the gas-over-oil system is to equip the drill or planter with a sensing device that measures the hardness of the soil as the opener travels through it. This signal is relayed to the hydraulic valving so that, as the soil hardness changes (which would otherwise alter the penetration depth of the opener), the oil pressure is automatically adjusted on the move to ensure that the openers get the correct amount of downforce to correctly maintain seeding depth in each metre of the field. This sophistication provides a fully automatic seeding depth-control capability, unparalleled with current technology.

Weights

One school of thought suggests that attaching weights to individual openers would be an effective way to ensure that each opener experiences the same downforce throughout its entire range of movement. But adding and removing individual weights for a multitude of openers on any one drill is impractical and would require the operator to carry surplus weights around in order to change the downforces for new conditions. It would also make changing the downforce on the move within a field impractical, but, then again, only the most sophisticated gas-over-oil systems with ADF allow this to be done.

Another downside to the use of weights is that, when an opener rises or falls, the inertia of the weight alters the effective downforce and that this inertia is highly dependent on the forward speed of the machine, which determines the speed of the rise and fall. For the technically minded, inertia is proportional to the square of speed in the direction of movement.

Where weights have their greatest use is for single-row drills, since many of the disadvantages above apply less to a single opener than to multiple openers on a larger drill and weights are often the cheapest and most effective option available where limited budgets apply (see Chapter 14).

Drag-arm design

The design and configuration of the drag arms that attach the seed opener to the drill frame are an important feature of drills or planters that have an impact on seed placement. A drill that has drag arms pivotally attached to the drill or planter frame will be designed to move the openers upwards and downwards to accommodate changes in the ground surface. This motion is provided either by a hinged attachment to the drill frame or by flexure of the drag arms themselves.

In the case of flexed drag arms, the whole drag arm must be constructed of spring steel. There are advantages in that this eliminates wearing joints, which, under the high forces involved in no-tillage, can become a maintenance problem. Such a desirable arrangement, however, must be balanced against the disadvantages of using mechanical springs as the downforce system in the first place and the difficulty in preventing the openers from also flexing sideways, which interferes with accurate row spacing.

With fully articulated (hinged) drag arms, the most common arrangement with conventional drills is to use a single arm pivoting on a simple unlubricated joint, as shown in Fig. 8.9. Because large forces are required to push openers into and drag them through the soil, there are quite large forces acting on the pivot, especially if the source of downforce is located close to the pivot itself. As a result, the wear rate within the pivoting mechanisms can be substantial.

This is an important issue with many seemingly advanced no-tillage machines. As new machines, they might appear to be of sound design. But as the pivoting joints wear, such machines soon provide poor seeding accuracy and become unserviceable, which creates an unforeseen cost penalty against no-tillage.

More sophisticated no-tillage drill designs provide pivots with lubricated and

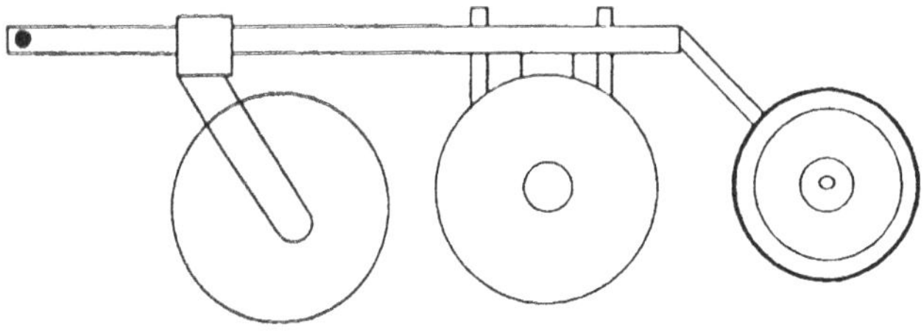

Fig. 8.9. A simple single-pivot drag-arm arrangement (slightly modified from Baumer *et al.*, 1994).

sealed bearings or heavy-duty bushings. While this adds to the initial cost, it can extend their service life to near that of the tractors that pull them.

Parallel linkages

To ensure correct functioning, some no-tillage openers must be maintained at a set angle to the horizontal in the direction of travel. Winged openers are a case in point. Such openers often employ two drag arms (upper and lower) arranged as a parallelogram in such a way that the horizontal angle of the opener remains unchanged throughout the entire range of its vertical travel.

The disadvantages of such an arrangement are the cost of the arms and pivots and the fact that four pivots have greater potential to create diagonal instability of the openers than one or two pivots if they become worn. To compensate, parallelogram drag arms are usually wider and more robust than single drag arms and utilize better-quality bushings or bearings in the pivots. Undoubtedly, they go another step towards perfecting precision seed placement in no-tillage, but to date they have only been included on advanced planter and drill designs.

Figure 8.10 shows a no-tillage opener mounted on parallelogram drag arms and the extraordinary range of travel provided by its gas-over-oil downforce system. The hydraulic cylinder is difficult to see but can be located from the position of the supply hoses (top right).

A variation on parallelogram drag arms is one where the parallelogram is designed to be deliberately imperfect (i.e. a trapezium). It is designed for operation with winged openers that are pushed into the ground with mechanical springs (Fig. 8.11).

Fig. 8.10. A winged no-tillage opener mounted on parallelogram drag arms and pushed down with a gas-over-oil system, showing its extraordinarily large range of vertical travel, which is important in no-tillage.

The objective has been to reverse the geometrical changes that occur with single-pivot drag arms, in which the angle of the wings normally becomes less in hollows and increases over humps. The effect is usually to accentuate the change in mechanical spring forces by drawing the wings into the ground more on humps than in hollows. But, in this design, the wing angle increases when the openers are in hollows and decreases when they go over humps. Since the steeper wing angle assists the opener to pull itself into the ground, the arrangement

Fig. 8.11. A no-tillage opener in which a deliberately imperfect 'parallel' (trapezium) linkage is utilized.

goes some way towards countering the disadvantages of variable downforces with mechanical springs.

Comparisons

The authors compared the capabilities of two different no-tillage drills (both of which featured gas-over-oil downforce systems) in terms of their abilities to ignore surface irregularities (Baker and Saxton, 1988). Three types of tillage tool were used to cause surface roughness in an otherwise smooth untilled soil that had been chemically fallowed. The roughness treatments were: (i) chiselled with a shank chisel at 380 mm centres operating 200 mm deep, which left the roughest finish; (ii) cultivated with 250 mm wide sweeps operating 100 mm deep (the next roughest finish); (iii) disced once with a heavy double disc (the next roughest finish); and (iv) no tillage at all, which left a smooth surface finish. The drills used are labelled in the diagrams as 'Cross Slot' (disc version of winged openers that created inverted-T-shaped slots) and 'Double Disc' (vertical double disc openers that created V-shaped slots).

The plant stands from the two drills and four surface roughnesses are shown in Fig. 8.12, and the resulting yields of winter wheat are shown in Fig. 8.13. The 'Cross Slot' drill had higher plant counts and yields than the 'Double Disc' drill for all surfaces, but significantly more so for the rougher surfaces. The much heavier 'Double Disc' drill had difficulty maintaining depth control in the more loosely tilled, rougher surfaces. The no-tillage surface was easily penetrated by both drills, but the double disc openers 'tucked' considerable residue into the seed slot, which probably contributed to the lower stands with that drill in the very dry seeding conditions that were experienced (see Chapters 6 and 10).

Seed metering and delivery

With small seeds sown on a mass basis, such as grasses, legumes, brassicas and small-grained cereals, the seed metering devices on drills are designed to distribute a continuous trickle of seeds with no attempt to single, or handle individual seeds separately.

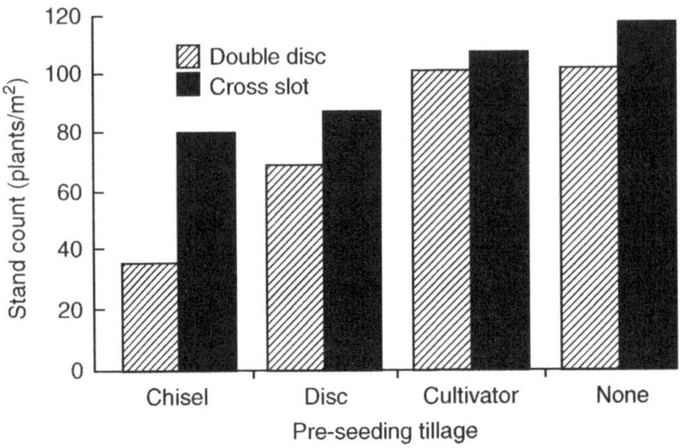

Fig. 8.12. The effects of surface roughness on wheat seedling emergence using two contrasting drills (from Baker and Saxton, 1988).

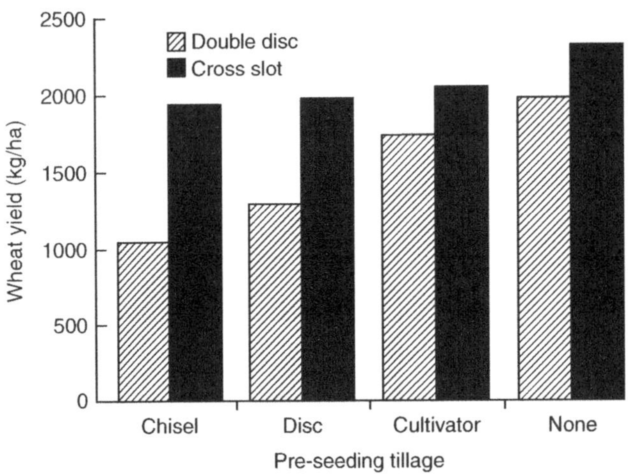

Fig. 8.13. The effects of surface roughness on yields of wheat using two contrasting drills (from Baker and Saxton, 1988).

As a result, such a trickle of seeds is largely unaffected by the length or shape of the delivery tubes that transport them from the seeder to the opener, so long as there is sufficient slope on the tubes for gravity to keep the trickle moving consistently or a stream of air to blow them along. Gravity delivery can be a problem when drilling up and down hillsides, where the drop tubes become too flat to maintain the seed flow. With air seeders, which substitute air flow for gravity, the air flow transports the seeds in a consistent manner to the openers and gravity plays only a minor role.

Seed metering and delivery are generally similar for no-tillage drills and drills used in tilled soils, with only minor differences. The seed metering mechanisms and delivery tubes can be expected to be common to both; however, the openers of no-tillage drills are often spaced further apart to clear residues and their vertical travel may be greater than for tilled soils. As a result, the seed delivery tubes may be longer and have further to span from the metering boxes to the openers, which may cause them to lie at flatter angles. Compensation for this loss of fall may involve raising the seed boxes higher on the drill or the use of multiple sets of seed boxes. Air delivery becomes an attractive option, since gravitational fall is then assisted by the air flow (see Chapter 13). An example of an advanced no-tillage drill with air-assisted seed and fertilizer delivery is shown in Fig. 8.14.

Fig. 8.14. An advanced design of no-tillage air drill.

Precision seeders that select single seeds at regular intervals, such as maize, cotton, beet and vegetable planters, provide a different situation. Ritchie (1982) and Carter (1986) showed that, once a single seed is released from the metering mechanism into a tube, its pathway through that tube may be somewhat random. It will have a tendency to bounce from wall to wall and at each bounce it will lose an unpredictable portion of its drop velocity. Consequently, any two seeds seldom arrive at their destinations at exactly the same time intervals from when they were released from the metering mechanism.

Thus, even if a precision metering mechanism selects individual seeds at precise intervals, the precision of the intervals at which consecutive seeds reach the ground will depend on the pathways each follows after leaving the metering mechanism. It is even possible for a seed that took a more direct route down a delivery tube to catch up with and pass an earlier seed that bounced on its way down the same tube.

For this reason, precision seed metering mechanisms in tilled soils are located as close to the soil as possible so that the seeds have only a short drop, often without touching the sides of any tubes at all. Commonly, the distance of drop is about 50 mm and often less. This free-drop approach is possible only because tilled soils are prepared so as to have no surface residues and are as smooth and fine as possible, allowing the bulky seeding mechanism to pass close to the ground surface without the risk of blockage or damage.

In no-tillage, however, surface residues often protrude 300–500 mm above the ground, are variable in their nature and extent and are often quite woody. Vertical clearance is therefore necessary to avoid blockage. Further, there is little or no opportunity to smooth the surface of the soil. Consequently, no-tillage openers are larger and more robust than their tillage counterparts and the metering mechanisms have to operate higher above the ground. This necessitates seeds having to be delivered up to 600 mm from the metering mechanisms.

Free drop of seed is not an option over such a distance in no-tillage because of the effects of wind, slope and machine bounce.

The result is that, although the same precision metering mechanisms are used for tillage and no-tillage planters and the same numbers of seeds need to reach the ground in a given length of row in both cases, precise spacing between individual seeds under no-tillage is more difficult to achieve than in tillage.

Opener bounce is likely to be greater under no-tillage. Attempts to qualify the effect of openers' bounce were reported in 2004 (Anon., 2004). The tests found that four conventional vacuum-type precision seed metering devices of European origin were all adversely affected by shifting from a tilled soil surface to an untilled surface and that the adverse effects increased with increasing forward speed.

The key question of whether or not these sources of inaccuracy have a measurable effect on the final yield of large, compensatory-growth plants, such as maize, will continue to be debated (for example, there is mounting evidence that precision seeding depth may be more important than precision spacing, due to inter-plant competition) but the fact remains that precision spacing has become an important marketing objective for machines designed for tilled seedbeds. Since there is no known agronomic downside to precision spacing, it makes sense for designers of no-tillage planters to attempt to duplicate these levels of precision spacing as closely as possible if they want to persuade farmers to make the switch from tillage to no-tillage.

Summary of Seed Depth, Placement and Metering

1. Wheat seedling emergence in no-tillage may decline by approximately 4% for every 10 mm increase in drilling depth below 20 mm and even more beyond 50 mm.
2. Lupin seedling emergence in no-tillage may decline by approximately 17% for every 10 mm increase in drilling depth below 20 mm.
3. Red clover seedling emergence in no-tillage will decline markedly at drilling depths above and below 10–15 mm.
4. The ability of no-tillage openers to maintain a constant seeding depth is very important but very demanding.
5. Harder ground, rougher surfaces and the presence of residues on the surface accentuate the depth-control challenge under no-tillage.
6. Because of the large opener downforces required in no-tillage, seeding depth control often uses one or more gauge wheels on each opener.
7. Press wheels are often also used on each opener to cover the slot.
8. Few no-tillage openers have both gauge wheels and press wheels, and even fewer have combined gauge/press wheels.
9. Zero-pressure tyres are a useful adjunct to gauge wheels.
10. Walking beams are also a useful adjunct to gauge wheels.
11. Mechanical springs are a poor means of providing downforce for no-tillage openers because their forces change with length.
12. Compressed-air cylinders are sometimes used to provide downforce but are seldom a practical option.
13. Removable weights are useful on single-row no-tillage drills but are not practical for multi-row machines.
14. Gas-over-oil systems offer advantages by using hydraulic cylinders to both apply the downforce and lift the openers for transport.
15. Automatic downforce control systems offer further refinement to gas-over-oil systems by changing the downforces on the move in response to changes in soil hardness.
16. No-tillage openers should provide up to 500 mm vertical travel compared with a maximum of 150 mm for tilled soils.
17. Single-pivot drag arms on drills and planters are less useful in no-tillage than in tillage.
18. Parallelogram drag arms maintain the opener angle but are mechanically more demanding.
19. Lubricated bearings or bushes used for the pivots on no-tillage openers contribute to a realistic service life of machines that operate under difficult conditions.

20. The function of no-tillage openers in depositing seed consistently in an uninterrupted horizontal band in the soil is important.
21. The function of no-tillage openers depositing fertilizer in a separate band is also important, as discussed in Chapter 9.
22. The delivery of bulk-metered seeds to no-tillage openers is made more demanding by their large horizontal and vertical spacing.
23. Air delivery of bulk seeds to no-tillage openers offers advantages.
24. Single-seed spacing along the row from precision planters may be compromised in no-tillage because of seed bounce down long delivery tubes.
25. No-tillage openers may have special problems, such as seed flick, seed sticking to the disc, soil turbulence, residue 'hair-pinning', opener bounce, seed bounce and slot closure.

9 Fertilizer Placement

C. John Baker

Simultaneous banding of seed and fertilizer by the openers is more important in no-tillage than for tilled soil and follows somewhat different principles.

It is especially important in no-tillage to sow fertilizers at the same time as the seed, but only if the fertilizer can be placed in a separate band from the seed. Much recent experience has documented the growth and yield advantages from fertilizers banded near the seed at the time of seeding. For autumn seeding this is often only a 'starter' amount of fertilizer, while for spring seeding it is usually the total seasonal requirements.

Crop responses to banded fertilizer at the time of seeding are nearly always larger in no-tillage than in tillage. There are several reasons for this.

- Tillage mineralizes organic matter to release nitrogen and this becomes readily available to the newly establishing plants. The downside is that, because no fertilizer is actually added to the system, the nitrogen is from 'mined', mineralized, SOM, which depletes this precious resource cumulatively. Because mineralization and nitrogen release are minimal under no-tillage, young no-tilled plants can appear nitrogen-deficient, particularly during early growth. Banding nitrogen fertilizer alongside the seed during no-tillage seeding cures the problem.
- Surface residues are often decomposing about the same time as seeding in no-tillage. The microorganisms responsible for residue decomposition temporarily utilize ('lock up') nitrogen during this process. Even though the nitrogen they demand may become available again later in the growth cycle as the microorganisms themselves die, it is temporarily made unavailable to young no-tilled plants.
- Soluble nutrients, nitrogen in particular, broadcast on to a no-tilled soil surface (as is common practice in tilled fields) are often preferentially carried by water flow down earthworm channels and other bio-channels (e.g. old root channels), which largely bypass young plant roots. In tilled soils, these bio-channels are destroyed and replaced by a smaller, more evenly dispersed pore system, which provides a more uniform infiltration of water and broadcast fertilizers.
- Under repeated no-tillage, surface-applied nutrients that readily attach to soil particles, such as phosphorus, accumulate in a narrow layer near the ground surface and may not be readily available to young plants.

Many of these factors often combine under no-tillage regimes to make nutrients less readily available to both seedlings and growing crops. Thus banding of fertilizers simultaneously at seeding becomes all the more important.

Numerous experiments and field observations have confirmed that the broadcasting of fertilizers during no-tillage often results in poor crop responses. Figure 9.1 illustrates a typical field response. A contractor (custom driller) had been sowing pasture species in New Zealand with winged openers into an otherwise fertile field while simultaneously banding 300 kg/ha of an N : P : K fertilizer mix alongside (but not touching) the seed. Near the end of the field the contractor ran out of fertilizer. The farmer asked him to carry on sowing seed alone while he (the farmer) broadcast the same rate of fertilizer on the remaining area, which he did. Inadvertently the farmer had set up a comparison of banded versus broadcast fertilizer. Figure 9.1 clearly shows the difference in plant response 8 weeks after drilling.

Nor are such responses restricted to grasses. In fact, responses to placed fertilizer under no-tillage were first identified with wheat in the USA in the 1980s (Hyde *et al.*, 1979). Almost every crop and soil have the potential to show a similar response to that illustrated in Fig. 9.1. Both narrow-leaved (monocotyledonous) and broadleaved (dicotyledonous) plants have regularly shown similar responses.

Figure 9.2 shows a marked response to banded fertilizer in France with maize. The four rows in the centre and left of centre in the photograph had broadcast fertilizer applied at the same rate as the placed fertilizer in all other rows. The differences are remarkable.

There are two important considerations when applying fertilizer by banded placement:

1. Possible toxicity of the fertilizer to the seeds and seedlings, often referred to as 'seed burn'.
2. Yield responses of the growing plants to the placed fertilizer.

We shall discuss these two aspects separately.

Toxicity

There are three options for applying fertilizer under no-tillage: (i) broadcasting on

Fig. 9.1. Pasture established by no-tillage in New Zealand with broadcast fertilizer in the foreground and banded fertilizer in the background at 8 weeks' growth.

Fig. 9.2. The difference between broadcast fertilizer (left-of-centre four rows) and banded fertilizer (all other rows) in no-tilled maize (France).

the surface; (ii) mixing with the seed; or (iii) banding separately from the seed at the same time as the seed is sown.

Since broadcasting of fertilizer is a separate operation either before or after seeding and not a function of the no-tillage drill or planter, we shall not consider it further here.

Mixing of fertilizer with seed is a risky undertaking at any time because of potential toxic chemical damage to the seed and seedlings. In tilled soils, a measure of dilution of the fertilizer with loose soil will often reduce the risk of 'seed burn'. But in an untilled soil, particularly one that is damp, soil dilution by mixing becomes minimal.

In general, fertilizer–seed toxicity will be affected by the following:

- The formulation of the fertilizer. Most forms of nitrogenous and potassic fertilizers are likely to 'burn' seeds, as well as some forms of phosphatic fertilizers. Secondary nutrients such as boron and sulphur can be particularly toxic.
- The form of the fertilizer. Dry granular fertilizers are more often placed directly with the seed than liquid fertilizers. While it is easier to direct the liquid placement away from the seed than the granular, either form will cause toxicity.
- The age of the fertilizer. 'Fresh' superphosphate may contain free sulphuric acid, although this dissipates over time in storage.
- The moisture content of the soil. Dry soils concentrate the fertilizer salts in the limited soil solution, which may damage or kill the seeds by the effects of reverse osmosis.

Mixing seed and fertilizer and sowing them together or alternatively allowing them to mix in the opener or the soil is therefore a very unsatisfactory way to provide nutrients for young no-tilled plants. At best, small amounts of starter fertilizer might be applied in this manner. Usual upper limits are considered to be at about 15–20 kg/ha of nitrogen. But a higher level of risk must be accepted compared with separate banding of seed and fertilizer.

Banded fertilizer

For separate banding of seed and fertilizer, the seed and fertilizer must be placed in different positions in the soil and remain in these positions after the opener has passed and the slot has been closed.

There are three realistic geometric options. The fertilizer can be placed directly below, to one side of or diagonally below and to one side of the seed. Placing fertilizer above the seed is not a logical option because this is very similar to broadcasting.

The ability of no-tillage drills and planters to simultaneously band seed and fertilizer without the two coming into contact with one another is widely recognized as one of their most essential functions. Indeed, an informal survey of no-tillage experts in the USA in the 1980s revealed that separate banding of seed and fertilizer was unanimously regarded to be the single most important design improvement that should be made to no-tillage openers. Unfortunately, providing this function has proved to be an elusive capability for many machinery manufactures.

Some no-tillage drills and planters employ two separate openers, one for seed and another for fertilizer. Others combine the two openers together into one (often complicated) 'hybrid' opener, while still others use one dedicated fertilizer opener between each pair of seed openers. But there are also modern openers designed specifically for no-tillage that band seed and fertilizer in the same slot without compromising seeding accuracy, row spacing or residue handling for a wide range of forward speeds, soils and residue conditions.

Vertical banding versus horizontal banding

The absence of friable soil makes vertical separation of seed and fertilizer more difficult in no-tillage than in tilled soils, even by successive openers or duplicated components.

Some drills and most planters in loose or tilled soils use a leading opener to place fertilizer at a given depth and then follow that with a scraper that fills the slot with loose soil. This in turn is followed by the seeding opener which opens a new slot that is either shallower and/or to one side of the fertilizer slot. Such repeated manipulation of loose soil is generally not possible or desirable under no-tillage, so the choice is to either broadcast or inject the fertilizer as a separate operation before seeding or simultaneously seed and place (band) the fertilizer to one side of the seed by a separate opener.

Experience with tilled soils suggests that vertical separation of seed and fertilizer should be at least 50 mm (known as 'deep banding'). Experience with no-tillage, however, shows that extrapolation of results from tilled soils requires adjustment for the nature of the soils and the machine performance.

The disc version of winged openers provides a physical barrier between the two sides of a horizontal slot in the soil, thus allowing seed to be deposited on one side and fertilizer on the other to provide adequate horizontal separation or banding. As the disc withdraws from the soil it tends to draw the soil up a little, resulting in a final horizontal separation distance of 10–20 mm. Figure 9.3 shows the horizontal separation of seed and fertilizer in an inverted-T-shaped slot created by a winged opener.

It is also possible to separate the seed and fertilizer vertically with this opener by arranging a long and short blade on the same side of the disc. Figure 9.4 shows a prototype winged opener with long and short blades to provide vertical separation of seed and fertilizer.

Yet another option exists with this opener using a long and short blade on opposite sides of the disc, thus creating diagonal separation (i.e. both vertical and horizontal). Figure 9.5 shows an excavated slot created by a winged opener in which there is a distinct step down from the seed shelf to the fertilizer shelf (i.e. diagonal banding). Figure 9.6 is a diagrammatic representation of diagonal banding using two separate disc openers. Similar placement patterns have recently been achieved with modified hoe-style openers using configurations to introduce the seed and fertilizer at different depths of penetration.

Baker and Afzal (1986) compared the effects of vertical and horizontal separation distances of ammonium sulphate (21 : 0 : 0 : 24) fertilizer from canola (rape, *Brassica napus*) seed in an untilled silt-loam soil using a winged opener. Canola seed is known to be particularly sensitive to the presence of ammonium sulphate fertilizer. Figure 9.7 shows seed damage determined by counts of seedling emergence, and Table 9.1 shows the seedling growth.

Figure 9.7 shows that horizontal separation by as little as 10 mm was equivalent to vertical separation by twice that distance

Fig. 9.3. A cross-section of an inverted-T-shaped slot showing the horizontal banding of seed (left) and fertilizer (right) (from Baker and Afzal, 1986).

Fig. 9.4. A prototype winged opener with long and short blades for vertical separation of seed and fertilizer (from Baker and Afzal, 1986).

Fig. 9.5. Diagonal separation of seed and fertilizer in the soil (fertilizer below the seed towards bottom of photo) using a winged no-tillage opener with an elongated blade on one side.

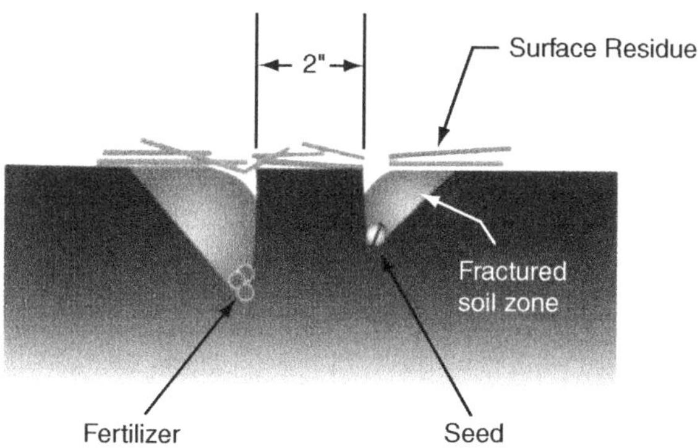

Fig. 9.6. A diagrammatic representation of diagonal fertilizer banding with two angled disc openers.

(20 mm) for reduced germination and emergence.

Table 9.1 shows that not only was there less seed damage from 20 mm horizontal separation, there was also a significant growth advantage for the 20 mm horizontal separation option compared with mixing of the seed and fertilizer together or separating the two by 10 mm either horizontally or vertically. Neither the horizontal nor the vertical separation by 20 mm was significantly different from where no fertilizer had been applied, which confirmed that no seed damage had occurred.

Afzal (1981) also compared the effectiveness of horizontal separation by a winged opener in tilled and untilled soils, to gauge the extent to which results from tilled soils could be safely extrapolated to untilled soils. Table 9.2 shows the results. At all three sampling dates (10, 15 and 20 days after sowing), the no-tilled soil contained more plants than the tilled soil, indicating that some seeds in the tilled plots had either been killed by the fertilizer, or had failed to germinate for other reasons.

An explanation for the effects in Table 9.2 seems to lie in the fact that, with

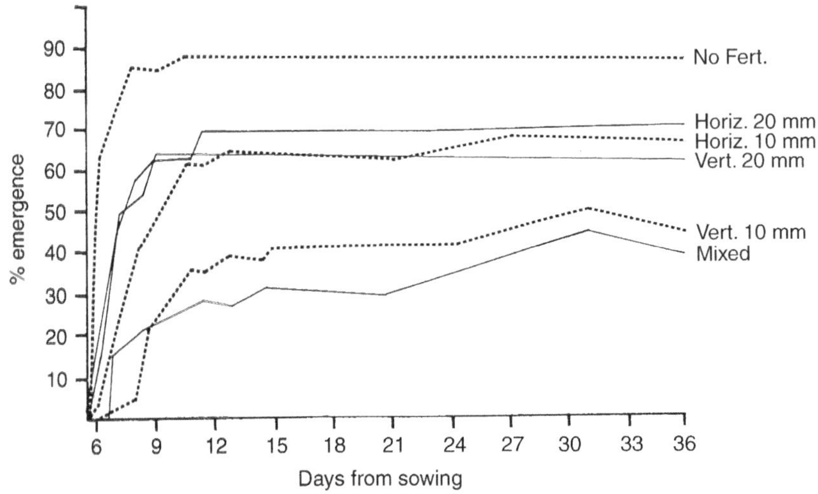

Fig. 9.7. Effects of the position of fertilizer placement, relative to the seed, on seedling emergence of no-tilled canola (from Baker and Afzal, 1986).

Table 9.1. Effects of method of fertilizer placement on seedling performance of no-tilled canola.

	Number of true leaves	Plant height (mm)	Plant weight (g)
No fertilizer	4.1 ab	63 ab	46 ab
Seed and fert. mixed	3.3 b	36 b	22 b
Horizontal separation by			
10 mm	3.3 b	34 b	19 b
20 mm	4.3 a	71 a	80 a
Vertical separation by			
10 mm	3.3 b	38 b	25 b
20 mm	4.2 ab	60 ab	54 ab

Unlike letters in a column denote significant differences ($P < 0.05$).

Table 9.2. Effects of tillage and no-tillage on horizontal separation of canola seed and fertilizer in the slot.

	Days after sowing		
Establishment method	10	15	20
No-tillage (plants/square metre)	25.1 a	50.7 a	55.2 a
Conventional tillage (plants/ square metre)	19.4 b	41.6 b	44.8 b
Increase of no-tillage over conventional tillage	29%	22%	23%

Unlike letters in a column denote significant differences ($P < 0.05$).

this particular opener design, the central disc cuts a thin vertical slot in the soil 50–75 mm deeper than the horizontal shelves on which the seed and fertilizer are placed. In an untilled soil, the integrity of this disc cut remains more distinct than in a tilled soil, where the friable nature of the soil allows soil to collapse into the disc-cut zone as the disc withdraws from the soil.

It is thought that this disc cut, in an untilled soil, effectively interrupts solute movement from the fertilizer, which might otherwise reach and damage the seed or seedling roots. It is also possible that the high humidity in the inverted-T slot in an untilled soil helps prevent reverse osmosis, which is one of the mechanisms by which seeds are damaged by high salt concentrations in dry tilled soils (see Chapters 5 and 6). Because the general humidity of a tilled soil is lower than that of an untilled soil, due to the artificially high porosity and the absence of surface residues, even the inverted-T-shaped slot is unable to maintain a high humidity zone around the seed when operating in a tilled soil.

Another important point in the tilled/no-tilled soil comparison is that the effects of separating the seed from the fertilizer are most apparent as the soil became drier. Collis-George and Lloyd (1979) had earlier noted that, in tilled soils, dryness tended to result in more fertilizer damage to seeds than where the soil was moist. Baker and Afzal (1986) examined whether or not this trend extended to untilled soils, using a winged opener.

Their results, shown in Table 9.3, indicate that plants suffered with both vertical separation and mixing together when the soil became dry, but these were equivalent to the other treatments in the moist soil. The only treatment that almost ignored the moisture status of the soil was the horizontal separation within an inverted-T-shaped slot. This may have been partly the result of the high humidity this slot maintains and partly the result of the disc cut. The result is that the optimum horizontal separation distance within an inverted-T-shaped slot was less than the distance commonly recommended for vertical separation by other openers and for tilled soils.

Field experience has shown that the particular disc version of the winged opener used in these experiments is equally well suited to separating seed from liquid or gaseous fertilizers as it is to separating it from dry powdered and granulated forms of fertilizer.

In two separate experiments (C.J. Baker, unpublished data), the author found that the upper limit of dry urea (46 : 0 : 0 : 0) application with this opener, sowing maize in 750 mm spaced rows, was about 200 kg/ha of urea (92 kg/ha/N), equivalent to 15 g urea per metre of sown row, before seed damage was detectable. Field applications of 780 kg/ha of 30% potassic superphosphate (0 : 6 : 15 : 8) with peas in 150 mm rows (117 kg/ha/K) have also been achieved with this no-tillage opener with no measurable toxicity damage to seed germination when compared with no fertilizer.

K.E. Saxton (unpublished data) also tested the ability of the same winged opener to effectively separate wheat seed from toxicity damage arising from the use of a range of rates and two forms of nitrogenous fertilizers sown in 250 mm rows in the USA. He found no detrimental effect on the seed from applying either dry urea (46 : 0 : 0 : 0) or liquid 'aqua' (ammonium hydroxide

Table 9.3. Effects of position of fertilizer placement and soil moisture status on germination of no-tilled canola (germination %).

Horizontal separation by 20 mm		Vertical separation by 20 mm		Mixed together	
Dry soil	Damp soil	Dry soil	Damp soil	Dry soil	Damp soil
89	81	64	90	58	85

solution in water: 40 : 0 : 0 : 0) at concentrations of up to 140 kg/ha of nitrogen.

Operators in New Zealand commonly apply up to 400 kg/ha of high-analysis fertilizer mixes (which sometimes include boron and/or elemental sulphur) in the field with this opener with no measurable effect from 'seed burn' but with substantial positive growth and yield responses (Baker *et al.*, 2001).

Although horizontal separation appears to be somewhat more beneficial than vertical separation in most instances, a range of vertical separation systems have been designed. Hyde *et al.* (1979, 1987) reported attempts to separate seed and fertilizer vertically with a single opener by modifying a hoe opener so that it deflected soil back over the fertilizer before the seed exited the opener. The deflecting action, however, was dependent on forward speed and soil moisture conditions, especially plasticity. In favourable conditions, its crop yield performance was comparable to horizontal separation by winged openers.

One solution that allows vertical separation of seed and fertilizer in no-tillage to be largely independent of soil moisture conditions is the use of slanted double disc openers. The leading (fertilizer) opener cuts a slanted slot and places the fertilizer at its target depth. The seed opener, which follows, is positioned either vertically or at the opposite slant and shallower, thereby placing the seed in the undisturbed soil above the fertilizer. This option appears to be effective but the downforces required to make two double disc openers penetrate the soil for each row limits it to reasonably soft soils. Figure 4.8 shows two slanted double disc openers so configured.

Another, more laborious but effective, method is to pre-drill the fertilizer as a separate operation to drilling of the seed at a shallower depth, and this can be achieved with virtually any design of opener.

Retention of gaseous fertilizers

Inverted-T-shaped slots are known to retain water vapour in the slot (see Chapters 5 and 6). It is possible that this slot also retains volatile gases from nitrogenous fertilizers (especially ammonia) within the slot in a similar manner to water vapour. It is well known that soil injection of both organic (animal waste) and inorganic forms of nitrogen as gas or liquid leads to problems with ammonia gas volatilizing and escaping into the atmosphere. With disposal of animal waste using knife-type openers (U-shaped slots), this is often overcome by deep (0.5 m) injection. Inverted-T-shaped slots also offer the option of shallow injection of this material (Choudhary *et al.*, 1988b).

During the no-tillage drilling of seeds, simultaneous deep injection of inorganic nitrogen is impractical because of the limitations on depth of placement and available tractor power. The result of simultaneous shallow placement has usually been a noticeable smell of ammonia at drilling as it escapes from the sown slots.

With the winged opener, less ammonia smell is evident, indicating entrapment of the valuable fertilizer within the slots. This was first noticed in the field in the USA by farmers using a winged opener. They were intrigued by the fact that the farm dogs ran along behind the drill. This apparently did not occur with other drills because the escape of ammonia from the soil immediately behind the drill made an unpleasant environment for the dogs.

Crop Yield

As previously discussed, broadcast fertilizers on no-tilled fields are often infiltrated by water moving into preferential flow paths and bypassing the early plant roots, or those constituents that bind to the soil remain on the soil surface. In contrast, tilled soils have more diverse flow paths through their microporosity and blend those binding constituents within the tilled zone. As a result, while broadcasting of fertilizers has been practised successfully for years with crops grown in tilled seedbeds, under no-tillage the same crop responses to broadcast fertilizer cannot be relied upon. Hyde *et al.* (1979) highlighted the problem in the Pacific Northwest of the USA, and a long-term experiment conducted by the authors over a 6-year period in New Zealand also illustrated the problem (Baker and Afzal, 1981).

In the New Zealand experiment, the scientists compared the continuous growing of summer maize, sown with a winged opener, on the one hand, into untilled soil and, on the other hand, into a conventionally tilled seedbed. It also coincided with some important technological developments of winged openers, which had an impact on the experiment.

Figure 9.8 illustrates the first 5 years of the maize yield results. To eliminate seasonal variations in yield, conventional tillage was given the arbitrary value of 100% each year and no-tillage was compared with it on a percentage basis. The seed was sown into inverted-T-shaped slots on all occasions with Class IV cover.

In year 1 no fertilizer was applied, either at planting or after the crop became established. The crop relied solely on the already high fertility of the soil, which had been under intensive pasture for 20 years. The maize yield under no-tillage was not significantly different from that under tillage.

In year 2 again no fertilizer was used. By this time, however, the advantages of mineralization, which is enhanced by the tillage process, had become evident. Only slow mineralization rates occur under no-tillage because of the absence of soil disturbance. As a result, the no-tillage maize yield was only 35% of that under tillage.

In year 3 a comprehensive NPK starter fertilizer (10 : 18 : 8 : 0) was surface-applied at 300 kg/ha by broadcasting on to all plots. At that time, simultaneous banding of seed and fertilizer by winged openers was not possible without risk of seed damage. The seed was sown with the simple original winged opener and mixing of seed and fertilizer together was not considered a viable option.

The disc version of the winged opener, which allows simultaneous banding, had not by then been invented. None the less, the surface-applied fertilizer lifted the yield under no-tillage to 60% of that under tillage.

In year 4 it was decided to apply a greater amount of broadcast NPK fertilizer than in year 3 (400 kg/ha) to both treatments to try to raise the no-tillage yield still further. Doing so had the opposite effect, however, and the no-tillage yield of maize fell to an all-time low of only 30% of the yield under tillage.

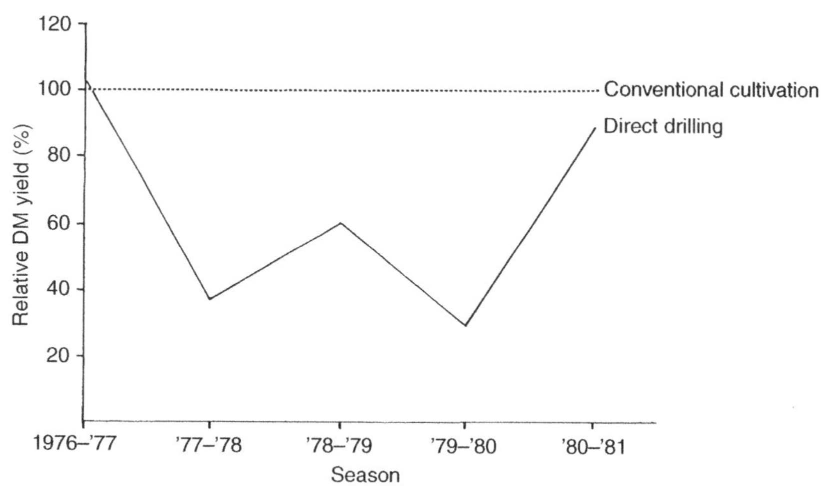

Fig. 9.8. Relative dry matter (DM) yield of no-tillage compared with tillage as affected by fertilizer application on no-tillage maize yields over a 5-year period (from Baker and Afzal, 1981).

Table 9.4. Effects of fertilizer placement on yield of maize (DM yield kg/ha) in the sixth year of a 6-year experiment.

	Fertilizer placed	Fertilizer broadcast	No fertilizer applied
No-tillage	10,914	4,523	1,199
Tillage	10,163	5,877	2,999

Year 5 coincided with the development of the disc version of the winged opener concept, which, amongst other things, allowed seed and fertilizer to be banded simultaneously with 20 mm horizontal separation in inverted-T-shaped slots.

The effect on the yield of no-tilled maize was immediate and spectacular. It raised the yield to again be not significantly different from the tilled yield.

In year 6 the experiment was altered to directly compare banded and broadcast fertilizer application under tillage and no-tillage and to check if the year 5 results were repeatable. Indeed, they were.

Table 9.4 presents the results for year 6. Clearly, the no-tilled soil benefited more from banding of fertilizer than the tilled soil. The final yields of the two methods with banded fertilizer were not significantly different.

Perhaps just as important were the yields of maize obtained from plots that had not received any fertilizer in the entire 6-year period. Although the unfertilized yields from both the tilled and untilled soils were poor in comparison with the fertilized plots, the enhanced mineralization that had occurred in the tilled soil each year produced plants almost three times as big as those under no-tillage. This mineralization, however, represents a 'burning out' of the SOM, with associated loss of soil quality, and is the reason why tillage is no substitute for no-tillage where fertilizers are applied correctly, in terms of both sustainability and crop yield.

An on-farm comparison was made in 2004 by a New Zealand farmer. He chose 11 fields and sowed a forage brassica crop into a randomly chosen selection of the fields over a 17-day period with two different no-tillage drills (M. Hamilton-Manns, 2004, unpublished data).

One drill was equipped with vertical triple disc openers. The triple disc openers had wavy-edged leading discs, which reduce the compacting effects normally associated with such openers. But they were not capable of banding fertilizer, so diammonium phosphate (DAP) fertilizer was broadcast at 300 kg/ha. The other drill was equipped with the disc version of winged openers, which banded the same amount of fertilizer 20 mm to one side of the seed at the time of seeding. Soil moisture conditions were not limiting and seedling germination was adequate with both drills.

The fields drilled with triple disc openers and broadcast fertilizer yielded, on average, 7069 kg dry matter (DM)/ha. The fields drilled with winged openers and banded fertilizer yielded, on average, 10,672 kg DM/ha.

While it cannot be said with certainty that the entire 51% average difference was the result of banded fertilizer alone (there may also have been opener differences), there is little doubt that most of the difference was due to fertilizer banding, and the heavier crops were worth, on average, US$468/ha more than the smaller crops.

Banding options

We have already seen that the need to band fertilizer beneath the soil without 'burning' the seed is greater under no-tillage than with tilled soils. Mixing of seed and fertilizer risks 'seed burn'.

Recourse to 'skip-row' seeding, in which every third opener sows only fertilizer in order to fertilize the two seeded rows either side of it (Little, 1987), has not been a feasible alternative either, although certainly better than broadcasting. Choudhary *et al.* (1988a) showed only mixed success with the 'skip-row' option, even when sown in narrow (150 mm) rows. Table 9.5 shows their results.

The 'skip-row' treatment produced the lowest fertilized barley yield (2072 kg DM/ha) but was equal to all other treatments when

Table 9.5. Effects of fertilizer application method on yield of two no-tilled crops.

	Barley grain DM yield (kg/ha)	Fodder radish[a] (whole plant) DM yield (kg/ha)
No fertilizer	1889 b	3240 ab
Horizontal separation by 20 mm	2580 a	3763 a
Fertilizer and seed mixed	2538 a	2809 b
Broadcast fertilizer	2432 a	3543 a
Skip-row separation	2072 b	3526 a

Unlike letters in a column denote significant differences ($P < 0.05$).
[a]*Brassica napus* L.

fodder radish was sown. In the latter case mixing of the seed with the fertilizer gave the poorest yield (2809 kg DM/ha). All other treatments were not significantly different.

Two other important points are evident in Table 9.5. The results are the mean of two soils, one of which was a fine sand, in which few, if any, preferential flow channels were present because of the exceedingly friable nature of the soil. Thus, even in its untilled state, surface-applied nitrogen fertilizer would have flowed more or less evenly through such a profile as if it had been tilled and showed less difference in favour of banding than where the soil was more structured.

The other point is that one of the fertilizer/seed combinations used in this experiment (DAP and barley) was not particularly damaging to barley seed. Consequently, mixing of barley seed and fertilizer together showed no disadvantage. On the other hand, mixing of the DAP fertilizer with the more susceptible brassica crop showed results similar to those of Afzal (1981) and Baker and Afzal (1986), who had used an even less compatible mix (canola and ammonium sulphate).

There is no evidence from any experiments conducted by the authors that greater amounts of fertilizer are needed under no-tillage. That which is applied just needs to be used more effectively by banding it alongside the seed. In fact, data from seven different experiments involving wheat (*Triticum aestivum*), drilled with double disc openers in a skip-row configuration (where every third row was sown with fertilizer only, at 100 mm depth), compared with horizontal separation by 20 mm with a drill equipped with winged openers, showed that fertilizer rates could actually be reduced with the latter openers (Saxton and Baker, 1990). Figure 9.9 shows the results.

On average, the winged openers showed a 13% increase in wheat yield compared with the skip-row drilling with double disc openers. Until then, that particular skip-row configuration had out-yielded all other methods with which it had been compared in the USA.

Not only did the plants sown with horizontal banding out-yield those sown with the skip-row method, but further measurements showed that the plants had been more vigorous from the outset. The improved vigour is likely to have been partly because of the positioning of the fertilizer and partly because of the high-humidity environment in which the seedlings developed beneath the ground in the horizontal (inverted-T-shaped) slots.

Table 9.6 shows analyses of the carbon and nitrogen contents of seedlings grown by these two fertilizer banding methods. Figure 3.1 had earlier shown the contrasting development of the seedlings in which the heavier and more fibrous nature of the root systems (more root hairs) from the horizontal banding and inverted-T-shaped slot was clear. Apparently, both the carbon and nitrogen levels were higher in the plants sown by the winged openers with horizontal banding of fertilizer compared with

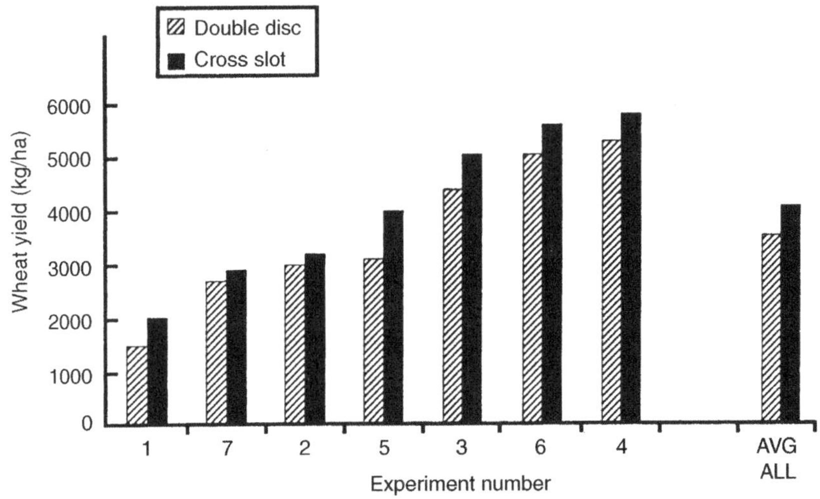

Fig. 9.9. Wheat yield comparisons from no-tillage using two different fertilizer banding options (from Saxton and Baker, 1990).

Table 9.6. Carbon and nitrogen contents of no-tilled wheat seedlings sown with two different openers.

Opener type	Field no.	Carbon (% DM)	Nitrogen (% DM)
Winged opener (inverted-T,	1	38.00	4.16
horizontal banding)	2	38.60	4.70
	Mean	38.30	4.43
Double disc opener (V-shaped	1	36.50	4.00
slot, skip-row application)	2	34.69	3.83
	Mean	35.60	3.92

those sown with the double disc opener and 'skip-row' fertilizer application.

Even where vertical banding of seed and fertilizer has been accomplished using a single opener, no clear advantage has yet been shown for this option.

Further, the technical difficulty of achieving satisfactory vertical banding in a wide range of conditions with a single opener makes implementation on a field scale unreliable. The problem is that, to achieve vertical separation, the fertilizer is usually drilled first at a greater depth than the target depth for the seed. In tilled soils it is relatively easy to induce soil to fall on to the fertilizer before the seed is sown. But in untilled soils this is much more difficult to achieve, particularly when the soil is damp and 'plastic'. For this reason, horizontal separation has become a more 'fail-safe' alternative since effective separation is not affected by soil looseness, surface cover or operating speed.

A comparison of horizontal banding (winged opener) and vertical banding (prototype hoe opener with a deflector to scoop soil on to the fertilizer prior to deposit of the seed) was made over several years by the authors. The results are shown in Fig. 9.10.

The figure shows that the winged opener with horizontal banding produced a greater yield in the first year of spring wheat (SW 87) and perhaps in the final year of winter wheat (WW 89), but there were no differences in yield in the other three seasons.

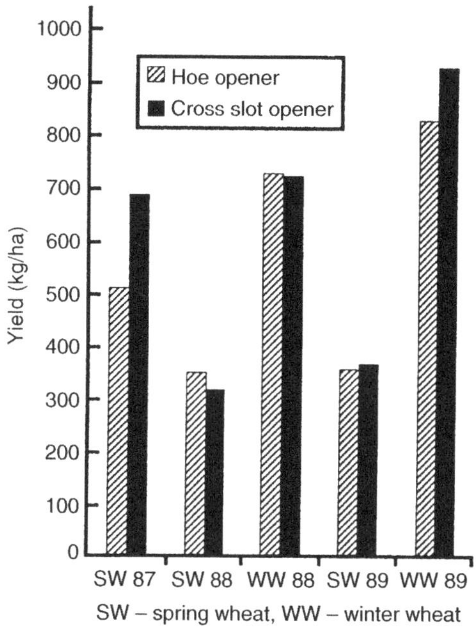

Fig. 9.10. No-tillage wheat yields from vertical banding of fertilizer with a hoe opener and horizontal banding with a winged opener.

A long-term double cropping experiment in Australia compared yields of soybean crops sown under no-tillage and tillage for 14 years using winged openers (Grabski *et al.*, 1995). For the first 2 years (1981/82 and 1982/83) the conventional tillage yields were superior, presumably because of the previous history of tillage. But for the following 12 years the no-tillage treatment was never bettered and averaged 30% higher yield of soybean than conventional tillage.

How close should banded fertilizer be to the seed?

Ferrie attempted to answer this question in Illinois, USA, in 2000. His results were reported by Fick (2000). Ferrie compared several diagonal distances of separation of starter fertilizer from maize seed sown with double disc openers, ranging from 90 mm deeper than and 50 mm to one side of the seed to 15 mm deeper than and 20 mm to one side of the seed. He concluded that, in terms of crop responses, 'the closer the starter was to the seed the better', provided that the fertilizer was not actually mixed with the seed and the action of banding the fertilizer did not disturb the accurate placement of the seed. The treatment with the greatest separation distance actually produced no measurable yield response to the starter fertilizer at all.

Ferrie also pointed out that slot wall compaction could have an effect on the ability of juvenile roots to access the fertilizer, especially in clay soils. He felt that, in such soils, even a narrow knife opener could cause problems.

Dianxion Cai (1992, unpublished data) tested two options for placing dry and liquid nitrogenous fertilizers at increasing rates of applied N, using winged openers and drilling wheat seeds 25 mm deep. The two options were: (i) standard horizontal banding 20 mm to one side of the seed (i.e. the fertilizer was also drilled 25 mm deep); and (ii) diagonal banding in which the fertilizer was banded 20 mm to one side of and 13 mm deeper than the seed (i.e. the fertilizer was drilled 38 mm deep). Figure 9.11 shows the effect on plant stand and Fig. 9.12 shows the resultant crop yields.

From Figs 9.11 and 9.12, it is apparent that the effects on seedling emergence (stand) were similar to the effects on yield, demonstrating the importance of initial plant population for final yield. In both experiments the horizontal banding (25 mm) produced more plants and heavier crops than the diagonal banding (38 mm) with both urea and aqua. These differences became most pronounced at an application rate of about 120 kg N/ha. At higher application rates, while the differences remained largely unaltered, both the plant stands and crop yields began to decline, possibly because of fertilizer toxicity. The decline of both plant stand and crop yield at the high application rates (160 kg N/ha) used in these experiments was considered to be of no consequence because these application rates were well in excess of normal application rates of nitrogen in any form (160 kg N/ha is equivalent to 350 kg urea or 400 kg aqua/ha).

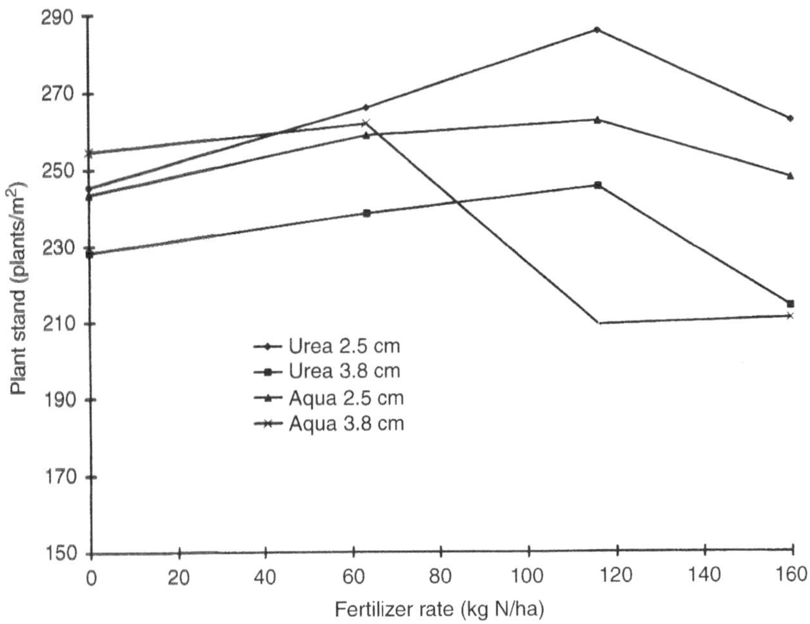

Fig. 9.11. Wheat stand responses to horizontal and diagonal banding of two forms of nitrogenous fertilizer separated from inverted-T-shaped slots.

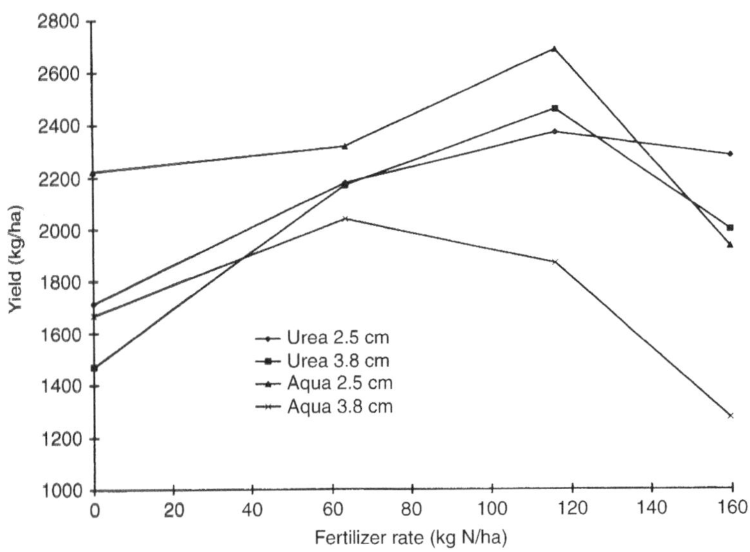

Fig. 9.12. Wheat yield responses to horizontal and diagonal banding of two forms of nitrogenous fertilizer separated from inverted-T-shaped slots.

Conclusion

One of the more noteworthy advances in no-tillage technology has been to develop a machine with the capability to separate fertilizer from the seed in horizontal bands and effectively entrap volatile forms of nitrogen in the slot. At the same time, these openers maintain the effectiveness of the separation function without being materially altered by forward speed, soil type, soil moisture content or the presence or absence of surface residues. From a field perspective, farmers find it easier to identify with this single factor, among all others, when assessing the performance of no-tillage versus tillage, and even when assessing the merits of competing no-tillage systems and machines.

It is interesting to speculate how many experiments and field observations showing poor yields for no-tillage crops have been the result of opener inability to adequately band the fertilizers.

Summary of Fertilizer Placement

1. Less nitrogen is available by organic matter mineralization under no-tillage than under tillage, making nitrogen application particularly important at drilling under no-tillage.
2. Some temporary nitrogen 'lock-up' may also occur under no-tillage as soil bacteria decompose organic residues.
3. Broadcast fertilizers are less effective in no-tillage than in tillage because soluble nutrients often bypass roots by infiltration occurring in preferential channels created by earthworms and decayed roots.
4. 'Deep-banding' of fertilizers at drilling is less effective or necessary in untilled soils than in tilled soils.
5. Fertilizer close to the seed is better than at a distance, so long as the two are not mixed.
6. Horizontal separation between seeds and fertilizer at distances as small as 20 mm have been more effective in no-tillage than vertical separation by any distance.
7. Relatively few no-tillage openers provide effective seed and fertilizer banding with a proper distance or direction.
8. Of those no-tillage openers that do provide effective separation, horizontal separation is preferable to vertical separation.
9. Where no-tillage openers are incapable of separating seed from fertilizer, other options include:

- Drilling every third row with only fertilizer ('skip-row' planting).
- Mixing seed and fertilizer together in the slot.
- Doubling the number of openers on a drill so as to provide separate fertilizer-only openers in addition to seed-only openers.
- Surface broadcasting of the fertilizer.
- Drilling seeds and fertilizer as two separate field operations at different depths.

10. Most double disc openers are incapable of banding fertilizer separately from the seed with a single opener.
11. Some angled disc openers have provided a fertilizer-banding capability.
12. One version of winged openers with a single disc effectively separates seed and fertilizer horizontally or diagonally.
13. Crop yields with winged openers have been good when using horizontal separation of seed and fertilizer, due to improved seed/seedling micro-environment and fertilizer response.
14. Only recently designed hoe openers separate seed and fertilizer in any direction.
15. Two disc openers (double or angled) slanted in opposite directions may be capable of providing vertical separation of seed and fertilizer.

10 Residue Handling

C. John Baker, Fatima Ribeiro and Keith E. Saxton

Successful no-tillage openers not only handle surface residues without blockage but also micro-manage these residues so that they benefit the germination and seedling emergence processes.

The second most valuable resource in no-tillage is the residue left on the ground surface after harvest of the previous crop. The only resource more valuable than residue is the soil itself – in its untilled state.

Unfortunately, the history of tillage is littered with descriptions of methods for disposing of residues so that they do not interfere with the operation of machinery. In tillage, surface residues have been regarded as a major nuisance and therefore have often been referred to as 'trash'. Those who take no-tillage seriously have dispensed with the term 'trash' in favour of the term residue. Trash is something unwanted. Residue is something left over, but in this case wanted and useful.

Before considering how well various openers and machines handle or manipulate surface residues, it is necessary to identify the various forms that residue can take (Baker *et al.*, 1979a). Then it will be appropriate to look at how the residues should be macro-managed on a field scale (Saxton, 1988; Saxton *et al.*, 1988a, b; Veseth *et al.*, 1993) and finally at the options for micro-managing residues in, around and over the slot zone (Baker and Choudhary, 1988; Baker, 1995).

The Forms that Residues can Take

Short root-anchored standing vegetation

Pasture (either growing or recently killed by herbicide)

Short root-anchored pasture is commonly encountered by no-tillage drills designed for pasture renovation or renewal in intensive animal grazing agricultural systems and for crop establishment in integrated crop/animal systems. In such systems, animal management can usually be sufficiently controlled to allow deliberate intensive grazing of selected fields prior to drilling, thus reducing the length of grass and therefore the residue-handling demands on such machines. This allows relatively inexpensive drills to be used for such conditions.

Short standing pasture usually presents few residue-handling problems as the vigorous root anchorage and firm soil beneath the plants allows even a 'rigid' tine or shank, without a pre-disc, to burst reasonably cleanly through it. If the pasture has been recently killed, the time-interval between spraying and drilling can have a profound effect on the handling properties of this residue. As decomposition starts soon after death of the plant, the material becomes progressively weaker and more likely to break away from its anchorage. At an advanced stage of decay, it may break away from the soil anchorage altogether and start to behave more like loose-lying residue than short anchored residue and therefore be more prone to causing blockage. Sometimes it pulls free in large pieces.

Pasture plants that have stoloniferous or rhizomatous growth habits (i.e. with horizontal and/or underground connecting stems), even though they might be grazed short by animals, present a different problem, since their creeping habit makes them likely to become entangled in non-disc-type openers. At least a pre-disc is essential for satisfactory handling of such residues with tine or chisel openers.

Short clean crop stubble after direct-heading with a combine harvester and baling of the straw

Clean crop stubble that has negligible loose straw lying on or amongst it offers only moderate residue-handling problems because the standing plants can usually be pushed aside by relatively unsophisticated no-tillage openers. In common with pasture plants, the key element is the anchorage offered by the root systems. The time interval between harvesting and drilling and the intervening weather will also influence the level of decay that has set in by the time drilling takes place. In the case of crop stubble, however, because harvesting normally takes place at a dry time of the year, the onset of decomposition may be slower than with pasture plants.

Standing stubble has important additional functions in no-tillage systems that experience snow and freezing winters or in which the crop is swathed prior to harvesting.

Where swathing takes place, long stubble, especially in narrow drill rows, will hold the cut swathe off the ground, which aids drying and makes harvesting easier as it aids the pickup mechanisms on combine harvesters compared with when the swathe lies close to the ground.

Where snow is expected, stubble holds the snow from blowing away. Snow, in turn, provides effective thermal insulation of the soil beneath and may be responsible for maintaining soil temperatures some 10° to 15°C higher than in soils that have no snow cover and are allowed instead to freeze (Flerchinger and Saxton, 1989a, b). In this respect, long stubble is better than short stubble (see below).

In either case, at the end of a cold winter, when such soils are drilled, stubble that has endured the cold months is usually brittle, though often it has not actually decayed much. It may break off at ground level, but due to its shortness will seldom present major residue-handling problems for no-tillage drills. On the other hand, no-tillage systems increasingly require that the full amount of residue disgorged from combine harvesters (including the threshed straw as well as the standing stubble) remains on the ground over the winter in such climates. This combination presents quite another problem as far as residue handling is concerned, which will be discussed later.

Standing stubble also has an important function in dry climates, by reducing wind velocity at the soil surface, which significantly reduces drying and soil movement. In windy conditions, standing stubble may protect young seedlings sown between the stubble rows from being blasted by wind-blown sand and other soil particles. In Australia, for example, planting between the rows of tall stubble offers wind protection to the new plants, while, in England, long stubble has another value, that of camouflaging wildlife, such as pheasants. Since many farmers in that country rate the commercial shooting of pheasants as an important source of farm income, no-tillage offers an opportunity through stubble retention for an extended game-shooting period that was not possible with tillage.

In tropical climates, tall standing stubble can result in etiolation of the new crop. But short standing residues lead to more vegetative material entering the combine harvester, resulting in a higher power requirement, more fuel consumption or decreased field capacity.

For all of these reasons, there has been recent interest in the use of stripper headers in association with no-tillage because such harvesting devices maximize the length of the standing stubble.

Tall root-anchored standing vegetation

Tall grass, sprayed-off cover crops and tall clean stubble (300 mm and longer), together with bushy weeds, present somewhat greater problems than short vegetation, even with root anchorage, but less than lying straw. There is a critical height above which each of these plants will collapse in the pathway of no-tillage openers (or simply over a period of time), at which point the residue behaves more like lying straw than standing stubble. Taller material may also trap a more humid microenvironment within, with the result that decay of the bases of the straw may be initiated more quickly than with short stubble and breakage is more likely.

Figure 10.1 shows the effect of drilling with the disc version of a winged opener through a partially standing matted legume crop 0.75 m high that had been sprayed. It is not common to drill into such very tall residue; not only because of the spatial constraints, but because it is difficult for seedlings to obtain sufficient light during early development to emerge satisfactorily.

Lying straw or stover

Detached stalk material, of any length, presents the most difficult residue-handling problems for no-tillage drills but is also a very valuable biological resource unique to no-tillage. Where such residues lie on firm ground (e.g. after a no-tilled crop has been harvested, or even when hay has been fed directly on to an established pasture and not fully consumed by animals), there will be less tendency to block no-tillage openers than where the residues lie on softer ground. Similarly, if the residues remain dry and brittle, they will be easier to handle and cut than where they have

Fig. 10.1. The effects of drilling with the disc version of winged openers into heavy partially standing residue.

become damp. Often dampness is a function of both the amount of straw (yield of the crop) and the weather. Heavy residues may generate their own dampness and increase in temperature from bacterial action.

The immediate history of the field may also be important. If the previous annual crop was established into tilled soil, for example, the soil background against which disc components of no-tillage openers will need to push to shear the straw, will be softer than if the previous seedbed had been untilled. This 'anvil effect', of course, will be influenced by soil type, which has an important influence on the effectiveness of some residue-handling mechanisms and presents farmers with some difficult choices when converting from tillage to no-tillage.

For example, a no-tillage machine that is good at drilling into residues previously established in a tilled seedbed (during the changeover period) may not be the best machine for drilling into residues previously grown in an untilled seedbed. Further, some farmers believe (usually erroneously) that they will still need to occasionally till their soil even under a predominantly no-tillage regime. There may be little logical basis for this belief, but it will none the less influence the farmer's choice of machine, perhaps to the detriment of the true no-tillage phase. The problem seldom exists when drilling into pasture because it is unusual for pasture to have been established for less than 12 months, during which time even a previously tilled soil will have consolidated again.

Fortunately, some no-tillage openers are equally well suited to soft and firm (or even hard) soils. The function of most tine- or shank-type, power till and winged openers is relatively unaffected by soil softness or firmness (except for downforce or power requirements), but those that tend to hairpin residue into the slot (double disc, angled flat disc and angled dished discs) have their hairpinning tendencies accentuated by softer soils. On firmer soils, they are more likely to shear the straw (which is desirable) than to push it bent over into the slot (which is undesirable). In firm soils, however, some openers are also more likely to compact the soil in the slot zone.

Lying residues have no anchorage to the ground and are therefore very easily gathered up to become entangled in 'rigid' machine components. Firmer ground provides greater friction (traction) for discs that may operate in conjunction with rigid components, ensuring that they keep revolving when they encounter lying residues. Some discs are especially shaped to further assist traction. Wavy-edged discs and notched or scalloped discs are cases in point. Even so, if the height of the lying straw is above the axle height of an approaching disc, it is likely to stall the disc, causing sledging and blockage. This is accentuated by dampness under the straw, especially if such dampness results in partial decay close to the ground. The decaying straw can become quite slippery on the ground and will often slide ahead of a disc, rather than allow the disc to grip and ride over or cut through it. Straw lying amongst standing stubble is less likely to slip than where it is lying on bare ground.

This sliding tendency is dependent to some extent on plant species. It is also soil-dependent and obviously weather-dependent. For example, pea straw becomes particularly slippery when partially decayed, especially on firm untilled soil, while most cereal straws do not. Sparse straw, such as soybean, canola, cotton or lupin, is less likely to remain damp long enough to promote decay close to the ground than crops that produce heavier vegetative growth. Further, the rigidity of the cut stubble of these somewhat woody crops helps prevent sliding of the lying residue.

Numerous methods have been devised to handle lying straw. Some of these are summarized below. The successful methods almost invariably involve openers where discs are used, either simply as the opener itself or where the discs assist the operation of other rigid components, such as winged blades, chisels or tines. In both cases, discs have become a common, though not exclusive, component of no-tillage openers designed for the handling of residues.

Management of Residues on a Field Scale

Macro-management refers to the way in which the residues are managed on a field scale. Their management is discussed separately for: (i) large field-scale no-tillage; and (ii) small-scale no-tillage. But in either case, surface biomass, whether from killed cover crops or harvested residues, plays a key role in no-tillage systems. For any no-tillage system (large or small), the handling of residues should:

1. Assist (or at least not hinder) the passage of no-tillage openers.
2. If possible, contribute to the biological functions of the openers.
3. Ensure that the residues decompose and add to soil carbon but at the same time remain on the soil surface long enough to protect the soil from erosion, keep the soil cool in tropical climates, retain soil moisture and suppress weeds;
4. Ensure that the residues do not compete with the sown crop.

These are demanding and sometimes competitive requirements, and compromises are often necessary. For example, tine (shank)- or knife-type openers do not handle residues well, so some farmers resort to burning or otherwise removing the residues to avoid blockages when drilling a field. But this compromises some of the other listed functions. For this and other reasons, the burning of residues is banned in several countries, although up to 45% of the biomass will be in the roots that remain even after burning.

In this respect, it is interesting to note that it makes little difference whether harvested residues are baled, burned or buried in terms of the amount of carbon they provide for the soil (see Chapter 2). Unless they are left to decompose on the soil surface, much of the carbon content of the above-ground plant residues will be lost from the system (oxidized and lost as carbon dioxide to the atmosphere). Therefore, to get the best out of a no-tillage system, the challenge for machinery designers is to provide no-tillage openers that can cope with any amount and type of surface residue without blockage. But, more than that, as is explained in Chapter 5, an opportunity exists for openers to harness the surface residues as an important resource to aid germination and emergence of the new crop.

Large field-scale no-tillage

Weed control and management of cover-crop residues

In larger field-scale no-tillage, weeds and cover crops are normally killed by herbicides. Indeed, the very feasibility of the modern concept of no-tillage owes its existence to the development of 'non-residual' herbicides in the 1960s and 1970s. This contrasts with small-scale agriculture (see below), which is more dependent on mechanical means of plant competition control.

No attempt is made here to analyse the pros and cons of specialist spraying machinery or different herbicides. Suffice to say that the control of existing competition is the first step in any no-tillage programme and that, unless this is achieved effectively, all other steps will be compromised. Effective chemical weed control is a function of understanding the biology of the plants to be killed and the efficacy of the herbicide(s) to be used and the mechanical performance of sprayers. Some herbicides (e.g. glyphosate) work best on actively growing unstressed plants, while others (e.g. paraquat) are more effective when plants are stressed. And, of course, there are species differences (and sometimes varietal differences) in the resistance of plants to different herbicides.

Management of harvested residues

CHOPPED OR LONG? The first and most important opportunity to correctly manage residues on a field scale occurs at harvesting. Once crops have been threshed and the residues ejected from a combine harvester in discrete windrows, they are very difficult to spread out again.

Modern combine harvesters gather together the material from cut widths of 5–10 m and process it in such a way that, unless spreading devices are added to the combine harvester discharge, the residues are ejected out of the back in the form of a windrow of light fluffy straw 2–3 m wide. Underlying this windrow will be the chaff from the separation processes, which consists of very short pieces of straw, awns, leaf material, empty glumes, chaff, dust and weed seeds. The chaff row forms a dense surface covering, somewhat narrower than the straw windrow covering it.

In contrast to these somewhat concentrated zones of residue, good no-tillage requires that the residue be spread evenly over the entire field. There are no-tillage openers that can physically cope with the concentrated windrows and tailings, but this capability is somewhat academic since the effect of surface residues on germination, emergence and crop growth is so vital that an uneven crop will almost certainly result from grossly uneven chaff and straw distribution. Uneven spreading can also affect the efficacy of herbicide applications.

Most combine harvesters have optional straw spreaders. These are different from straw choppers in that spreaders do not chop the straw into shorter lengths. They spread the straw with beaters rather than with wind assistance (see Fig. 10.3). Most straw choppers spread as well as chop. Straw spreaders are not high power-demanding additions and are easily fitted and operated. They are essential standard equipment on all combine harvesters for no-tillage systems, as indeed they already are on some makes and models.

Whether or not a chopper is also needed will depend on the residue-handling capabilities of the no-tillage drill or planter to follow. Straw choppers are unpopular in some respects because they consume up to 20% of the total power requirement of the combine harvester (Green and Eliason, 1999). Chopping damp straw requires more power than chopping dry straw, although the distribution of damp straw on the soil surface may be more even than that of dry straw.

Generally, if the straw needs chopping to avoid the no-tillage openers blocking, this reflects inadequate performance on the part of the openers.

CHAFF. Another area of concern is the tailings or chaff. With some openers, this thick mat of fine material is more troublesome than thick straw. Fortunately, in recognition of this, many combine harvesters now offer chaff spreaders (or tailings spreaders) as well as straw choppers or spreaders (see Fig. 10.2).

Most straw choppers/spreaders can be adjusted to produce longer or shorter cuts and to spread the residues different distances through adjustments of the deflector, the vertical positions of the knives and the speed of the chopper (Siqueira and Casão, 2004).

Some modern straw choppers use improved cutting principles and blower support for spreading. For example, auger types can be applied to both straw and chaff with spreading widths up to 10 m in either direction without visible separation of different fractions (Lücke and von Hörsten, 2004).

SPREADING AFTER HARVEST. There are limited residue management options available where it is not possible to spread the residue with the combine harvester. Re-spreading of the residues evenly after harvest has been only partially successful because most straw is light and fluffy, making it difficult to throw or blow any distance. One way of handling the situation after harvesting is to pass the material through a large fan or forage harvester and blow it as high into the air as possible on a mildly windy day. In this manner the wind will spread it reasonably evenly, but it requires a tractor with cab and good air filtration system or an operator who can tolerate dusty conditions. Variations on this have been attached to combine harvesters to create ‘straw storms’.

Another way is to use straw harrows, which consist of feely rotating angled spikes that are pulled at an angle and flick the residues more evenly across the field.

Fig. 10.2. A straw and chaff (tailings) spreader on a combine harvester. Note the dust associated with spreading chaff.

Fig. 10.3. A pair of simple beater-type straw spreaders on the rear of a combine harvester. No attempt is made to spread chaff (tailings) with such a device.

They also double as a convenient way to disturb weeds seeds and induce them to germinate so they can be killed with a herbicide before drilling the next crop (referred to as 'chitting' in Europe).

Small-scale no-tillage

The killing of cover crops on a small scale is not as dominated by herbicides as is the case for large-scale no-tillage.

Mechanical destruction is frequently used, or a combination of mechanical and chemical methods. Mechanical destruction is favoured because it results in lower repetitive cash outlays and less exposure by small farmers and their families to chemicals, although chemicals such as glyphosate have a high level of safety associated with their use. But other herbicides (e.g. paraquat) are less safe and more difficult for farmers operating on small fields to take proper protective measures against than in larger operations, where fully enclosed vehicle cabs with filtered air supplies are common. Mechanical methods for cover-crop handling in small-scale agriculture are therefore being widely promoted.

Mechanical destruction of growing plants is achieved by slashing, chopping, crushing, spreading or bending the plants. Each method is suited to different conditions and results in different amounts of plant material being left on the soil surface.

Manual slashing

Manual slashing is a very labour-intensive operation. Schimitz *et al.* (1991) reported that labour requirements of 70 man-days/ha for manual slashing have been measured when managing a 3-year-old grass-residue field yielding 10 t/ha dry matter.

Knife roller

Knife rollers are amongst the more useful residue-management tools to achieve evenly distributed plant material on the soil surface. Figures 10.4, 10.5 and 10.6 show examples of

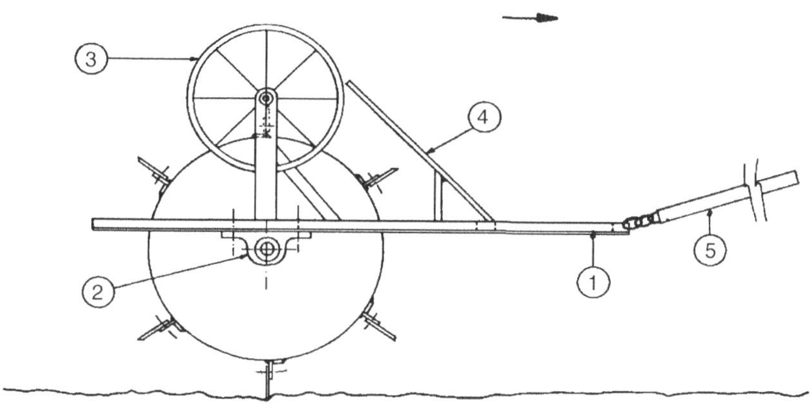

Fig. 10.4. Side view of a knife roller: (1) frame; (2) bearings; (3) transport wheel; (4) protection structure; (5) shaft (from Araújo, 1993).

Fig. 10.5. Animal-drawn knife rollers: (left) with full-width knives and (right) with short knives.

Fig. 10.6. A tractor-pulled knife roller operating in oats.

typical knife rollers. They have the advantage of allowing non-chemical organic production methods to be combined with no-tillage. For example, such implements are in common use for no-tilled organic soybeans in southern Brazil (Bernardi and Lazaretti, 2004) and are available for both animal and tractor power.

Knife rollers have flat metal knives mounted on a roller with a frame for support, wheels for transport and a protective structure. The knives are mounted on the roller in various patterns, most commonly perpendicular to the direction of travel. The effect of the knives is to bend, crush and chop off plant material. Their effectiveness depends upon the width, diameter and weight of the roller, the number, height, mounting angle and sharpness of the knives, speed of operation and the fibre and moisture content of the plants (Schimitz *et al.*, 1991; Araújo *et al.*, 1993).

Rollers are constructed from either steel or wood. Steel rollers are often filled with sand, so that their weight can be adjusted according to the condition of the plant material and the desired result of chopping, crushing or bending. But on slopes the sand can move to one side of the roller and affect the evenness of performance and stability. Monegat (1991) recommended roller widths between 1 and 1.2 m as a compromise between stability on hillsides and an ability to stay in contact with irregular surfaces.

Knives may be the same width as the roller (Fig. 10.5 – left) or in short sections (Fig. 10.5 – right). Shorter sections increase the pressure exerted as each knife impacts the ground and spreads the impact forces more evenly, which is important for draught animals in particular. For a given diameter of roller, the effectiveness decreases as the number of the knives increases because the pressure on each knife is reduced (Schimitz *et al.*, 1991). For the best cutting action, the knives should be perpendicular (i.e. not angled) to the surface of the roller (Siqueira and Araújo, 1999).

Tables 10.1 and 10.2 show recommendations for the construction of knife rollers for draught animals and tractors, respectively (Araújo, 1993).

Table 10.1. Recommendations for the construction of animal-drawn knife rollers (1 m wide) operating at 1 m/s (3.6 km/h) (from Araújo, 1993).

Roller		Diameter	Height of	Number of
Material	Density (kgf/m^3)	(cm)	knives (cm)	knives
Eucalyptus wood	1040	60	5	5
			10	6
			15	6
Steel + sand	2000	40	10	4
		60	5	10
			10	10

Table 10.2. Recommendations for the construction of tractor-mounted knife rollers (1 m wide) (from Araújo, 1993).

Roller		Speed,	Diameter	Height of	Number of
Material	Density (kgf/m^3)	m/s (km/h)	(cm)	knives (cm)	knives
Eucalyptus wood	1040	2 (7.2)	40	5	4
				10	4
				15	6
Steel + sand	1500	2 (7.2)	30	15	12
		3 (10.8)	25	8	4

The design, construction and operation of knife rollers must also take safety considerations into account. When working on slopes, it is advisable to use a fixed shaft instead of chains, so that the shaft will work as a brake for the roller. Other considerations are manoeuvrability, including reversing (Schimitz *et al.*, 1991), and the use of protective shields. Figure 10.7 shows a protective shield, which is important to both the draught animal and the operator.

The force required to pull a knife roller in black oats at the milky seed stage (sown at a density of 100 kg/ha) was measured at approximately 3430 N (350 kgf) per metre of width (Araújo, 1993).

Time requirements for handling black oats with a knife roller are about 3 h/ha for animal-drawn and 0.9 h/ha for tractor-pulled (Fundação ABC, 1993; Ribeiro *et al.*, 1993), although Schimitz *et al.* (1991) reported requirements as high as 6 days/ha with animal-drawn units.

The crushing action of knife rollers interrupts the flow of sap through the plant, which will kill many annual plants if the timing is correct (see Fig. 10.8). In this regard, it is best if the cover crop is uniform and rolling is undertaken at the beginning of the reproductive stage, when seeds are not yet viable. This is at full flowering for leguminous species and at the milky stage for cereals (Calegari, 1990). In some environments, such as sub-Saharan Africa, it is desirable that the cover crop remains green as long as possible to avoid burning during the dry season. In this situation, a knife roller should be used at the beginning of the rainy season, prior to planting.

Different methods of cover-crop residue handling will result in different rates of biomass decomposition. Araújo and Rodrigues (2000) compared the decomposition rates of black oats (*Avena strigosa*) as a function of mechanical treatment. They found that after 68 days the residues

Fig. 10.7. A protective structure for the draught animal and operator alike.

Fig. 10.8. Black oats killed with a knife roller.

remaining in relation to the initial amount were 59% for a knife roller, 48% for a flail mower and 39% for herbicide application. A similar study carried out by Gamero *et al.* (1997) indicated that after 75 days the amount of black oat dry matter was 68% for a knife roller and 48% for a flail mower. The authors also found a lower weed population when the knife roller was used compared with a flail mower.

Yano and Mello (2000) evaluated the distribution of various cut lengths of pigeon peas (*Cajanus cajan*) as a result of different mechanical treatments of cover-crop residues. A flail mower resulted in 70% of the cut lengths being 100 mm or less compared with 45% for a rotary mower and 22% for a knife roller.

Another advantage of mechanical treatment of heavy cover-crop residues is that, if the crop is sprayed with herbicide before mechanical treatment, the main canopy may prevent the herbicide getting to lower-growing weeds beneath the canopy. Alternatively, the cover crop can be treated with a knife roller and then sprayed, provided that sufficient time is allowed for the weeds to appear through the bent-over canopy so that they can be targeted by the spray. This option is best suited to heavy cover crops. Spraying options are most effective where the cover crop is not heavy.

Can a knife roller substitute for herbicides?

Knife rollers are not designed for weed control, even though the mulch they produce may contribute to weed suppression. But one purpose of growing a cover crop is to pre-suppress the weeds with a dominant monoculture, which can itself be killed by a knife roller at the appropriate time prior to planting the main cash crop. If the cover crop is vigorous and the weed incidence is low, a knife roller alone may be sufficient to prepare the field. In Tanzania, for example, Schimitz *et al.* (1991) reported that a knife roller had been effective for weed control in grass up to 3 m high after a fallow. The factors that make such a totally mechanical option viable are:

1. Perform the planting operation as close as possible to the destruction of the cover crop.
2. Use planters with minimal slot disturbance.
3. For planters that create substantial slot disturbance, plant before the cover crop is treated so that the residue will cover the slot opened by the planter.

Management of Residues by Openers, Drills and Planters: Micro-management of Crop Residues

Micro-management refers to how the residues are handled by the openers themselves and the role the residues play in the opener functions. It is a sad fact that the designers of many no-tillage openers still treat residues as an unwanted nuisance. While recognizing the macro-value of residues to no-tillage, these designers often show little sign of recognizing the micro-value of residues for opener function and seeding results. As explained in Chapter 5, the highly desirable Class IV slot cover is only possible if the ground is residue-covered in the first place and then only if the openers are designed in such a way as to retain that residue over the slot itself.

Opener handling of residues

Chopping (strip tillage)

All power till openers chop the surface residues with the soil. There is no practical way to avoid their doing this. Where the surface residues consist of undecomposed accumulated organic matter in colder climates, such incorporation may be of benefit, but, in all other circumstances, some of the value of no-tillage is lost when the residues are incorporated, even on a strip scale. Besides, strip tillage itself defeats some of the objectives of true no-tillage in the planting zone.

Sweeping aside

Hoe, knife, shank, angled flat disc and angled dished disc openers all push soil and some of the surface residues aside as they proceed through the soil. Disc openers may also push some of the residue into the soil to form hairpins in the seeding slot. With hoe- and shank-type openers, if the residue is reasonably thick and of some length, it will accumulate on the shank of the opener rather than be pushed aside, causing opener blockages. Angled disc-type openers do not have this problem, but, in either case, residue that is pushed aside will have negligible influence on the micro-environment within the slot that is being created.

On the other hand, because the residue is usually heaped to one or both sides of the slot (Fig. 10.9), careful choice and operation of a subsequent covering device may succeed in collecting some of this residue and guiding it back into the slot zones (Class III cover), although it is then likely to be mixed with soil. This process will occur if the soil remains dry and friable. If the soil becomes damp, the covering device is likely to create a smearing effect and the value of the residue will be lost by becoming smeared into the soil alongside but not over the slot.

Pushing down or through

All discs, to a greater or lesser extent, push down through surface residues. Double or triple disc openers mostly push down, whereas angled discs sweep aside as well as push through. The problem with pushing down is that, because it is impossible to cut all of the residue all of the time, a proportion of residue is doubled over and pushed (tucked) down into the slot in the form of a 'hairpin'.

The tendencies of different discs to hairpin depend on several factors:

1. Sharpness of the disc. Sharper discs are more likely to cut than to hairpin, but it is impossible to keep discs sharp all of the time.
2. Brittleness of the straw. Brittle straw is more likely to break than fibrous straw. Brittleness itself is a function of crop species, dampness and stage of decay.
3. Softness of the soil. Firm soil will assist shearing by a disc (the anvil effect) more than soft soil. More hairpinning will occur in soft soils.
4. Speed. Faster operating speeds generally reduce the incidence of hairpinning. The straw has less time to bend because of its inertia and is therefore more likely to be cut or broken.
5. The presence of chaff and tailings. Where straw is lying over a mat of fine tailings, as is often the case, the tailings provide a soft mat beneath the straw, which acts like a soft soil and encourages hairpinning. Worse, a portion of the tailings themselves may be pushed down into the slot, where they make the hairpinning problem worse by coming into contact with the seed.

Fig. 10.9. Residue swept to one side of a no-tillage slot.

6. Diameter of the disc. Smaller-diameter discs, because of their reduced footprint area, will put more pressure on the residue than larger discs and are therefore more likely to cut the residue than to hairpin it. But small discs are also more likely to sledge, since larger discs have a flatter cutting angle at the soil surface.
7. Disc design. Wavy-edged discs, because of their self-sharpening tendencies, will cut better than plain discs. Notched discs do not remain any sharper than plain discs but cut more residue because of the slicing action of the sides of the 'points' and the increased footprint pressure of the 'points'.

Folding up from beneath

The disc version of winged openers manipulates the surface residues by first pushing a notched disc down through the residues and then using the lateral wings of the side blades to fold the residue and soil upwards and outwards while the seed and fertilizer are deposited in the slot. A pair of following press/gauge wheels then fold the material back over the seeded slot. The end result is a horizontal slot covered with soil and residue (Class IV cover) in much the same layering as the soil and residues had been before seeding.

The limited amount of vertical hairpinning caused by the notched disc is of little consequence because, unlike with all other no-tillage openers, the seed is placed to one side of the central disc slot away from any hairpins. In this way the seed is effectively separated from any hairpinned material and instead benefits from the presence of residues over the slot (see Chapter 5).

Row cleaners

One method of assisting no-tillage openers to operate in residues is to clean the row of residues immediately ahead of the openers. The devices designed to achieve this are known as 'row cleaners' or 'residue managers'.

With small-scale no-tillage, it is often not feasible to use disc openers because of the weight required to push them into the ground compared with tine- or shank-type openers. 'Row cleaners' require little additional weight since most of them only work on the surface of the ground. But in such situations they may make the difference between being able to undertake no-tillage or not.

With large-scale no-tillage, where weight is less of a problem, 'row cleaners' are often used in springtime to remove residues from over the immediate row area so as to allow sunlight to warm the soil more quickly after a cold (and often freezing) winter.

Most 'row cleaners' consist of spiked rotating wheels, notched discs or rakes set at an angle to the direction of travel and operating ahead of the openers. The spikes just touch the ground, which causes them to rotate much like a finger-wheel rake for turning hay. In the process, they sweep the residues to one side or both sides while at the same time moving as little soil as possible.

With tougher residues, such as maize stover, two wheels may be set at opposite angles to one another and the spikes are synchronized at their fronts to reduce the side force on the whole device by sweeping residues to both sides of the row rather than to one side. Figure 10.10 shows a 'row cleaner' consisting of a pair of synchronized spiked wheels. Figure 10.11 shows unsynchronized notched discs designed to push residues aside.

Chopping of straw into short lengths

There is a critical length for most straws, above which they will bend and thus wrap around approaching rigid tools (e.g. tines). Chopping all straw into relatively short lengths allows short lengths to fall away from rigid tools rather than wrap around them. Other objectives for chopping straw trace their origins to tillage by making the straw easier to incorporate into the soil and enhancing the decomposition process.

To drill into maize stover with shank-type openers, Green and Eliason (1999)

Fig. 10.10. A pair of synchronized star wheels (row cleaners) for pushing residues aside.

Fig. 10.11. A pair of angled, notched, disc row cleaners designed to push residues to either side of the row ahead of the opener.

recommended that the cut lengths should be no longer than the shank spacing of the openers.

Chopped straw may also settle down on to the ground more easily and closely than long straw and may therefore provide a more effective mulch. On the other hand, an effective no-tillage opener will ensure that even long straw is replaced on the ground after its passage (see Fig. 10.1).

One of the most effective ways to obtain chopped straw is to fit a straw-chopper to the rear of a combine harvester. Such devices are not readily favoured by operators, however, because they consume considerable power and are yet another component that must be adjusted correctly on an already complicated machine. In any case, they seldom chop every single straw, with the result that the long straws that persist may eventually accumulate on openers not equipped to handle them.

Other methods produce chopped straw with a separate chopper. Some of these machines incorporate the straw into the soil as they chop it, which departs from true no-tillage because of the general soil disturbance. Others simply chop it and redistribute it back on the ground again.

Yet a third approach is 'vertical mulching', where the straw is chopped and then blown into a vertical slit created simultaneously in the soil by a large soil opener on the machine (Hyde *et al.*, 1989; Saxton, 1990). The result is a series of vertical slits filled with straw, thus solving a disposal problem as well as providing an entry zone for water infiltration.

Because no general tillage takes place, vertical mulching complements no-tillage, but the absence of a horizontal surface mulch reduces the options for maximizing the benefits of true no-tillage. Figure 10.12 shows a prototype vertical mulching machine in the USA.

Random cutting of straw in place

The most obvious way of handling long surface residues in place is to cut a pathway through them with some form of sharp tool. Generally, discs are the most commonly used device but other forms of tool have included rigid knives and powered rotating blades.

Rigid knives

These have been known to work for short periods only if their cutting edges remain smooth and very sharp, but extended use is impossible because of random damage and dulling by stones and soil abrasion. Figure 10.13 shows a knife-edge opener where the sweeping action of a tapered front was combined with a sharp edge in an attempt to slide past and/or cut residues. The sliding action was unsuccessful because small imperfections soon developed in the otherwise smooth edge from contact with stones,

Fig. 10.12. A prototype vertical mulching machine.

Fig. 10.13. A prototype knife-edge opener designed to sweep and cut residue (from Baker *et al.*, 1979a).

resulting in straw catching as it slid down the face. This led to deterioration of the cutting effect and blockage.

Rotating blades

These, such as on power till openers, are not always successful either. To be most effective as a soil pulverizer, power till blades are usually L-shaped. The horizontal portion of the L is important because it elevates and accelerates the soil upwards and throws it against the surrounding cowling, breaking it into smaller particles. Unfortunately, the horizontal L is also a perfect catch for wrapping of residue. As a consequence, backward-facing C-shaped blades are often used in residue situations because they allow the residue to be brushed off as they rotate. C-shaped blades, however, do not have a truly horizontal portion to the blade and the trade-off is that they are less effective as a soil pulverizer.

Discs

These can be most effective for breaking through or slicing straw, but, as explained previously, their action is highly dependent on the firmness of the background soil against which they must shear the straw and the brittleness of the straw itself. No matter what design of disc is used, no disc will cut all of the residue all of the time.

Cutting damp fibrous straw is particularly difficult. Cutting it against a soft soil background is even more so. One variation that has been tried is to power the disc so that it rotates faster than its peripheral forward speed. The aim is to create a slicing action as the disc presses the residue against the ground. Figure 10.14 shows a prototype powered disc. A further variation is to cause the disc to vibrate as it rotates by using a power drive on the disc hub. Both of the powered disc options, however, are disadvantaged by the cost and complexity of providing individual drives to a multiplicity of openers together with the interruption to residue flow between adjacent openers brought about by the bulkiness of such drives in the vicinity of the disc hubs. Besides, some unpowered designs have managed to achieve comparable results at a fraction of the cost.

The most appropriate diameter of discs for handling agricultural residues is always a matter for debate. Small-diameter discs have a smaller footprint and are therefore easier to push into the soil than larger discs. For this reason they also cut residues better than larger discs. However, the closer the disc axle is to the ground, the easier it is to stop the disc rotation (stall) when the

Fig. 10.14. A powered disc opener designed to rotate faster than the speed of travel through the soil.

thickness of residue exceeds the height of the disc axle. Also, a large-diameter disc has a flatter angle of approach between the leading edge of the disc and the ground, making it less likely to push ('bulldoze') the residue ahead of it and more likely to trap it in the 'pinch zone' and then roll over or cut through it. The most appropriate disc size is a compromise between getting sufficient penetration and avoiding disc stall. The most appropriate disc diameters used in agriculture seem to be between 450 mm (18 inches) and 560 mm (22 inches) and are used extensively on no-tillage openers.

DISC DESIGNS. Another debatable aspect is the design of the disc. Essentially discs can be of five designs.

PLAIN FLAT DISCS (FIG. 10.15). These are used on more no-tillage openers than any other form. They are the least expensive option to manufacture and have a sharpened edge, although experiments have shown that sharpening of the edge is not altogether necessary in all situations. They have the least traction of all alternate designs to ensure turning, which is not a disadvantage when used in short standing residue, but can be a disadvantage in long lying residue. When sharpened, their action is intended to be one of cutting the residue, but as the edge becomes dull they tend to trample rather than cut some of the residue. As such, they have a strong tendency to hairpin when configured as double discs, angled flat discs or a single vertical pre-disc.

One redeeming feature of flat discs is that they are able to handle large woody sticks better than most other discs. The smooth edge tends to push such sticks away, whereas other disc types may slice into and catch on the sticks without actually cutting them, which then prevents the disc from rotating.

WAVY-EDGED 'FLAT' DISCS (FIG. 10.16). These are designed to gain maximum traction by interrupting the smooth sides of plain discs with a series of ripples. These ripples are designed to 'gear' the disc to the soil, ensuring the disc will rotate in even the heaviest lying residue. For reasons that are not well understood, the ripples also result in the discs being self-sharpening. As such, their action is more one of cutting than with

Fig. 10.15. A plain flat disc.

Fig. 10.16. A wavy-edged 'flat' disc.

plain discs, making them somewhat less likely to hairpin. Penetration forces are similar to those of plain flat discs. Although wavy-edged discs are sharper than plain discs, making penetration easier, their waviness actually increases their footprint area and increases rather than decreases required penetration forces.

Their self-sharpening tendency also results in relatively high wear rates. The ripples also become collection zones for sticky soils, interrupting their effective functioning. Their most common use is as a single pre-disc ahead of rigid components such as hoe openers. They may also have the function of loosening the soil ahead of the double disc openers so as to counter the compacting tendencies of such discs. They are sometimes known as 'turbo-discs' for this reason.

NOTCHED, OR SCALLOPED, FLAT DISCS (FIGS 4.27 AND 8.10). These have semicircular notches cut from their peripheries, leaving about 50% of the periphery as 'points' (actually they are not points, as such, but simply a portion of the edge of the original plain disc left unaltered) and 50% as gullets. The objective is to reduce the footprint area on the ground, which aids penetration when compared with plain discs, and to 'gear' the disc to the soil to assist traction. The 'points' of the disc penetrate the soil first and present approximately half the footprint area of a plain disc of the same diameter, although the gullet zones of the notches also eventually penetrate the soil to a shallower depth. The net effect, therefore, is easier penetration than with plain or wavy-edged discs.

Furthermore, as the 'points' penetrate the soil, they change their angle of attack slightly as they progress further around the rolling circle. One important effect of this is that the near-vertical edges from the 'points' to the gullets slide into the soil at a range of angles and thus produce a slicing action against a portion of the residue. This cuts that portion of the residue more effectively than when it is simply pressed on from above, as with all other disc types.

DISHED, OR CONCAVE, DISCS (SEE FIGS 4.10 AND 4.11). These are nearly always angled to the direction of travel. As such, the friction against them is increased compared with plain discs travelling straight ahead. They therefore have good traction and are less likely to stall in heavy, flat residue than plain discs, but have all of the other attributes of plain discs, including power requirements and a tendency to hairpin residue.

One of the difficulties with all angled discs is delivery of the seed to the U-shaped slot created behind the lee (back) side of the disc. Usually, a boot is positioned close to the disc beneath the ground, but the gap between this boot and the disc is a collection point for random residue when used in no-tillage. Unless this gap is continually adjusted, blockage soon results.

One way of solving the problem is to spring-load the boot so that it rubs on the disc at this point. An advantage of dished discs is that their curvature provides considerable strength, allowing the discs to be made from thinner steel than is common for flat discs of any nature. This in turn has clear advantages as far as penetration and sharpness are concerned. For example, a 3 mm thick disc will require only 60% of the penetration force required for a 5 mm disc, although with dished discs this advantage is offset by the resistance to penetration of the convex (back) side of the disc.

NOTCHED DISHED DISCS. These combine the attributes of dished discs with those of notched discs. Although such designs have been used extensively in heavy residues for cultivating new land from native scrub and felled bush, there are no known no-tillage openers that use the principle in a more refined role. Similarly, there are no known no-tillage openers that use a wavy-edged dished disc.

Realigning residue on the ground

A novel approach to avoiding hairpinning with plain discs has been to use realigning fingers ahead of the discs. One drill of US origin had vertical spring tines designed to agitate and jostle the lying straw so as to cause each straw to lie end-to-end with the approaching disc. This was intended to avoid the tendency of discs to pass across straws, the starting point of all hairpinning. The tangled nature of many straw residues, however, ensured that this approach was never wholly successful.

Flicking

Another novel approach to the operation of single plain or wavy-edged discs ahead of rigid tines has been to attempt to flick off any residue that collects on the leading faces of the tines, since a single disc operating ahead of a rigid tine will not allow that tine to pass cleanly through lying residue all of the time. Clean-cutting ahead of a tine can sometimes be achieved with short cut straw and often with anchored residue, but long and lying residues are another problem. Regardless of how well the disc cuts the residue, there will always be some straw remaining uncut and passing the disc to collect on (or wrap around) the tine. Even when the disc is positioned close to, or even touching, the front edge of the tine, residue will collect on this front edge. Besides, it is most difficult to ensure that a disc remains permanently touching a tine when both are subject to normal wear.

Scottish designers created a self-flicking device (Fig. 10.17). Two spring-loaded fingers were attached to the hub of the disc in such a way that, as the disc rotated, the fingers became tensioned against the ground. At a certain point in the rotation, each of the

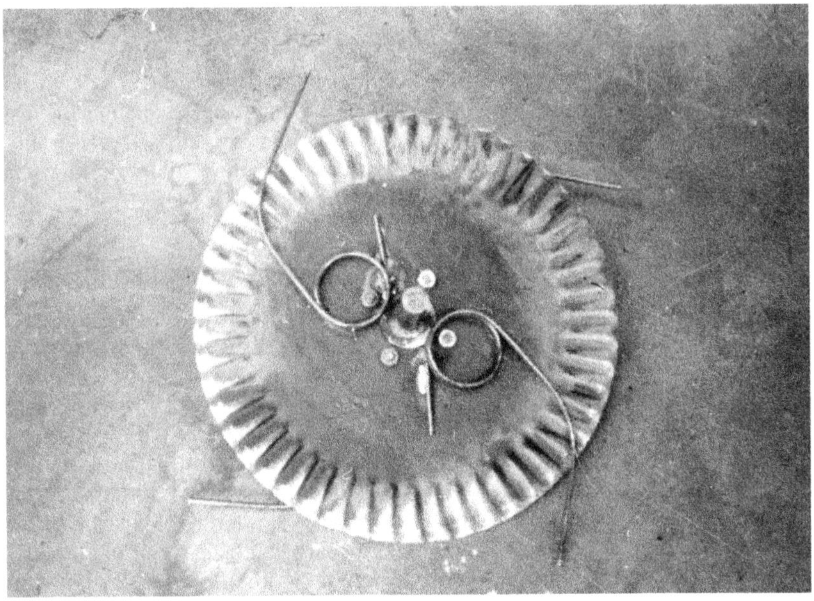

Fig. 10.17. A flicking device designed to self-clean stationary tines.

fingers was suddenly released from the ground, whereupon it flicked upwards at high speed past the front edge of the tine and dislodged residue that had collected there. Similar devices have been used by the authors, but these were attached to separate wheels that ran alongside the tine.

While the flicking devices worked in light and dry residues, heavy residues, especially when wet, tended to interfere with the flicking action. Failure to dislodge all of the straw from the tine with any one flick became a cumulative problem, leading eventually to total blockage of the tine.

Treading on residues

To overcome the 'hit-and-miss' nature of flicking, recourse to more predictable treading has been tried with mixed success. To achieve this, wheels are located alongside the tines so that they continuously roll on to one side of the residues wrapped around the leading edges of the tines. The intention is to cause the residue to be pulled off to one side. Even though they may achieve this objective, the presence of the wheels themselves generally interferes with free passage of other lying residue between the openers.

Self-clearance by free fall of residues off tines

Provided that sufficient space can be provided around each tine, most accumulated residue on the front of tines will eventually fall off, simply as a function of its own accumulated weight. Unfortunately, this does not always occur, especially with wet residue, necessitating irregular stops to clear what can become a sizeable mound of accumulated debris. Not only do these mounds of debris on the ground interfere with subsequent operations, their clearing is invariably a source of annoyance for operators at seeding time.

The most serious disadvantage of this principle, however, is the spatial demands on the drill required for clearance between individual tines. Drills of this type are limited to relatively wide row spacing (250 mm or greater) and the extended area occupied by the tines interferes with accurate surface following by individual tines and seed delivery.

Unfortunately, there are some designers and operators who are willing to widen the row spacing of their drills beyond what is agronomically desirable, expressly to provide more clearance for residue. But, if anything, no-tillage, by conserving soil moisture, should allow closer row spacing to be used than for tilled soils, with resultant higher potential crop yields. An example of a drill with wide row spacing is shown in Fig. 4.15.

Combining rotating and non-rotating components

An important new residue-handling principle was designed in 1979 (Baker *et al.*, 1979b). This involved rubbing the entire leading edge of a rigid component (such as a tine, shank or blade) against the vertical face of a revolving flat disc. For the rubbing action to be self-adjusting (so as to accommodate wear of components), the stationary component needs to be wedge-shaped so that it presents a sharp leading edge against the disc but tapers outwards away from the disc towards the rear. In this way it is held against the disc by lateral soil forces as the soil flows past. If two such rubbing components are positioned one on either side of the disc, all of the soil forces become symmetrical, thus avoiding undesirable side-loading on the discs and their bearings.

The design is illustrated in Figs 4.27 and 8.2. In the design of the disc version of a winged opener, an opportunity was taken to deliver the seed to the base of the slot by directing it to fall between one such stationary blade and the corresponding face of the disc. By directing fertilizer in an identical manner down the other side of the disc, an effective method of horizontal separation of seed and fertilizer in the slot was achieved (see Chapter 9).

There are four important principles involved in the rubbing action:

1. The intimate contact between the stationary blades and the revolving disc allows any residue passing the disc to also pass the whole assembly, thus making openers involving a rigid tine or blade at least as able to handle residues as a pure disc opener. This combination of disc and rigid component has achieved a remarkable residue-handling ability. This is important because all pure disc openers compromise at least some of the slot-shape functions to achieve residue handling. The best micro-environments for seeds in no-tillage are generally created by horizontal slots formed by a rigid tine (see Chapter 4).
2. The contact between the rigid component and the revolving disc is lubricated by a thin film of soil (Brown, 1982). This means that the rigid component can be manufactured from material that is much harder (and therefore more wear-resistant) than the disc, without cutting into the face of the disc to any appreciable extent.
3. There needs to be a small amount of pre-load between the rigid component and the disc, even though, in operation, the soil continually presses the two together. As the device enters the soil, and before the soil forces have pressed the two components together, a single piece of straw may occasionally become wedged between the two components if there is no pre-load between them. This residue will hold them fractionally apart for a short period. Other pieces of straw are then likely to enter the gap, with the result that blockage eventually occurs.
4. There is a disc-braking effect on the disc from the rubbing of the rigid components. For this reason, traction of the disc must be maximized. Notched flat discs are most commonly used for this type of opener, although plain flat discs have also been used. Wavy-edged discs are unsuitable because a flat surface is necessary for the blade and disc contact to be effective.

Wet versus dry straw

The action of most openers is affected by the brittleness of the straw, which is itself a function of dampness or dryness as well as other physical attributes, such as fibre content. After spraying or physical killing of growing material, the residue will lose water and become increasingly ‘stringy’. Sometimes best results will come from waiting 10–15 days so that the residues will dry completely and be more easily cut by discs. This also allows root material to begin decaying, which makes the soil more crumbly and usually leads to better slot formation. In other situations, drilling might be undertaken before or immediately after the residues are killed, provided that competition for soil water does not take place between the crop and the cover crop before the latter is killed.

On the other hand, harvested straw will normally be at its most brittle shortly after harvest. Discs are most effective when operating on brittle straw in warm dry weather and when the background soil is firm. Standing residue often becomes increasingly brittle as it 'ages' over winter, making spring no-tillage seeding more easily accomplished.

The case for and against scrapers

A natural reaction to problems of accumulation of sticky soil and/or residues on rotating components of openers is to strategically place scrapers and deflectors to remove the unwanted material. Such scrapers and deflectors can range from those designed to deflect residue from ever coming near the opener (e.g. Fig. 10.18) to those designed to protect a specific part of the opener. Figure 10.19 shows a circular scraper designed in Canada to remove soil from the inside of double disc openers.

However, most scrapers create more problems than they solve. Often, they simply present yet another point on which unwanted material can accumulate. While they may remove the original problem from interfering with a critical part of the opener, they seldom result in a total cure from accumulated debris. With the disc version of winged openers, both the side blades and the disc-cleaning scrapers (Fig. 10.20) operate beneath the ground and are therefore self-cleaning.

Clearance between openers

Even if individual openers are designed to freely handle surface residues without blockage, arranging multiples of such openers to handle residues in narrow rows is often a difficult problem in its own right. The main principles generally involve lateral spacing. To provide sufficient lateral space between adjacent openers for residues to pass through, the openers need to have a minimum of 250 mm clearance. Even then, the actions of different openers

Fig. 10.18. Residue deflectors on a maize planter.

may interfere with their neighbours and therefore require greater clearance.

For example, while 250 mm might be sufficient for openers that create minimal disturbance of the soil (e.g. double disc), greater distances may be required for those that either throw the soil (e.g. angled flat discs and angled dished discs), push it aside (hoe) or fold it back (winged). In such circumstances, each alternate opener needs at least to be offset forwards or rearwards of its neighbours so as to give diagonal as well as lateral clearance (referred to as staggering). An alternative to staggering is to create greater lateral distances between openers, but this usually means increasing row spacing, which may be agronomically undesirable.

The problem is further complicated by the greater downforces required for opener penetration under no-tillage, which is usually applied to the drag arm connecting the opener to the drill frame. The strength required of drag arms to transmit these large downforces discourages the use of alternately long and short drag arms to create staggering, especially if such drag arms are also of the parallelogram type with multiple pivots. In contrast, long and short drag arms are common on drills designed for tilled soils because the forces are small in comparison.

One way of overcoming this problem has been to operate the openers from two separate tool bars, one in front of the other.

Fig. 10.19. Circular scrapers for double disc openers.

Fig. 10.20. Underground scrapers on the disc version of a winged no-tillage opener.

This allows the openers on each tool bar to be spaced twice the distance apart as the row spacing. If used, it also allows the longer drag arms for a stagger arrangement to be of robust construction without interfering unduly with the between-opener spacing.

The problem of lateral spacing largely applies to drills and not to planters, because drills may have row spacing as close as 75 mm, while planters seldom require row spacing closer than 375 mm.

Summary of Residue Handling

1. The most serious physical problem relating to the handling of surface residues is mechanical blockage.
2. The most serious biological problem relating to the handling of surface residues is hairpinning (or tucking) of residues into the seed slot.
3. Macro (whole field)-management of surface residues starts with the combine harvester and is important for soil and resource management of no-tillage in general.
4. Macro-management should aim at even spreading of both straw and tailings over the entire field. Chopping of straw is optional.
5. Micro-management of surface residues is a function of no-tillage openers and is important for controlling the micro-environment of the seed slots.
6. Micro-management should strive to return the residue over, but not into, the seed slot (Class IV cover).
7. Large-scale no-tillage almost invariably involves the use of herbicides to kill existing vegetation.
8. Small-scale no-tillage relies predominantly on mechanical or manual residue management.
9. Knife rollers are a useful tool for management of residues in small-scale no-tillage.
10. Residue can be classified as 'short root-anchored'; 'tall root-anchored'; 'short flat'; or 'long flat'.
11. 'Long flat' residue is the most difficult to handle.
12. Relying solely on the cutting of residues is seldom effective. No device will cut all of the residue all of the time.
13. Most pure disc-type openers handle residues well, but also tend to hairpin (or tuck) straw into the slot, which is undesirable.
14. Most rigid component openers (hoe- or shank-type) handle residues poorly with regard to blockage but do not hairpin.
15. Most power till openers handle residues poorly except when the blades are C-shaped.
16. Wavy-edged and notched discs handle residues better than plain discs.
17. Small-diameter discs penetrate soil and residues more easily than larger discs, but are more likely to form blockages in heavy residues.
18. Firm soils provide a better medium for residue handling and cutting by openers than soft soils, thus reducing hairpinning.
19. Small no-tillage machines often have poor performance by having tined openers (due to cost) but benefit from the manual attention that can be given to residue management by operators.
20. Wet soil and/or wet residue are more difficult to handle than dry soil and/or dry residue.
21. Unless operating beneath the ground, most scrapers are of limited value because they accumulate residue on themselves while they are removing it from elsewhere.
22. Vertical mulching consists of disposing of straw into deep vertical slits in the soil.
23. Any rigid opener component, such as a tine or shank, will accumulate residue regardless of the design or positioning of a disc ahead of it.
24. Only when the leading edge of a rigid tine is forced to rub in intimate contact with the side face of a revolving flat disc will a tine/disc combination handle residues as well as a disc alone.
25. The minimum distance between adjacent openers for self-clearance of residues is approximately 250 mm, either laterally or diagonally, or both.

11 Comparing Surface Disturbance and Low-disturbance Disc Openers

C. John Baker

The surface disturbance of soil and residues often represents the most visible difference between no-tillage openers; yet the real effects may lie underground.

The passage of a seeding drill over a no-tilled field causes a wide variety of soil and residue disturbances, largely depending on the opener design, soil condition and operation speed. These disturbances are quite visible and yet the impacts on crop establishment and subsequent yields may only become obvious in stress conditions.

In the first part of this chapter, we revisit the seeding principles of the previous chapters to relate the disturbance effects to the effectiveness of common no-tillage slot shapes. In the second part, we compare the design features of common disc-type openers, since it is mainly disc openers that create minimum-disturbance slots.

Minimum versus Maximum Slot Disturbance – How Much Disturbance Is Too Much?

The largest concern centres on openers that create significantly different amounts of disturbance, such as a single, straight-running disc versus a broad hoe or chisel opener. These results are recognized as minimum versus maximum opener disturbance drills. Minimum disturbance creates just enough soil movement for the seed insertion with a single cut through the overlying residue while maximum disturbance moves a significant volume of soil to create a seed slot and allows the soil to fall or be moved back over the slot with the residue moved well away from the seed row.

Crop residues are the lifeblood of no-tillage. Indeed, they are the lifeblood of sustainable agriculture itself. In the past, debates about surface residues have mostly centred on their macro-management: the percentage of ground that is covered by residues in relation to erosion control, surface sealing, shading and the ability of machines to physically handle them. Recent emphasis has been to reduce the amount of residue disturbance during drilling for the erosion protection that greater amounts of ground cover offer.

Micro-management of residues centres on the influence that residues have on seed, seedling and plant performance in individual rows, all of which ultimately affect crop yield.

One aspect relates to soil erosion. The other to crop yield. Is one more important than the other?

Unless crop yield is maintained, few will undertake no-tillage and the soil erosion benefits become irrelevant. Therefore it could be said that micro-management of surface residues should be the first objective in any no-tillage system. But, sadly, history shows that that has seldom been the case.

Then again, minimum slot disturbance means different things to different people. For example, an allowable limit of 30% slot disturbance means that the disturbed zone in 150 mm spaced wheat rows can only be 45 mm wide – a difficult but achievable expectation for many no-tillage openers. But 30% disturbance in rows of maize or cotton sown in 750 mm–1 m rows represents 225–300 mm of disturbance – a much more generous objective.

So the development of no-tillage openers for wheat and other narrow-row crops may take a very different course from that for wide-row crops. But, since there is twice as much wheat sown in the world as the next most common crop, the constraints on openers for narrow-row crops provides the greatest challenge for machinery designers.

Minimum-disturbance no-tillage is created by openers that disturb the surface of the ground as little as possible, retaining at least 70% of the surface residues intact after their passage, with residues evenly distributed over the surface of the ground. Minimum-disturbance openers include double and triple disc (so long as the soil is not sticky); the disc version of winged openers; some narrow knife openers operating in low-residue conditions; and some angled disc openers operating at slow speeds on flat ground and in non-friable soils.

Maximum-disturbance no-tillage is created by openers that either burst the soil aside or deliberately till a strip at least 50 mm wide. Maximum-disturbance openers include most hoe, shank and sweep types; angled discs operated at high speed and/or on hills; double or triple disc openers in sticky soils; dished disc type openers; and powered till-type openers.

Disturbance effects

No-tillage opener design has the biggest influence on the amount of slot disturbance that occurs, and this in turn can have a direct influence on multiple factors directly related to the effectiveness of no-tillage seeding. Each will be discussed using many of the principles previously introduced, but more specifically related to the amount of visual soil and residue disturbance as the seeding is accomplished.

Slot cover

In tilled seedbeds, it is relatively easy to cover the seeds with loose soil. Therefore, aiming to create localized tilled strips during no-tillage has been an obvious objective of some no-tillage machinery designers. But no one has ever advanced a good biological reason for regularly tilling or disturbing the soil in the slot zone other than to compensate for the inadequacies of the openers that place the seed.

Many low-disturbance no-tillage openers cut a vertical slot in the soil. Even though this creates minimal surface disturbance (which is desirable), unless the soil is dry and crumbly at the time, closure of such slots is difficult and is worst in damp and 'plastic' soils. No-tillage slots that remain open dry out and attract birds, insects and slugs, which may cause crop failures even before the plants emerge from the ground. This problem has probably been responsible for more crop failures in no-tillage than any other single factor.

Covering problems can largely be solved while still retaining minimal residue disturbance by creating horizontal or inverted-T-shaped slots (winged openers). The seed is located on a horizontal soil shelf on one side of these slots and with advanced designs fertilizer is banded on an identical shelf on the other side. Horizontal flaps of soil with residue covering the soil are folded back to cover both. Even if the central slit dries and cracks open, as is inevitable in some untilled soils, neither the seed nor the fertilizer becomes exposed.

Viewed from the surface, inverted-T-shaped slots may appear similar to vertical V-shaped slots. Both are usually classified as minimal disturbance. The difference is beneath the ground. Vertical V-shaped slots may have compacted near-vertical side walls and get narrower towards their bases. It is often difficult to push a finger into them. They usually provide class I or, at best, II cover. On the other hand, inverted-T-shaped slots are loosened beneath the surface, get wider with depth and are usually very easy to push a finger into, providing Class IV cover.

In-slot micro-environment

Minimum slot disturbance does not always equate with a beneficial slot micro-environment. But nor does maximum slot disturbance. In fact, the best in-slot micro-environment that maximum-disturbance slots can provide is seldom better than a tilled soil, but may be better than poorly made and covered V-shaped slots (Class I cover).

Within the various minimally disturbed slots, horizontal slots (inverted-T-shaped with Class IV cover) create about as favourable a micro-environment as possible by trapping vapour-phase soil water in the slot (see Chapter 5). Seeds will germinate on the equilibrium relative humidity (RH) contained within the soil air, so long as this RH remains above 90%. Tilled soils seldom contain an equilibrium RH greater than 90% due to air exchange with the atmosphere, whereas untilled soils nearly always have an equilibrium RH between 99% and 100%. The problem is that unless the seed slot created in an untilled soil has sufficient coverage to trap the air (which usually means residues overlying soil), the potentially superior micro-environment in an untilled soil will be lost and seeds must then rely on a slot micro-environment that is no better than a tilled soil.

Vertical V-shaped slots (Class I or II cover) do not trap in-slot RH and are therefore about the least tolerant of all no-tillage slots of dry conditions.

All slots that involve strip tillage of some nature (Class III cover) fall into the maximum-disturbance category. They are likely to be more tolerant of adverse conditions than vertical V-shaped slots, simply as a function of the friable soil within the slot, but will still be inferior to horizontal inverted-T-shaped slots, which contain RH as well as liquid-phase water.

Slots created by angled discs fall between the extremes. As a general rule of thumb, if a slot made by an angled disc results in minimal surface disturbance, it will contain a better slot micro-environment than where such slots are more disturbed.

Carbon dioxide loss

Slot shape and residue retention may affect the ability of no-tillage slots to retain carbon dioxide. There is no doubt that all no-tillage offers major advantages over tillage in this regard (Reicosky, 1996; Reicosky *et al.*, 1996), but differences in no-tillage slot disturbance may also affect the amount of carbon dioxide that is lost from the slot zone.

In-slot moisture and temperature

Some studies have shown that slot shape and residue retention have only minimal short-term effects on liquid-phase soil water content and temperature within the sown slots, even though they are both affected on a macro-scale by residue retention (Baker, 1976a, b, c). On the other hand, the practice of removing residues from over the slot to raise the soil temperature in the slot zone in spring has a measurable effect.

The objective of this process is to expose the slot zone to direct sunlight when soil is warming up (such as in springtime), which in turn causes drying, thereby raising the soil temperature in the row. This begs the question whether seeds sown shallow beneath a residue canopy (Class IV cover) experience any lower soil temperature regimes than seeds sown deeper in uncovered slots, because the former option provides water for germination at shallow sowing depths and involves minimal-disturbance of the residues.

Seed germination

Chapters 5 and 6 showed that, while most minimum-disturbance slots promote high germination counts in dry soils, not all such slots perform well in wet soils, even though some do, such as the inverted-T shape. Nor does good germination always translate into good seedling emergence in dry soils (see below).

Maximum-disturbance slots are neither the best nor the worst for promoting germination. They attempt to emulate tilled soils and as a result usually perform similarly to tilled soils.

Seedling survival and emergence

The most critical time for no-tillage seedlings is the time between germination and emergence, as discussed in Chapter 5. Retaining surface residues over the slot (inverted-T-shaped slots, Class IV cover) sustains seedlings beneath the surface of the soil awaiting emergence better than loose soil (Class II–III cover), which is better than no cover (Class I). In addition, the retained residues are desirable from a soil erosion point of view. Not all minimum-disturbance slots create Class IV cover, depending on the residue amount and condition. Some may even be as poor as Class I cover. Most maximum-disturbance slots create Class II–III cover.

Soil-to-seed contact, smearing and compaction

The amount of slot disturbance visible from the ground surface is not always a good indicator of what is taking place beneath the soil in terms of soil-to-seed contact. For example, V-shaped slots in heavy soils (minimal-disturbance) may create neatly cut, smeared (if wet) and even compacted slot walls but still have adequate soil-to-seed contact, even in dry soils, because the seeds become wedged between the near-vertical slot walls. But such seeds may germinate and die (see Chapter 6), even with adequate soil-to-seed contact. In other cases, seeds sown into highly disturbed dry slots may have good soil-to-seed contact but fail to germinate because loose soil conducts liquid-phase soil water poorly.

In inverted-T-shaped slots (minimal-disturbance) without vertical slot walls, soil-to-seed contact may be little different from highly disturbed U-shaped slots, but, because inverted-T-shaped slots are covered with residue (Class IV cover), the presence of water vapour will ensure germination and emergence take place.

Root development

Vertical V-shaped slots create minimal surface disturbance but may restrict root development more than other openers, especially in heavy damp soils. The use of wavy-edged pre-discs with such openers reduces root restrictions but increases surface disturbance.

Most maximum-disturbance openers, together with most winged openers, present little, if any, restrictions to root growth.

Infiltration into the slot zone

Slot disturbance has a direct effect on infiltration. Earthworms and other soil fauna that feed on surface residues create channels that have a positive effect on infiltration. Earthworms, in turn, respond to the positioning of the residues. Minimum-disturbance slots that leave or replace the residues over the slot encourage earthworms to colonize the slot zone, which increases infiltration.

Maximum-disturbance slots may kill earthworms in the immediate vicinity. The wider and more severe the disturbance (especially if a power till mechanism is involved), the greater the earthworm mortality. But nearby earthworms will recolonize the disturbed zone shortly afterwards.

Other factors may also contribute. Minimum-disturbance slots created by vertical double or triple disc openers compact the side walls of the slot. This has a direct negative effect on infiltration from sealing as well as an indirect negative effect because earthworms avoid the compacted zone.

Hairpinning of residues

The most quoted negative effect from residues positioned close to the slot zone has been tucking of residues into the slot, termed 'hairpinning' (see Chapter 6). Decomposing residues in wet (and especially anaerobic) soils produce acetic acid, which can kill seeds or seedlings that are touching the residue. In dry soils, seeds suspended in hairpins have difficulty accessing liquid-phase water.

All disc-type no-tillage openers hairpin residues at least some of the time. But no one has yet designed an opener that can physically handle surface residues in closely spaced rows without the assistance of discs. The disc version of the winged opener physically separates seeds from direct contact with hairpinned residues and thus avoids the problem. Acetic acid is rapidly broken down in the soil by bacteria, so small separation distances are effective. But all double disc and angled disc openers (whether slanted or upright) experience hairpinning problems because the seeds remain embedded in the residues.

Fertilizer banding

Banding of fertilizer close to, but not touching, the seeds at seeding is vital to maximize crop yields (Baker *et al.*, 1996; Fick, 2000). Some designers achieve this by combining two openers together, which increases inter-row spacing and surface disturbance, or by using 'skip-row' planting (one row of fertilizer between every two rows of seed). Others use altogether separate fertilizer openers, which increase slot disturbance even more. But there are other 'double-shoot' openers (e.g. disc version of winged openers) that have been purpose-designed with no sacrifice of row spacing or surface disturbance (Baker *et al.*, 1979b).

Soil erosion

Since retention of surface residues is the most effective mechanism for controlling soil erosion, the more of the surface that remains covered with residues after seeding, the better.

Pests, diseases and allelopathy

Early predictions of uncontrollable residue-related pest and disease problems attributable to no-tillage and residue retention have proved to be exaggerated and, in most cases, groundless. In early trials with no-tillage, poor crop results were often attributed to toxic exudates from dying residues (allelopathy). But, as scientists have come to understand what really affects seed germination and seedling emergence during no-tillage (particularly the role of residues in improving the slot micro-environment), examples of true allelopathy have become difficult to find.

In any case, the advantages of residue retention are so great that they far outweigh any minor residue-related disease or pest problems that might occur from time to time.

Disc opener feature comparisons

No comparison would be complete without examining the designs of a selection of mainstream openers and/or machines. In this case, we have compared three different designs of disc openers: the disc version of a winged opener; angled vertical discs; and double discs.

Comparisons in Table 3.2 showed that the risk of impaired crop performance with the disc version of the winged opener was rated at 11%, while the angled vertical disc opener was 30% and the double disc opener 53%. Table 11.1 lists the causes of these differences. Shank-, hoe- and tine-type openers were not compared because the designs and performance of such openers vary widely and are affected by soil conditions and operating speed; thus the results are difficult to generalize.

Summary of Comparing Surface Disturbance and Low-disturbance Disc Openers

1. The dual objective of minimizing the amount of disturbance to surface residues while at the same time maximizing seed, seedling and plant performance is possible to achieve with modern no-tillage techniques and equipment.

2. Not all minimum-disturbance openers will create optimum crop yields, but all maximum-disturbance openers will reduce

Table 11.1. Comparison of selected features of three disc-type no-tillage openers.

Opener features	Disc version of winged opener	Angled vertical disc opener	Double disc opener
Risk of impaired performance	11%	30%	53%
Opener description	Opener comprises a vertical notched disc with two blades with horizontal wings and scrapers in intimate contact with either side of the disc	Single-shoot form of opener has a single vertical disc arranged at an angle to the direction of travel. Double-shoot option has duplicate fertilizer-only openers	Single-shoot form has two discs arranged at about 10° vertical angle to one another. Double-shoot option has duplicate fertilizer-only openers
Effects of forward speed	Functions are not greatly affected by forward speed	Functions are affected by forward speed because of the angle (7°) of the disc(s) to the direction of travel	Some functions (e.g. seed flick) can be affected by high forward speeds
Maximum speed	Capable of operating at speeds up to 10 mph (16 kph)	Forward speed is limited by conditions, but maximum is less than 10 mph (16 kph)	With adequate deflectors against seed flick, capable of speeds up to 10 mph (16 kph)
Seed covering	Seed covering (Class IV) is residue layered over soil, which scientific experiments have shown to be superior to all other forms of cover	Seed covering is mostly loose soil but the covering function is very speed-dependent (Classes I–III)	Often covering is very difficult to achieve at all because of the wedged V-shaped slot. Rippled pre-disc helps in this respect (Classes I–III)
Slot compaction	No in-slot compaction occurs	In-slot compaction occurs on one side only	Severe in-slot compaction occurs on both sides
Slot smearing	Any in-slot smearing that occurs remains moist and therefore of no consequence	Most in-slot smearing that occurs remains moist and therefore of no consequence	In-slot smearing is common and difficult to prevent from drying, when it forms in-slot crusting, which is even worse
Surface residues over slot	Retains 70–90% surface residue cover	Tends to push residues aside rather than replacing them. Higher speeds push residues further aside	May retain 70% residue ground cover except in sticky soils, when residue retention declines
Vapour moisture	Retains maximum vapour moisture at seed zone	Medium vapour moisture retention if loose soil covers slot (Class III)	Poor vapour moisture retention
Residue hairpinning	Will create hairpins but the seeds are effectively separated from the hairpins so they are of no consequence	Creates hairpins and, because seeds remain embedded in the hairpins, this affects germination in both dry and wet soils	Creates hairpins and because seeds remain embedded in the hairpins this affects germination in both dry and wet soils
Seedling emergence	High seedling emergence, almost regardless of soil or climatic conditions	Seedling emergence is dependent on favourable soil and climatic conditions	Seedling emergence is highly dependent on favourable soil and climatic conditions
Crop failure	Low crop failure rate	Medium crop failure rate	High crop failure rate

Crop yields	Regularly produces superior crop yields to both tillage and other no-tillage openers	Requires double-shoot version to produce best crop yields, but farmer confidence is not high	Produces acceptable crops in favourable conditions, but yields are restricted by inability to band fertilizer
Emergence retarding	Seedling emergence is not retarded by soil flaps in wet plastic soils because seedlings follow pathway to the surface provided by the vertical disc slit	Seedling emergence is not affected by soil flaps, but exposure of seeds can be a problem	Seedling emergence is not restricted, but desiccation and bird damage of exposed seedlings can be a problem
Banding fertilizer	Simultaneous and separate seed and fertilizer banding occurs on either side of a single disc on the same opener	No separate banding of seed and fertilizer by each opener. Requires duplicate openers for double-shooting	No separate banding of seed and fertilizer by each opener. Requires duplicate openers for double-shooting
Fertilizer placement	Therefore one compact opener is used to sow both seed and fertilizer	Double-shooting of seed and fertilizer requires either a complex doubled-up opener or duplication of two openers. This increases the complexity and space occupied by each composite opener and encourages farmers to only do half the job by purchasing the cheaper single-shoot option	No known double-shooting openers exist. Therefore duplicate openers are the only option, with the same problems that angled disc openers have
Row spacing	Can be arranged in close row spacings down to 5.5 inches (140 mm)	Minimum row spacing of double-shoot version is 7.5 inches (190 mm)	No double-shoot option. Minimum row spacing of single shoot is 4.72 inches (120 mm)
Hillside operation	Effective on hillsides	Not effective on hillsides because of the angle of the disc	Effective on hillsides
Prior tillage	Effective in tilled and minimum-tilled soils	Not very effective in tilled and minimum-tilled soils	Effective in tilled and minimum-tilled soils
Service wear	All moving pivots use sealed bearings, leading to long trouble-free service	Most versions use many simple pivot arrangements with limited service life	Most versions use many simple pivot arrangements with limited service life
Penetration force	Individual hydraulic rams on each opener ensure consistent downforce that can be varied infinitely on the move from the driver's seat and automated to adjust to soil hardness on the move	Most have downforce springs with stepwise settings and each setting changes its downforce with elongation and contraction	Most have downforce springs with stepwise settings, and each setting changes its downforce with elongation and contraction

Continued

Table 11.1 *Continued*

Opener features	Disc version of winged opener	Angled vertical disc opener	Double disc opener
Depth gauge	Press/gauge wheels located close to seed zone	On some models gauge wheels located right alongside seed zone	Gauge wheels sometimes behind seed zone, but other versions are right alongside seed zone
Depth control	This results in superb opener depth control and makes drilling over contour banks and variable soils effective	Springs limit opener depth control, especially when drilling over contour banks and surface changes, but good location of gauge wheels helps	Limited opener depth control, especially when drilling over contour banks and surface changes
Downforce control	Electronic monitoring of press wheel footprint allows the downforce to be altered on the move in response to changing soil hardness	Downforce cannot be changed on the move	Downforce cannot be changed on the move
Downforce range	Normal downforce range is 0–500 kg	Most designs have a downforce range 0–250 kg	Most designs have a downforce range 0–250 kg
Maximum downforce	High-downforce rams (1100 kg) are available for openers operating in the wheel tracks of compactable soils	No high-downforce openers available	No high-downforce openers available
Opener adjustments	There are only two operational adjustments with each opener, neither of which changes with wear and one of which is made in the tractor cab	Many operational adjustments are necessary, most of which are affected by wear and require dismounting from the tractor cab	Main operational adjustments are closing wheels, downforce
Disc adjustment for penetration	Disc axle can be located in three different positions, which minimizes the penetration forces required, especially in hard soils and stones	Disc(s) are located in one fixed position	Discs are located in one fixed position
Operation skills	Medium skill and training level required. The electronic monitoring system protects the finished job from the effects of inexperienced operators	Skill level is dependent upon operators gaining experience with every condition likely to be encountered so as to make the correct adjustments. Therefore a good finished job is highly skills-dependent	Medium skills dependency. Simple design creates fewer adjustments than angled disc but also reduces adaptability compared with either other opener
Scientific evaluation	Extensive scientific validation of manufacturer's claims	Little or no scientific validation of manufacturer's claims	Scientific tests have mostly been negative about this opener

the effectiveness of erosion control and soil improvements offered by no-tillage.

3. Minimum slot disturbance is an optimum no-tillage objective, with full consideration given to the several other seeding requirements for stand establishment.

4. Horizontal (inverted-T-shaped) slots provide good slot coverage with minimal residue disturbance (Class IV). V-shaped slots provide poor slot coverage and residue cover (Class I).

5. Minimum-disturbance slots do not necessarily create favourable slot micro-environments unless adequately covered with soil and residue. Horizontal minimum-disturbance slots most readily create favourable slot micro-environments while vertical minimum-disturbance slots do not. Maximum-disturbance slots create slot micro-environments similar to tilled soil.

6. Minimum-disturbance slots are likely to lose somewhat less carbon dioxide than maximum-disturbance slots.

7. The amount of residue cover over the slot has minimal long-term effect on liquid moisture content. Minimum-disturbance slots trap water vapour, while residue-free slots warm more quickly in spring.

8. It is possible to have both minimum residue disturbance and maximum seed germination.

9. It is not always desirable or necessary to sacrifice residue disturbance to encourage seedling emergence. Depending on the design of opener and the climatic condition, it may, in fact, have the opposite effect.

10. Slot disturbance by itself is not necessarily a good indicator of soil-to-seed contact. The amount of residue disturbance may have little effect on soil-to-seed contact.

11. Some, but not all, maximum-disturbance openers may enhance early root growth. Restrictions by some minimum-disturbance openers may occur with unfavourable soil conditions.

12. Provided that compaction is not a factor, most minimum-disturbance slots encourage earthworm activity, and thus increase infiltration compared with maximum-disturbance slots. In the absence of earthworms, maximum-disturbance slots may have greater infiltration than the best of minimum-disturbance slots.

13. All non-disc openers, especially those associated with maximum residue disturbance, avoid hairpinning problems but experience residue-handling problems. Most disc openers, except those that create horizontal slots, have hairpinning problems.

14. Some no-tillage drills that band fertilizer are less capable of minimizing residue disturbance or close row spacing, or both. But there are notable exceptions, such as the disc version of winged openers.

15. No-tillers should ensure that surface residues are well distributed and minimally disturbed.

16. Disturbing surface residues as little as possible in the slot zone will have more positive effects on seed, seedling and crop performance than harmful effects by pathogens or allelopathy.

17. Disc-type openers vary widely in their specific designs, which in turn affect their biological functions, including slot disturbance.

12 No-tillage for Forage Production

C. John Baker and W. (Bill) R. Ritchie

The establishment and/or renovation of forage species is a special no-tillage case requiring additional techniques and management.

Pastures and other forage crops provide food for foraging animals in countries, regions or seasons in which animal production is profitable. In some situations animals are grazed outdoors, often all year round. In other situations, forage crops are harvested for storage or transport to the animals housed indoors for at least part of the year. Many of the world's pasture plants are self-sown native species on rangeland, which have survived in the ecosystems to which they are adapted. Most of these species, however, produce relatively poor feed for domestic animals in terms of quality and quantity.

In the improved pastures of temperate countries, genetically superior species have been sown into the rangelands and, together with the judicious use of fertilizers and rotational grazing management, have led to vastly improved animal productive capacity. Over time, however, some of these improved pastures have slowly reverted to the original, less productive species, requiring intermittent renewal or renovation with improved species. In other situations, the continual genetic improvement of pasture species dictates their introduction into otherwise 'permanent' grazing systems to improve animal performance, regulate seasonal production or repair damage from pests, flood, drought or natural mortality.

We shall discuss the drilling of forage species and pastures separately, although in reality they are as often integrated into a single system as they are dealt with in isolation.

Forage Species

Forage crops are similar to arable crops for their establishment requirements by no-tillage except that small-seeded species are often involved, which require very accurate depth control from the openers. Many brassica species are used for forage cropping, along with grasses, legumes and herbs, all of which require shallow seeding. But a wide range of cereal species are also used, often for whole-crop silage, which have a greater tolerance of the depth of seeding.

One problem is that farmers often value their forage crops lower than arable crops, presumably because the cash returns from forage crops are derived by indirect animal harvesting rather than directly through machine harvesting of seed, fibre or grain. When a forage crop fails, there will often be an alternative forage crop nearby that can be used to compensate, or, at worst, animals can be sold to reduce demand. In contrast, when an arable crop fails, that source of income is lost for ever and cannot be replaced. For this reason, there seems to be more acceptance by animal farmers of sub-standard no-tilled forage crops than is the case with arable farmers. Even those people who farm integrated animal and arable systems put less value on the forage crops than on the arable crops, possibly because the latter usually comprise the major part of the farm income.

Further, because pastures are regularly grazed or mowed, differences in individual plant performances are more difficult to detect by eye. As a consequence, rather than attracting greater precision at drilling, much pasture establishment actually attracts less precision.

But this situation is changing. Animal farmers in New Zealand, for example, are finding that a new level of animal intensification is possible using 'fail-safe' no-tillage that rivals arable cropping, in terms of both returns per hectare and risks.

Animals are often grown on 'permanent' forage species (usually pastures), which are characterized by uneven growth cycles during the year. Maximum forage production and quality of feed occur in warm moist months, while minimum production and quality occur in cold and/or dry months. Management of animal systems that rely on such feed supplies is constantly restricted by the lowest-productivity months. Often this involves the use of feed supplements, either purchased in or saved as silage or hay during the more productive months.

But a new level of productivity can be achieved by replacing 'permanent' forage species with highly productive, short-rotation, speciality forage species, which are re-established at least once (and often twice) per year and are chosen according to their suitability for specific growth periods or animal requirements during the year. Some are cold-tolerant; others are dry-tolerant; still others produce a quality of feed suited to particular stages of growth of the animals. The possible combinations are virtually endless and can be regularly changed.

But they all depend on the availability of 'fail-safe' no-tillage techniques and systems. Such systems of forage production cannot be accomplished using tillage because few productive soils can stand being tilled continuously once or twice per year. The soils would quickly deteriorate to unmanageable conditions, and utilization by animals would become almost impossible.

Although the quality and quantity (and therefore productivity) of short-rotation forage species grown under continuous no-tillage regimes are superior to those obtainable from 'permanent' pastures, the new system puts much pressure on the ability of the no-tillage system and equipment to deliver maximum crop yields for each successive crop.

Because the forage crops are established at least once and often twice per year, it is a 'high-input' and 'high-output' system, but some of those practising it have reported tripling the numbers of stock grown to slaughter weight per year on the same area. Outdoor weight gains from lambs and beef cattle in the order of 400 and 1000 g per day, respectively, have been common when the animals are fed *in situ* on a continuing supply of short-rotation no-tilled forage crops.

A variation on this system is to grow continuous short-rotation silage and hay crops for cash sale rather than grazing by animals. In some systems, animals never enter the fields. This restricts the choice of forage species to those that can be converted into hay or silage, but again the system is totally dependent on no-tillage.

Integrated Systems

The optimum diversification is to integrate animal and arable crop production systems. This is a common practice in countries where climate allows animals to graze outdoors all year. Typically one or more arable crops are grown during the most productive time(s) of the year and forage crops are grown between the arable crops and either fed directly to animals or mechanically harvested and fed indirectly to them. Up to three integrated crops can be grown per year in some climates.

Where no-tillage is not available, such intensive cropping will not sustain soil structure and tillage delays planting. For this reason, typical tillage-based rotations have included a period in permanent pasture with the objective of repairing the damage to the soil structure by previous tillage and readying the soil for further destructive cropping processes to follow.

No-tillage changes all of that by allowing continuous cropping (forage and/or arable) to take place almost indefinitely without significant damage to soil structure. Crop rotations are then not constrained by the need for a remedial pasture phase and can be selected for the relative values of crops at any one time.

Figure 12.1 is an example of where two crops of summer turnips (*Brassica* spp.), one established by tillage and the other by no-tillage on a light organic soil, have been grazed by dairy cattle *in situ*. The difference in soil damage is clear.

Of course, severely wet weather and heavy concentrations of stock will damage even untilled ground eventually. The question then becomes: 'How serious must the damage be before some form of tillage is justified?' Figure 12.2 shows severely damaged soil from repetitive hoof treading in a gateway when the soil was wet. The soil damage in Fig. 12.2 is about the upper limit that winged, hoe or angled disc no-tillage openers can be expected to repair without assistance from tillage tools. Indeed, the result from a single pass with a disc-version winged opener drill can be seen on the left. Double or triple disc openers would not cope well with such surface damage because they have only a minor smoothing effect as they travel through the soil.

Damage beyond that shown in Fig. 12.2 is best repaired with a shallow tined implement or rotating spiked harrow, the actions

Fig. 12.1. Two turnip crops sown with angled disc openers and fed *in situ* to dairy cattle in New Zealand. The upper photograph shows a wide view of the treading damage to the tilled soil on the right compared with the untilled soil on the left. The lower photograph shows close-up views of the respective soil surfaces. Note that the soil surface under no-tillage (left) has not been broken open at all, whereas the tilled soil (right) is heavily scuffed.

Fig. 12.2. Severely hoof-damaged soil that is about the upper limit that can be repaired by a single pass with the disc version of winged openers (seen on left).

of which are to drag or flick surface soil into the hollows rather than invert the soil. More severe treading damage may also compact the surface layers (to about 300 mm) of the soil profile. This is best relieved with a shallow subsoiler with narrow vertical tines or sweeps that leave the soil surface reasonably smooth so that no-tillage may take place again without further smoothing being required.

Where the integration of animal and arable enterprises is practised, it is common for last-minute decisions to be made between the growing of one or more arable crops or forage crops based on expected relative returns. Such flexibility is only possible if last-minute crop-establishment decisions are based on no-tillage. No-tillage provides the flexibility that allows truly integrated animal and arable systems to develop to new heights.

Arable crops are sometimes rotated with pastures where land is retired or 'set aside' to allow it to revert to native grasses and/or scrub for periods of 10 years or more to protect the soil from erosion or reduce agricultural production. With the world's demand for food continuing to expand, however, such land is likely to be returned to arable farming in due course. When it is, it will be more important than ever to retain the sustainability of the soil health, which will, in most cases, have reached new heights from the retirement process, by adopting no-tillage from the outset. This means learning how to drill or plant into heavy 'unmanaged' sod.

No-tillage of Pasture Species

In some circumstances when drilling pasture species, it is not appropriate to kill all of the competing species. If the competing species are other desirable grasses and not destroyed, this is known as 'pasture renovation'. In other circumstances, it is necessary to kill all of the existing species. If the new species to be sown are also pasture plants, this is known as 'pasture renewal'.

Pasture renewal

One-quarter of the world's surface, some 3000 million hectares, is grassland (Kim, 1971; Brougham and Hodgson, 1992). Renewal and establishment of this valuable resource are a major effort, which can be enhanced with no-tillage practices.

Pastures are traditionally renewed either to improve the productivity of existing vegetation (e.g. bush, scrub, native grasses or introduced swards) or to replace a harvested annual crop with grazable pasture.

The objective can be to establish a long-term 'permanent' sward of monoculture (single species) pasture, including lucerne, or a mixed sward of several grass species and/or compatible legumes, such as clovers or lotus species. A further objective can be to establish a short-term (usually single species) temporary pasture to utilize land between successive arable crops.

Not all pastures will be grazed directly by animals. Many are harvested by machine or hand and fed to animals either directly as grain or as silage or hay, or are regularly mowed to keep them short (e.g. sports turf). This has some importance for the establishment method used. For example, if a pasture is to be directly grazed by cattle, the young plants may be damaged by treading or pulling. No-tillage offers clear advantages over tillage in this respect because the stability of the untilled soil resists treading damage and provides better root anchorage than tilled soils. No differences between no-tillage openers have so far been found in the pulling resistance during grazing (Thom *et al.*, 1986).

Where pasture plants are mechanically cut, pulling damage is minimized. Surface damage to the soil may result from heavy vehicle traffic under wet conditions. In this respect, the improved soil structure by no-tillage offers significant advantages over tillage.

The largest problem with pasture renewal by no-tillage is meeting the requirements of many pasture seeds for depth of sowing and germination micro-environment. The more rapidly establishing grass species, such as ryegrasses, are usually tolerant of sowing depths from 5 to 30 mm, but suffer reduced germination outside this range. The more weakly establishing species, such as lucerne, clovers and some grasses, are much less tolerant of improper depth, preferring the narrower range of 5–15 mm.

In a tilled soil, a narrow depth-tolerance range is relatively easy to achieve because the soil has been previously prepared to a uniform physical consistency and is easily penetrated by drill openers. Accurate sowing depth in a tilled soil favours large flotation-type openers (such as on V-ring roller drills), which are unable to operate in no-tillage because of the more dense untilled soil.

No-tillage openers for pasture renewal therefore need mechanisms for depth control and surface following and to be capable of creating a desirable slot micro-environment within the top 10–15 mm of soil. These are demanding requirements.

The choice of drilling pasture plants in rows compared with random scattering of seeds followed by harrowing has been discussed because the objective is to utilize all the available ground space. With no-tillage, random scattering (oversowing or broadcasting) almost invariably results in poor establishment because untilled soils offer little loose soil or debris to cover undrilled seeds by harrowing. Trampling of the seed into the ground by stock is no substitute for positive placement by a drill opener. None the less, where the operation of drills has been impossible (such as on steep hillsides and on some sports turfs), the practice of oversowing by aircraft, hand or light machine has been undertaken, with acceptable establishment by pelleting the seed and/or increasing the seeding rate to compensate for mortality.

Row spacing

Where no-tillage drilling can be successfully undertaken (i.e. tractors can access the land), the debate shifts to the most desirable row spacing and drilling times. Common design and space limitations of drills provide a narrowest practical row spacing of about 75 mm, with wider spacing up to 300 mm used in dry climates for pasture species with surface creeping habits or for forage seed production.

Research in New Zealand using a rapidly establishing ryegrass species (Inwood, 1990; Thom and Ritchie, 1993; Praat, 1995) showed little or no production differences between: (i) single-pass drilling with winged openers in 150 mm rows; (ii) single-pass drilling with the same openers in 75 mm rows; and (iii) cross-drilling in 150 mm rows with the same openers. In the last case, two passes were made at approximately 30° to one another, sowing half the intended application rate with each of the two passes (Thom and Ritchie, 1993).

Results in Table 12.1 (Praat, 1995) show that a slowly establishing species (tall fescue, *Festuca arundinacea*) initially benefited from narrow (75 mm) row spacing as a result of reduced weed population. Cross-drilling had no long-term benefits over 150 mm spacing, possibly because the gains of closer plant spacing were offset by greater stimulation of weed seed germination by the second pass of the drill.

Single-pass drilling of tall fescue in 75 mm rows produced greater 5-month growth than single-pass drilling in 150 mm rows, but was not significantly different from the cross-drilling in 150 mm rows. The latter two treatments were themselves not significantly different from one another. The advantage of the 75 mm rows at 5 months was not repeated with ryegrass. By 23 months there were no significant differences among any of the drilling or species treatments.

Since the only differences were during the early stages of pasture growth, and then only with a slowly establishing species, the single-pass 150 mm row option is preferred because it is less expensive, both in terms of drill design and operational costs. An added advantage is that most no-tillage drills can also be used for drilling small-grained cereals, pulse crops, oil seed crops and forage crops.

In mild, wet winter climates, pastures and sports turfs are often renewed by tillage, in autumn because weeds are more easily controlled than in the spring and post-drilling soil moisture levels are likely to be more reliable than in the hotter summer periods. With no-tillage, however, the availability of herbicides and reduction in physical stimulation of dormant weed seeds largely eliminates the disadvantage of spring weed germination.

Further, the moisture conserved with no-tillage reduces the risk to new pasture and turf seedlings in dry summer soils. These factors have led to more spring drilling of new pastures and sports turfs using no-tillage than with tillage, but the majority of such swards are still established in the autumn.

Even in the autumn, the debate about row spacing has centred on the ability of pasture plants to quickly tiller and spread to occupy otherwise bare ground so as to compete with expected natural weed germination between the rows. Data in Table 12.2 (Praat, 1995) show the results of autumn no-tillage drilling of ryegrass and tall fescue into a recent alluvial soil containing a high weed seed population.

Only the two-pass cross-drilling treatment using winged openers increased weed seed germination and growth compared with single-pass drilling in 75 mm and 150 mm rows. Even then, these differences (approximately 20%) occurred only within the first 5 months after sowing. Thereafter there were no significant differences between drilling methods.

On the other hand, data in Table 12.3 (Hamilton-Manns, 1994) show a clear trend of declining weed growth with date of sowing in autumn and early winter, using the same pasture species sown in a single pass with winged no-tillage openers in 150 mm rows in a similar soil.

At the earliest sowing there were twice as many weeds in the tall fescue pasture

Table 12.1. Pasture production from differently spaced no-tillage drill rows, kg/ha dry matter yield at the time of sampling after sowing.

	5 months after sowing		23 months after sowing	
Treatment	Ryegrass	Tall fescue	Ryegrass	Tall fescue
75 mm rows, single pass	1893 a	2066 a	1399 a	1827 a
150 mm rows, single pass	1911 a	1525 b	1449 a	1734 a
150 mm rows, cross-drilled	2196 a	1826 ab	1453 a	1711 a

Unlike letters following data in a column denote significant differences ($P = 0.05$).

Table 12.2. Effects of no-tillage drilling method on herbage composition of pasture species, kg/ha dry matter at time of sampling.

	5 months after sowing		23 months after sowing	
	Sown grasses and clover	Weeds	Sown grasses and clover	Weeds
75 mm rows, single pass	902 a	531 a	2086 a	119 a
150 mm rows, single pass	835 a	545 a	2146 a	125 a
150 mm rows, cross-drilled	796 a	675 b	2123 a	178 a

Unlike letters following data in a column denote significant differences ($P = 0.05$).

Table 12.3. The effects of time of sowing on the proportion of weeds in a no-tillage pasture.

Time of sowing	Pasture species sown	% weeds present at 70 days after sowing	Mean % weeds for both species
Early autumn	Ryegrass	4.5 b	7.6 a
	Tall fescue	10.7 a	
Early–mid-autumn	Ryegrass	4.8 b	4.8 b
	Tall fescue	4.9 b	
Mid-autumn	Ryegrass	3.4 b	3.7 b
	Tall fescue	3.6 b	
Late autumn	Ryegrass	0.6 c	1.3 c
	Tall fescue	2.0 c	
Early winter	Ryegrass	1.1 c	1.4 c
	Tall fescue	1.8 c	

Unlike letters following data in a column denote significant differences ($P = 0.05$).

(10.7%) at 70 days after sowing compared with the ryegrass pasture (4.5%), because the more slowly establishing tall fescue pasture took longer to colonize the inter-row spaces. Thereafter there were no differences between these two pasture species in terms of weeds. As the season became colder (from early autumn to early winter), the percentage of weeds in both pastures steadily declined from an average of 7.6% to 1.3–1.4%, reflecting increasingly less favourable conditions for weed seed germination.

For total pasture production in New Zealand, Hamilton-Manns (1994) also found greater yield potential from early autumn sowing (March) than later winter sowing (June) if sufficient soil moisture was available to sustain early seedling development. This held true for both rapidly establishing species (e.g. ryegrasses) and slowly establishing species (e.g. tall fescue). The earlier sowings and warmer temperatures favoured tiller development of the sown species, although there was also an increased (though manageable) weed problem.

The retention of crop residues from a harvested summer crop or the fallowing of ground in the spring by spraying the previous pasture will help offset potential problems of weeds and low soil moisture levels during early autumn no-tillage sowing of new pastures. During winter in temperate climates, the retention of residues may result in increased earthworm populations (Giles, 1994).

In drier climates, improved establishment of new pasture species in the autumn has been achieved by chemical fallowing of fields over the dry summer. Resident species are sprayed out during late spring when they are still actively growing and receptive to herbicides, after which the fields are left fallow for several dry months. If sufficient residue remains on the surface as a mulch, considerably less soil moisture is lost compared with unsprayed pastures because the spraying reduces moisture loss by transpiration and evaporation. Moisture gains as high as 12-fold have been reported (Anon., 1995).

The potential loss of pasture production over summer in dry climates is small and a more moist soil environment is maintained for early autumn establishment. Control of resident species is enhanced by using a more appropriate time of year for spraying, and there is also the opportunity for an autumn herbicide application prior to drilling, if required.

In climates with adequate summer rainfall, autumn establishment of new pasture species may be enhanced by drilling a forage crop the previous spring. This not only provides the opportunity for a double application of herbicide, but it also provides time and stock trampling to break down the intense root mats that exist with some native pasture species in low-fertility situations.

The emphasis with most of these techniques is to ensure effective long-term control of resident species to provide the greatest opportunity for a competition-free environment into which the new species can vigorously establish.

Pasture renovation

Pasture renovation, where at least partial recovery of the existing vegetation can be expected, adds another requirement to no-tillage seeding. The existing vegetation must be suppressed or managed such that it will not unduly compete with the introduced species. This renovation method is often referred to as overdrilling or sod seeding (see Chapter 1).

The renovation of existing pastures may be undertaken for several reasons:

1. To introduce a more productive long-term pasture species into an existing pasture.
2. To introduce a short-term pasture species that is more suited to a particular time of the year or animal performance than the existing species.
3. To repair damage from natural mortality, drought, flood, erosion, pests, physical damage or poor drainage.
4. To compensate for management or fertility limitations for particular fields, soils or climates.
5. To capitalize on nitrogen fixation brought about by previously introduced legume species.

No-tillage pasture renovation was accomplished before the modern concept of general no-tillage. Early reports show renovation of animal pastures began in the mid-1950s (Blackmore, 1955; Cross, 1957; Robinson, 1957; Cullen, 1966; Dangol, 1968; Kim, 1971). Sports turf renovation came later (Ritchie, 1988).

In the 1950s the dominant reason for overdrilling was to capitalize on nitrogen fixation (Robinson and Cross, 1957). Low-fertility hilly pastures and pastures sown into bush burns on recent volcanic ash soils tended to become clover-dominant because of their low natural fertility. With time, however, this legume base improved the fertility and organic matter levels of such soils to a stage where they could sustain productive grass growth from pasture plants, mainly ryegrasses. The problem was how best to introduce the new grasses without destroying the clover base, or tilling and burying the organic matter layer of such fragile soils.

Since herbicide use was new at that time and, in any case, all available herbicides had a residual action of several weeks in the soil, early overdrilling machines concentrated on mechanical destruction of existing plants in a limited-width track (up to 50 mm wide) centred on the seed row. The objective was to provide a competition-free habitat for the new seedlings that would remain so until regrowth eventually revegetated the strips. By that time the newly introduced species were expected to be competitive with the resident species.

Even today, several designs of no-tillage openers for pasture renovation, e.g. power till and furrow openers, rely on physical, rather than chemical destruction or suppression of the resident species to temporarily check the existing competition.

Band spraying

More recent research has shown that physical removal of vegetative matter from the slot cover zone has a negative effect on the micro-environment within the seed slot, thus seeding into less than optimal soil conditions. Fortunately, the advent of non-residual herbicides now permits selective spraying of a strip of existing vegetation (band spraying) at the same time as the seed is sown with openers. This creates a vegetative mulch, while at the same time suppressing the competing vegetation. Figure 12.3 shows an example of a band-sprayed and drilled pasture.

Figure 12.4 shows the trade-off effects of the various options for overdrilling of vegetation in comparison with the various slot shape options for promoting germination and seedling emergence. The top left illustration is of a slot left by a hoe- or shank-type opener without any attempt to cover the seed. The bursting effect of the opener has a positive effect on competition removal by physically pushing it aside. This is represented by a tick alongside the illustration. But the open (uncovered) slot has a negative effect on seedling survival (represented by a cross alongside the illustration). Therefore there is some risk of failure from this technique.

The centre left illustration shows that covering the slot with loose soil will improve seedling survival (tick and cross) while still having a positive effect on competition removal. The risk of failure is decreased.

The lower left illustration shows that an uncovered V-shaped slot created by a double disc opener has a negative effect on both seedling survival and competition removal. The risk of failure is high. In this case, however, the absence of physical bursting allows band spraying to be used to kill a strip of vegetation over the slot. The top right illustration shows that this has a positive effect on competition removal but does nothing to improve seedling survival. But the risk of failure decreases accordingly.

The centre right illustration shows a slot left by a winged opener. While such an opener might have a positive effect on seedling survival, the absence of physical

Fig. 12.3. The effects of band spraying and simultaneous drilling of pasture.

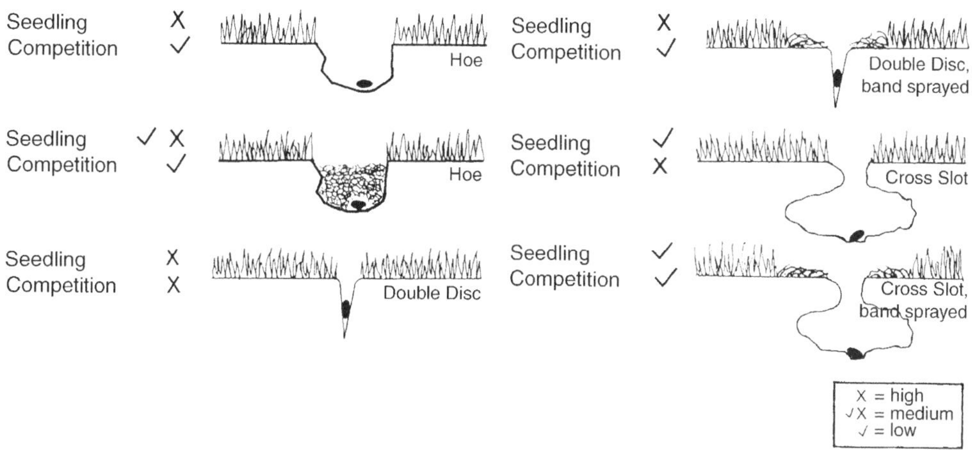

Fig. 12.4. The risk factors of vegetative control and slot micro-environmental control when overdrilling pastures.

bursting has a negative effect on competition removal (the risk of failure is medium).

Only when band spraying is used in conjunction with winged openers does the combination have a positive effect on both seedling survival and competition removal, as shown in the lower right illustration. The risk of failure is low.

There can be debate about how much suppression of existing vegetation is necessary or desirable to bring about the most productive pasture possible as a consequence of overdrilling an improved species into an existing sward. At one end of the scale is complete eradication of all existing species (blanket spraying by herbicide), producing a competition-free environment over the entire field, in which the new species can be expected to express its maximum yield potential. But, during the eradication and establishment period, production from the original sward is lost and must be deducted from the total pasture production for that year or season.

At the other end of the scale is no suppression at all, in which the new species is forced to compete with the existing species from the outset. Lost production from this option is only minor from damage to the existing sward, but the early and continuing competition adversely affects the yield and growth potential of the introduced species.

Between these two extremes is band or strip spraying, where a strip of existing vegetation is sprayed simultaneously as the new seeds are sown, or strip tillage. These are compromises and the loss of yield of the old species and realization of yield potential of the new species both reflect this by falling midway between the other two extremes.

Figures 12.5, 12.6 and 12.7 show the effects of the three spraying options with overdrilled ryegrass. With blanket spraying, the distinct rows of the new species are clear and vigorous. Where no spraying was undertaken, the new rows are less obvious, while band spraying lies in between. On the assumption that the new species has a greater yield potential than the existing species, any pasture that promotes vigorous growth of the new species is likely to have a greater long-term yield potential than the original sward.

To quantify the three options discussed above, scientists in New Zealand measured milk production from the yields of pastures renovated by the three methods (Lane *et al.*, 1993). They also took account of the relative costs of undertaking each practice and expressed their findings in terms of the time taken to recover those costs from the relative milk production figures for each of the options under the prevailing conditions. Their findings are shown in Table 12.4.

Fig. 12.5. Ryegrass establishment by overdrilling with blanket spraying.

Fig. 12.6. Ryegrass establishment by overdrilling with band spraying.

The blanket spraying option was the most expensive compared with band spraying and no spraying, but this option also created the best pasture, resulting in greater returns from milk fat per hectare. When the costs were offset against the returns, however, there was little difference between the three options and all repaid themselves within 8 months or a year. After the payback period, however, the extra pasture production becomes clear profit to the farmer, since the establishment costs would not be repeated annually. This clearly favours blanket spraying since returns from that technique are higher than from either of the other two options.

The technique of band spraying was first tried by L.W. Blackmore (1968, personal communication) and later developed by Collins (1970), Baker *et al.* (1979c) and Barr (1980, 1981). The most desirable band width was not obvious because the cost benefits described above suggest that band spraying is somewhat inferior to blanket spraying. Altering the band width is a simple matter of raising or lowering the spray nozzles. Collins (1970) and Barr (1980) therefore studied different band widths in terms of their effects on yield of both the introduced species and resident species during pasture renovation. Table 12.5 records the results of band spraying during overdrilling with winged openers in 150 mm spaced rows.

Clearly, the wider band (75 mm) reduced the competition from the resident species more than the narrower (50 and 25 mm) bands. This was reflected in lower yield of the resident species with the wider band.

The effects on yield of the introduced species (even at the early stage of 12 weeks) reflected the levels of competition within the bands. The wider band produced the greatest yield of juvenile plants. Since the drilling row spacing was 150 mm, an optimum sprayed band width of 75 mm bands represents 50% removal of the competing vegetation. The results shown in Table 12.4 did not involve bands as wide as 75 mm, so the band spraying treatment may have been

Fig. 12.7. Ryegrass establishment by overdrilling with no spraying.

disadvantaged somewhat in the analysis of Lane *et al.* (1993).

In Barr's (1980) experiments, there was also an effect from fertilizer placement, which at first seemed to be at odds with trends described earlier (see Chapter 9). On closer examination, however, the effects with overdrilling are predictable and logical. It appears that by placing fertilizer with the seed in these circumstances, those resident plants left alive are able to utilize the nutrients before the introduced species because of the mature root systems of the resident species. This disadvantages the young introduced plants through increased competition, as seen in Table 12.6.

The addition of fertilizer at drilling increased the yield of resident species by 25%, which in turn competed with the drilled species and reduced its yield at 12 weeks by 18%. Ryan *et al.* (1979) had earlier illustrated the relative superiority of blanket spraying by comparing blanket spraying, 50 mm band spraying and no spraying. They obtained 1413 kg/ha dry matter yield with blanket spraying, 930 kg/ha

Table 12.4. The cost–benefits of renovating dairy pasture by three different methods.

	Blanket spray	Band spray	No spray
Contract cost of renovation (US$/ha)	113	100	70
Extra pasture production (kg dry matter/ha, first year)	2049	1187	1146
Extra cows/ha able to utilize this pasture	0.43	0.26	0.24
Returns from extra cows (US$/ha)[a]	170	102	96
First-year return on investment (%)	150	98	137
Time to recover renovation costs (years)	0.7	1.0	0.7

[a]Assumes that 25 kg of extra pasture production in New Zealand results in 1 kg of extra milk fat, which sells for US$3.24/kg.

Table 12.5. Effects of spray band width on dry matter (DM) yield of overdrilled ryegrass 12 weeks after drilling (Barr, 1980).

Sprayed band width	DM yield of drilled species (kg/ha)	DM yield of resident species (kg/ha)
25 mm	130	1298
50 mm	143	1184
75 mm	196	776

Table 12.6. Effects of fertilizer application at drilling on overdrilled ryegrass plants 12 weeks after drilling (DM, dry matter).

	DM yield of drilled species (kg/ha)	DM yield of resident species (kg/ha)
With fertilizer	141	1207
Without fertilizer	172	966

with band spraying and 906 kg/ha using no spray at all.

It is recommended, therefore, that, with overdrilling where the resident species have not been totally killed, fertilizer application should be delayed until after emergence (or even after the first grazing) of the drilled species. This is the only no-tillage situation for which such a recommendation is made. If, for example, pasture is being established into an untilled seedbed in which all of the existing competition is dead, the recommendation would be to band fertilizer with the seed at drilling if the openers used are capable of separating the two in the drilled slot.

Although the parameters for optimum results with band spraying are now well defined as above, the practice represents another function from the drill, increasing the opportunity for error. Further, the total yield of the new pasture is seldom as high after 12 months as from blanket spraying (total kill), so the technique is not used as much as drilling into the weed-free environment offered by blanket spraying.

Band spraying represents a realistic option when total kill is not desirable; thus, the techniques and designs of equipment needed to undertake the technique are included. Situations where band spraying is appropriate include:

1. The rejuvenation of lucerne stands where the stand has become thin with age (as is typical) but the surviving plants are healthy and strong, favouring their retention along with newly introduced plants.
2. The temporary balance change of a pasture, e.g. where a legume pasture becomes semi-dormant over the winter months, the temporary injection of an annual ryegrass or winter forage cereal in autumn may increase winter production.
3. The repair of pest-, trampling- or drought-affected pastures where the surviving species are assumed to be resistant to the factors that killed many of the other plants in the pasture and are therefore considered to be a valuable resource worth retaining.
4. The introduction of a new species suited to the habitat created by a resident species, such as in the fertility build-up described earlier in this chapter.

Band spraying equipment

Early designs of band spraying devices centred on placing a spray nozzle ahead of the opener. The option of spraying behind the opener was quickly discarded for two reasons (Collins, 1970):

1. The herbage is often covered by soil after passage of the opener, which tends to deactivate paraquat or glyphosate herbicides.
2. Paraquat is phytotoxic to many seeds, which might be exposed in the slot before covering has had a chance to be completed.

For the nozzle to remain a constant distance above the ground, it has to either be mounted independently on its own height-gauging device (Fig. 12.8), or, if mounted directly on the opener, the latter has to have a positive height control of its own, which is necessary for adequate control of seeding depth anyway.

Even with adequate height control, there are other problems with spray nozzles. The application rates of water that the manufacturers of herbicides recommend be applied per sprayed area are difficult to achieve because the narrow bands mean water application becomes concentrated on to a very small area for each nozzle. This requires very fine nozzles, which require

Fig. 12.8. A band spraying nozzle mounted on separate skids to control spraying height.

micro-filtration to avoid blockage by water impurities that would otherwise be acceptable to farm boom sprayers. Further, because these nozzles operate close to the soil (50–75 mm), they are subject to blockage from random soil splash and debris and damage through contact with stones, etc.

Hollow cone nozzles are most suited to single-nozzle band application, although fan-type nozzles have been used successfully, largely because the inherent variations across the band are acceptable when the objective is only to suppress rather than kill all of the target plants. Hollow cone nozzles generally have a more uniform pattern from a single nozzle than fan-type nozzles.

An innovative method of applying banded herbicides has been used with the disc version of winged openers. Because this opener is equipped with two semi-pneumatic rubber gauge/press wheel tyres, the herbicide can be dripped on to the top of the tyres at low pressure and rolled on to the ground in much the same way as a lawn marker (Ritchie, 1986a, b). This avoids problems of blockage, micro-filtration, wind drift, the presence of tall plants and physical damage, common with small nozzles, and introduces the feasibility of ground-metering of the herbicide. Figure 12.9 shows a drip roller arrangement.

Ground metering involves using a positive displacement pump driven by the ground wheel of the drill in such a way that its output per metre of travel remains largely constant, regardless of travel speed or pressure. Such a system is impractical with pressurized nozzles because inevitable variations in ground speed cause variations in nozzle pressure, which in turn cause variations in band width because the width of the spray pattern from a nozzle is partly dependent on its operating pressure. With the drip roller system, the output pressure is very low and unimportant, since there is no pattern of spray to maintain and, even if there is, this is aimed at the top of the tyre, which then delivers the herbicide to the ground as a wet film on the bottom of the tyres, rather than directly as a jet.

In one respect, the rolling on of herbicide is a disadvantage because the tyres operate behind the opener and inevitably pick up soil, which quickly turns to mud on the wet tyres. Their use for this task is only possible with winged or double disc openers because of the minimal surface soil disturbance each of these openers creates.

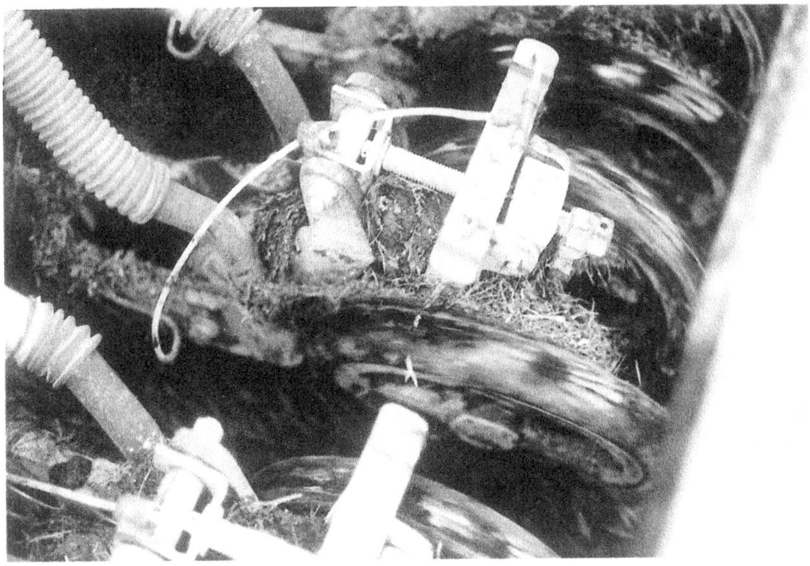

Fig. 12.9. Herbicide being dripped on to the press/gauge wheels of a no-tillage opener.

Any soil contamination that does occur is countered by the improved efficacy of uptake of most herbicides from being rolled rather than sprayed on to the leaves. The result from many thousands of hectares of field-testing is that the 75 mm wide bands created by rolling on of herbicides works as well as spraying the same width of band and has a greater tolerance of the conditions under which it can be used.

Depth control and slot formation

Control over the drilling depth of pasture, sports turf and many forage crop species is particularly demanding. Most drills designed expressly for pasture renovation have been promoted on a low-cost basis. Because of this, the control mechanisms for depth of seeding are generally primitive and sometimes non-existent.

For example, the simple low-cost drills that dominate the pasture drill markets in Australia and New Zealand are almost all equipped with 'Baker Boot' versions of simple winged openers (inverted-T-shaped slot). While the choice of slot shape is appropriate, the ability of these openers to follow the surface is limited by the simplistic drill designs to which they are attached, particularly the mechanisms for articulating each opener up and down. This causes the angles of the opener wings to change throughout their arcs of travel (see Chapter 4). To avoid complete loss of wing angle in hollows, the preset angle for level ground is about 10°. This relatively steep wing angle means that the shallowest this opener can drill and still maintain a true inverted-T-shaped slot without breaking through the covering surface mulch is about 25 mm.

In contrast, the more sophisticated disc version of winged openers is mounted on parallelogram drag arms, ensuring that the wing angle never changes. The preset wing angle is reduced to 5° to allow the wings to operate with integrity at depths as shallow as 15 mm. There is a major advantage, therefore, in being able to drill pasture with a machine equipped with similar technology demanded for the more highly valued arable crops.

While many drill designers consider pasture and sports turf to be the most difficult of media to drill, with winged openers the matted roots of pasture and turf plants provide a mulch medium of considerable elasticity and tensile strength, which can be readily folded back and replaced while retaining the integrity of the inverted-T-shaped slots (Ritchie, 1988).

Seed metering

Most pasture and forage crop seeds are small, light and/or fluffy. Many pasture seeds also have awns attached. This presents several handling and metering problems.

First, they are difficult to meter accurately. Small-grain metering devices that commonly dispense several hundred kilograms of seed per hectare are often not well adapted to dispensing less than 1 kilogram of small seeds per hectare. Further, if the seeds have large awns or are fluffy, they will have a tendency to bridge above the metering device, interrupting the feed. This requires an agitator to be fitted to the drill to continuously avoid bridging. Often, drills for sowing small and/or difficult seeds use an auxiliary hopper designed especially for such seeds.

Many pasture species are sown as blends of two or more species. Common blends are clovers and grasses. Clover seeds are generally round and dense. Grass seeds are generally elongated and often fluffy and light. A previously mixed blend of such different seeds may partially separate into its individual components within the seed hopper of a drill in response to the continual vibration of the machine. To reduce separation and aid metering, the small seeds are often mixed with inert filling material such as sawdust or rice hulls to bulk up the material and reduce settling. Separation can be a problem with these mixes as well, especially if they are metered and dispensed by an air-delivery system. In these circumstances the high-speed airstream may blow some lighter, fluffier seeds out of the seed slot altogether before it has been covered.

Summary of No-tillage for Forage Production

1. Farming systems that depend on an intensive forage supply demand maximum and consistent feed supplies, which favour the use of successive forage crops in preference to more traditional 'permanent' pastures.

2. Establishment of successive forage crops is only sustainable on a long-term basis using no-tillage.

3. The integration of forage and arable cropping systems is desirable in climates that permit economic utilization of forage crops by animals.

4. Farmers generally place lower values on forage crops than on arable crops and will more readily accept inferior results.

5. Drills for pasture and many forage crops need to have more accurate depth control and sow at shallower depths than equivalent machines for arable crops.

6. Drills for pasture and forage crop species need to be able to meter small seeds.

7. Forage crops should generally be treated with the same care and attention as arable crops, but they seldom are.

8. Drills for pasture need to handle tightly root-bound soil and also utilize this covering medium to advantage.

9. With pasture renovation by overdrilling, there may be a trade-off between providing a suitable environment for germination and emergence and reducing competition from the existing sward.

10. Because the drilling time of forage crops and pastures is usually not as critical as with arable crops, there is more opportunity to wait for suitable weather to offset substandard openers.

11. On a cost-recovery basis, blanket spraying of the existing competition will give a greater long-term return than band spraying, which gives a greater return than no spraying.

12. Cross-drilling slowly establishing pasture species may produce greater short-term weed infestation than single-pass drilling.

13. Drilling early in the autumn is likely to produce more pasture production than later drillings, provided adequate soil moisture exists at the time.

14. Early autumn drilling and spring drilling are likely to produce more weed problems than later autumn drilling, especially with slowly establishing pasture species.

15. Single-pass drilling in 75 mm rows may produce a short-term yield advantage with slowly establishing pasture species compared with single-pass drilling in 150 mm rows.

16. Neither single-pass drilling in 75 mm rows nor cross-drilling in 150 mm rows has any long-term agronomic advantage compared with single-pass drilling in 150 mm rows, or any short-term advantage with rapidly establishing pasture species.

17. With band spraying for overdrilling, 75 mm wide bands are preferred with 150 mm spaced rows.

18. With overdrilling for pasture renovation, fertilizer should not be applied at the time of drilling but should instead be applied about 3 weeks post-emergence.

19. With complete pasture renewal, the new pasture should be drilled and fertilizer applied during drilling, similar to an arable or forage crop.

13 No-tillage Drill and Planter Design – Large-scale Machines

C. John Baker

A no-tillage seed drill is no more nor less than a device designed to service the functions of its openers.

While most of the desirable functions of no-tillage drills and planters can indeed be related back to the desirable functions of their openers, other components and functions are also important. These will be examined in a general sense with no attempt to approve or disapprove design criteria for individual commercial drills or planters.

Manufacturers and designers who seriously consider the desirable functions of drills and planters and variations required to achieve these most often will present a range of design options. Consumers must then ascertain for themselves what represents the best value after having weighed the risk, performance and cost factors.

For example, drills for pasture renovation might not need to be as sophisticated as those to establish cash crops, because residue handling is seldom a high requirement with pasture establishment and there may be more time flexibility allowable in choosing an appropriate sowing date. This in turn permits a delay in drilling until favourable weather patterns arrive. The target sowing dates for cash crops, on the other hand, are often dictated by a narrow window of climatic opportunity or harvesting and seldom allow the luxury of being able to wait very long for favourable conditions. Cash-cropping drills and planters, therefore, must function to their maximum potential with less dependence on weather and therefore need to be more sophisticated than pasture renovation drills.

This chapter considers large field-scale and tractor-drawn machines. The following chapter considers small field-scale and animal-drawn machines. In both cases we consider drill and planter design under several headings:

- Operating width.
- Surface smoothness.
- Power requirements.
- Downforce application.
- Transport considerations.
- Matching to available power.
- Storage and metering of product.

Operating Width

The most important factors that should influence the design width of no-tillage drills and planters are the total time available to establish a given crop and the tractor power available to pull the machines. Unfortunately, many converts from tillage to no-tillage expect no-tillage to achieve the same rates of ground coverage as each of their previous tillage machines. Such expectations fail to account for the fact that no-tillage machines are only going to cover the field once and can therefore afford to operate at a slower rate of ground coverage. Because most no-tillage drills and planters are capable of operating at equivalent forward speeds to tillage machines, this means they can be narrower.

A sensible and practical comparison was made by an English farmer, who concluded that, so long as he could drill with his no-tillage machine at the same rate as he could previously plough, he would be gaining by adopting no-tillage. Despite such pragmatism, it is common to hear other farmers demanding that no-tillage machines must be the same width as conventional tillage machines. Some machinery designers accede to this request but in so doing are forced to select openers with low power demand. Almost invariably, the lower the power demand from no-tillage openers, the less work they do on the untilled soil and the greater will be the risk of biological failure.

For example, a farmer practising minimum tillage will cover the field at least twice and probably three times to establish a crop. If each of the machines used for minimum tillage (including the drill) was 4.5 metres wide, the effective working width would be 1.5 metres (4.5 ÷ 3). And yet many such farmers complain that a 3 metre wide no-tillage drill would be too narrow for them, even although once-over with a 3 metre wide drill would complete the whole job in half the time that three times over with 4.5 metre minimum-tillage machines could achieve. While seemingly simple, it is surprising how often this argument is voiced.

For 'diehard' tillage exponents, such an argument seems to be an excuse for avoiding the issue. For others already practising no-tillage with wide low-power-demanding drills, it reflects ignorance of the benefits that the more sophisticated no-tillage technologies offer (which are almost invariably accompanied by greater power demand).

While increases in both the power and downforce demand from openers translate into increases in tractor power and machine weight, these are relatively cheap and readily available inputs. Increases in biological reliability and crop yield from improvements in opener design are much more expensive and sophisticated inputs. Some operators choose to minimize power or weight requirements rather than maximize biological reliability. It is a matter of how individual operators approach the whole concept of no-tillage: whether they are yield-driven or cost-driven.

Those that see no-tillage as short-cutting tillage, but still regard tillage as the benchmark, will probably rate cheapness, maximizing working width and minimizing power and weight requirements as high priorities. Those that see no-tillage as the ultimate goal and regard tillage or minimum tillage as having been only interim learning steps (albeit practised for centuries) will take a different view. They will seek to maximize biological performance, almost regardless of cost, weight and width, and readily add the changes needed to their management practices. The world is full of people with both of these outlooks and is not likely to change in this respect.

The design and desirability of an operating width include a number of functions beyond that of the associated opener: power available, field topography, amount of product to be carried and field-to-field transport, to list a few. Each added function integrates into the overall design and machine width. Example machines shown in Figs 13.1 to 13.4 have a range of widths from 4 to 18 m, all outfitted with the same inverted-T opener but with widely varied configurations.

Surface Smoothing

The opportunity to smooth the ground prior to drilling is lost under a no-tillage regime. Thus, the drill or planter openers need to be able to faithfully follow significant changes in the surface of the soil without

Fig. 13.1. A 4.5 metre wide rigid-frame end-wheel no-tillage drill.

Fig. 13.2. A 12 metre trailed toolbar with folding wings for transport.

detriment to drilling depths or functions. This is a demanding requirement (see Chapter 8), but for general drill or planter design it places limitations on overall machine width and design considerations.

Six metres (20 feet) seems to be about the upper limit a machine can be expected to span in a single frame and allow the openers to rise and fall sufficiently to follow each hump and hollow. Even then, unless the openers are pushed in with a downforce device capable of exerting consistent force as the openers move vertically approximately 0.4 metre (16 inches), some

Fig. 13.3. A 4 metre rigid toolbar that is lifted clear of the ground for transport.

Fig. 13.4. An 18 metre toolbar that is end-towed for transport.

inconsistent seeding depth will result from a 6 metre wide drill or planter. Where widths greater than this are required, multiple units or folding wings from a central unit should be considered. Even a 6 metre width with good opener surface-following ability is feasible only on reasonably flat ground. A more universal size would be 4.5 metres.

Nor does it make any difference whether the openers are spaced 150 mm apart or up to 1 metre apart. Each individual opener must rise and fall in response to surface irregularities independently of its neighbours. Its inability to do so will result in a missed row, regardless of how many other rows there are.

Because the micro-contour of the ground surface remains undisturbed, the gauge/press wheels of no-tillage openers must operate on a rougher surface than with tillage. Cushioning of this roughness can be achieved by springing the gauge/ press wheels, but this virtually eliminates their gauging (or depth-control) function, since the relationship between the position of the wheels and the base of the slot (position of the seed) constantly changes when gauge wheels are sprung. Alternatively, mounting the wheels on walking beams effectively halves the magnitude of each surface irregularity, which will smooth the passage of an opener equipped with rigid or semi-pneumatic gauge wheels, without compromising their gauging function.

Then there is the question of speed. Obviously the faster the drill or planter is pulled, the rougher will be the ride. This is especially important with planters, because the accuracy of seed selection and final spacing is affected by the smoothness of ride. A speed that is acceptable for operating a given precision seeder on tilled soil may well be too fast when the same seeder is operated on untilled soil. This is a negative factor as far as no-tillage is concerned but must be balanced against the fact that several passes with tillage tools would have been necessary before planting was even attempted into a tilled soil. Therefore, if a slower planting speed is necessary for no-tillage planting, it will only reduce, not reverse, the advantages associated with no-tillage. And, with drilling of small seeds compared with precision planting of larger seeds, there are almost no speed restrictions. Indeed, some no-tillage drills operate at faster speeds than their tillage counterparts.

Power Requirements

No-tillage drills and planters require more power to pull them through untilled soils than do their tillage counterparts. This is partly due to the fact that the openers are designed to break untilled ground and partly because the machines are heavier. Typical power requirements are 3 to 9 tractor engine kilowatts (kW) (4 to 12 horsepower (hp)) per opener (see later in this chapter). This amount of power also requires an associated traction increase; thus four-wheel-drive and tracked tractors are used more with no-tillage drills than with drills used in tilled seedbeds.

This power requirement places constraints on the number of openers that can be pulled with any given tractor. For example, a 25-opener drill operating on flat, light soil might require a tractor engine of approximately 150 kW (200 hp), while the same drill operating on silty and/or hilly soils or in dense sod might require a tractor with 50% more power.

Power requirements are also related to drilling speed. Some openers can operate satisfactorily at relatively high speeds (up to 16 km/h). Others should not be used above 7 km/h. The tractor power requirement will increase at higher speeds, but this will be put to good use by covering the field more rapidly.

Planters gain an advantage over drills with respect to tractor power requirement. The smaller number of openers on planters, due to their wider row spacing of up to 1 metre, means that tractor size will seldom be the limiting factor to machine size. Generally, it will be the surface-following ability of the openers that will dictate the upper limit of planter size, whereas with drills available tractor power is often the limiting factor. As a rough guide, for any given width of operation, a planter will require half the tractor engine power of a similar-sized drill.

Finally, drill width will be determined by a combination of opener number and row spacing. In general, crops benefit from closer row spacing under no-tillage than under tillage because of the improved moisture availability of untilled soils. On the other hand, the physical limitations imposed by residue handling dictate that no-tillage rows are seldom spaced less than 150 mm apart on drills.

Weight and Opener Forces

Each design of no-tillage opener requires a different downforce to obtain its target seeding depth. Required downforce is determined by a number of variables:

1. Soil strength, which determines the soil's resistance to penetration.
2. Soil moisture and density, which affect soil strength.
3. The presence or absence of stones and their sizes.
4. The presence or absence of plant roots that directly resist penetration.
5. The decay stage of plant roots, which is affected by the interval between spraying or harvest and drilling.
6. Operating speed, because openers penetrate better at slower speeds than at higher speeds.
7. The draught of the openers (their resistance to moving through the soil).
8. The attachment geometry of the openers to the drill frame, because, as an opener moves downwards into a hollow, the vertical component of pull increases, acting upwards, opposing and reducing the downforce pushing the openers into the soil.

Mai (1978) measured both the downforces and the draught forces, at 38 mm seeding depth at very slow speeds, of vertical triple disc and simple winged no-tillage openers operating in sprayed turf in a silt loam soil at two moisture contents. The results are shown in Table 13.1.

Data of Table 13.1 show that, while the vertical triple disc opener required about four times as much force to penetrate to 38 mm depth as the simple winged opener, it required 50% less force to pull it through the soil. The penetration action of the triple disc opener is one of wedging the soil sideways and downwards, accounting for its high downforce requirement. The winged opener, on the other hand, tends to heave the soil upwards, reducing its penetration force. In fact, soil acting on the upper surfaces of the inclined wings tends to draw that portion of the winged opener into the ground, although this is more than countered by the resistance to penetration of the pre-disc, the vertical shank portion of the opener and the lower frontal edges of the wings.

The vertical triple disc opener is comprised entirely of rolling discs. Once it has attained operating depth, the forces required to pull it through the soil are smaller than with the winged opener, which cuts roots and shatters a wider zone of soil than the triple disc opener as it moves forward. This is reflected in the downforce : draught ratios for the two openers, which averaged 0.65 for the vertical triple disc opener and 0.11 for the simple winged opener.

Not surprisingly, the wetter soil required less downforce and draught force from both openers than the drier soil, but the downforce : draught ratios remained reasonably consistent, regardless of soil moisture content.

Table 13.1. Downforce and draught requirements of two no-tillage openers.

	Vertical triple disc opener[a]		Simple winged opener[b]	
Moisture content (g/g)	23%	28%	23%	28%
Downforce (N)	882	842	221	203
Draught (N)	1684	1210	2096	1852
Downforce : draught ratio	0.53	0.70	0.11	0.11

Conversion: N (newton) = 0.2 lb force.
[a]The vertical triple disc opener had a flat 3 mm thick pre-disc of 200 mm diameter; the double discs were 3 mm thick and 250 mm in diameter.
[b]The simple winged opener had a flat 3 mm thick pre-disc of 200 mm diameter; the wings of the tine measured 40 mm across.

Baker (1976a), in three separate experiments, measured the downforces required for 38 mm penetration by a range of openers into a dry, fine, sandy, loam soil covered with sprayed pasture residue and at moisture contents ranging from 14.1% to 18.2% (g/g). The results are shown in Table 13.2.

Data of Table 13.2 show that the difference in downforce between the vertical triple disc and simple winged openers is slightly less than in Table 13.1, probably because of the softer (sandier) soil. The hoe opener was similar to the winged opener, suggesting that the draw-in effect of the wings on the winged opener played only a small role, since hoe openers do not have wings.

The angled flat disc opener required the least downforce of all openers tested, but the angled dished disc opener required more downforce than all other openers except the vertical triple disc, possibly because of the resistance to penetration of the convex (back) side of the angled disc.

For a drill or planter to operate, its weight or downward drag component must be sufficient to provide the required combined downforces of all its openers when operating in the worst (usually driest) conditions in which its openers can obtain seedling emergence. This concept is particularly important and often confuses would-be purchasers of drills when faced with the claims and counterclaims of manufacturers. For example, vertical double or triple disc openers are known to perform poorly in terms of seedling emergence in dry soils (see Chapter 6). With few exceptions, drills and planters featuring such openers generally do not provide sufficient downforce (weight) for them to obtain drilling depth in dry soils. The drills therefore often appear to be relatively light in construction, giving the erroneous impression that they can penetrate the ground more easily than other drills, when in fact the reverse is true.

Winged openers, on the other hand, can tolerate very dry soils, in biological terms, so their drills and planters are often built to be heavy enough to force the openers into soils that might otherwise be biologically hostile. Thus, the overall weight of a drill or planter does not necessarily reflect the penetration requirements of its openers in any given soil. It may, in fact, reflect more the biological tolerance (or intolerance) of its openers to dry soils than anything else.

But there is more to forcing openers into the ground than just dead weight. Figure 13.5 shows four geometrically different arrangements for attaching openers to drill frames.

The first (and simplest) arrangement is to fix the openers rigidly to the drill frame, preventing articulation between the two. This gives the drill a very poor ability to follow ground surface changes, but the downforce provided for each opener will remain reasonably constant and largely predictable.

The second arrangement uses a length of heavy spring steel to: (i) introduce a separate drag arm between the drill chassis and the opener; and (ii) provide limited movement between it and the drill frame. To accomplish the second function, the upper

Table 13.2. Downforce requirements of a range of no-tillage openers.

	Vertical triple disc[a]	Simple winged[b]	Hoe[c]	Angled flat disc[d]	Angled dished disc[e]
Downforce (N)	770	281	263	133	445

Conversion: 1 N (newton) = 0.2 lbs force.

[a]Vertical triple disc design was as for Table 13.1. The value is the mean of three experiments.

[b]The simple winged design was as for Table 13.1. The value is the mean of three experiments.

[c]The hoe opener had a flat 3 mm thick pre-disc of 200 mm diameter, and the tine was 25 mm wide. The value is the mean of three experiments.

[d]The angled flat disc was 3 mm thick and 250 mm in diameter. The value is for a single experiment.

[e]The angled dished disc was 2 mm thick and 250 mm in diameter. The value is for a single experiment.

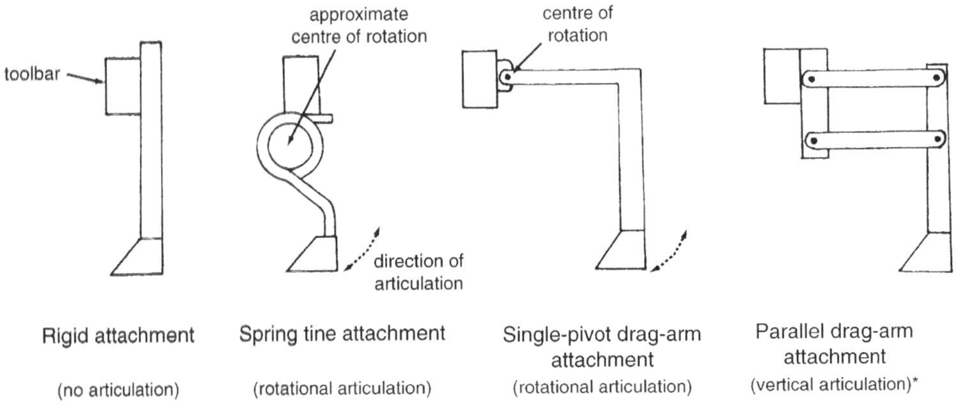

Fig. 13.5. The geometrical options for attachment of drag arms to a no-tillage seed drill. *Opener also moves forward, but since the whole machine is moving forward anyway this is ignored as it does not affect the function of the opener in any way.

portion is extended and often coiled several times to increase its flexibility. In operation the soil drag on the opener tends to cause the drag arm to pull backwards as well as deflect upwards, but the actual displacement in either direction is relatively small. This means that the point of action of the applied downforce in the soil remains relatively constant in relation to the drill frame, and there is therefore little change in downforce as the openers traverse undulations in the ground surface.

This design limits their ability to faithfully follow variations in the ground surface. In addition, many similar designs allow the openers to wander sideways, with the result that inter-row spacing varies somewhat, although this also gives them an ability to handle large surface stones with less blockage than either rigid openers or drag arms that move only in the vertical plane.

The third arrangement is commonly used for conventional drills for tilled seedbeds and has been simply transferred to many no-tillage drills with adjustments only for robustness and the magnitude of the applied downforces. It consists of a pivot-mounted single drag arm, which is pushed down from above or sometimes pulled down from beneath. The opener cannot deflect rearwards, only upwards and downwards in a limited arc about the pivot point between the drag arm and the drill frame. Because the force applied by the tractor to create forward movement (drag force) acts through this pivot point and is opposed by the resistance of the opener at the point of soil contact, these forces can be resolved into their vertical and horizontal components by triangulation.

Figure 13.6 shows the resulting force diagram. The drag, or draught force applied by the tractor is opposed by the soil resistance to forward movement (P) through the soil. This is shown as the horizontal component of pull (H) in the diagram. The vertical component of pull (V) is derived from the resultant line of pull (R) which passes through the point of attachment of the opener to the drill and the centre of resistance (X) of all soil forces, which is the point of equilibrium of all soil resistance forces on the opener and is located somewhere beneath the soil. The vertical component of pull (V) acts upwards and, together with the vertical force arising from the soil's resistance to penetration, has to be overcome by the net vertical downforce (D), which is applied separately by springs or other means on the drill (not the tractor) for the opener to remain in the ground.

All of these forces find an equilibrium, but a problem arises when the position of the opener changes relative to the drill frame. For example, as the opener passes into a slight hollow and moves downwards (relative to the pivot point or drill frame), the horizontal component of pull (H) may

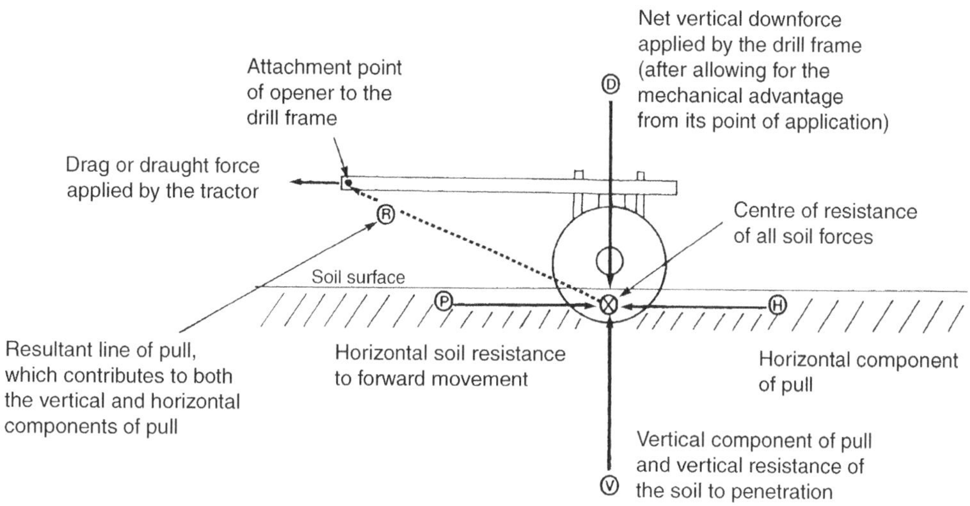

Fig. 13.6. The distribution of forces acting on a no-tillage opener as it is pulled through the soil.

not change, but the vertical component of pull (V) will increase because the resultant line of pull acting through the pivot point (R) will have become steeper.

This means there will then be a greater upward force opposing the net vertical downforce (D) on the opener, which at best remains constant, resulting in shallower drilling. It would be a big enough problem if the applied downforce did in fact remain constant, but, where the mechanism of downforce application on the drill is commonly a spring, the downforce will actually decrease somewhat as the opener moves downwards because the spring lengthens. The net effect is a significant reduction in the net vertical downforce (D) applied to the opener, resulting in shallower drilling for that portion of the field.

The opposite effect occurs when an opener passes over a hump. Characteristically, openers with this common geometrical arrangement drill ‘hollows’ too shallowly and ‘humps’ too deeply.

However, the problem does not stop there. If the soil resistance to forward movement (P) increases because the drill encounters an area of harder soil, the magnitude of the resultant line of pull (R) will increase, even though its slope may remain the same. This in turn will increase the vertical component of pull (V), which, unless it is compensated for by an increase in net vertical downforce (D), will also result in shallower drilling.

In reality, both the soil surface and resistance to forward movement of individual openers continually change under no-tillage. Therefore, so too does the vertical component of pull, causing penetration variation.

The fourth arrangement (Fig. 13.5) is common on precision planters and more sophisticated no-tillage drill designs. Here the single pivoting drag arm used in the third arrangement is replaced by two parallel drag arms of equal length arranged as a parallelogram, illustrated on the right of Fig. 13.5. The objectives of this configuration are fourfold:

1. To maintain a predictable relationship between several components on an opener assembly. Some planter openers, for example, have up to six separate components following one another in a fixed relationship. If the assembly were mounted on a single pivot drag arm (Figs 13.5 and 13.6) and moved in an arc as it rose and fell, the vertical relationship between the forward and rear components would alter appreciably as it travelled vertically.
2. To maintain a given approach angle of critical components to the soil, regardless of the vertical position of the opener assembly.

Winged openers, for example, have soil wings that slope downwards towards the front at an angle of 5–7° to the horizontal so that they can operate at shallow depths with the wings still beneath the ground. If the opener were mounted on a single-pivot drag arm, the preset wing angle would need to be increased to about 10° to ensure that a positive wing angle remained at the bottom of the arc of movement. But in the mid-position an angle of 10° would limit the shallowness of drilling because the wings would break through the surface of the soil.

3. To reduce the magnitude of the forces opposing the downforce. Although a parallelogram arrangement will have little or no beneficial effect on the vertical component of pull opposing the downforce, there is yet another force that also opposes the downforce on single-pivot drag arms. This is the rotational force arising from the horizontal soil drag acting rearwards on the base of the opener (which is always positioned lower than the pivot itself). With a single-pivot drag arm arrangement, this rotational force causes the opener to attempt to rotate upwards, regardless of the opener position or angle of the drag arm, and opposes the downforce. The actual magnitude of this opposing force is somewhat self-cancelling because, if the opener rotated upwards, the soil drag would then be reduced because the opener would be drilling more shallowly. On the other hand, when the arms are horizontal in a parallelogram arrangement, this rotational force is eliminated altogether and has no effect on the downforce. Most drills and planters are designed so that the drag arms are nearly horizontal in the normal drilling position.

4. To facilitate the design of long and short drag arms without changing the position or geometry of the downforce application. The force mechanics of parallelograms is such that, if a downforce is applied part-way along one of the horizontal arms, there will be a resulting vertical downforce at the rear pivots. Further, if a rigid horizontal frame is attached to these rear pivots, the same downforce will be applied at any point along this rigid frame. Since an opener attached to the rear pivots of a parallelogram acts as a rigid horizontal frame, this principle applies to openers mounted on parallelogram arms.

In drill designs, this allows openers of difference lengths to be attached to parallelogram linkages in order to create stagger for residue-clearance purposes, and each opener will experience the same downforce as its neighbour.

Although the best of the innovations and geometric arrangements discussed above go a long way towards ensuring that no-tillage openers receive constant downforces throughout their extended ranges of travel, it should be emphasized that the magnitude and direction of the main opposing forces (i.e. the upward vertical components of pull and soil resistance) vary with soil conditions and the position of the opener at any one point in time and are therefore seldom constant. Thus, no geometrical arrangement so far devised has the ability to maintain a truly consistent net penetration downforce on an opener.

Re-establishing Downforce

An adjunct to the general downforce requirements of no-tillage drills and planters is the range of methods used to ensure that a drill or planter re-establishes the downforce to its preselected level after the openers have been raised from the ground for transportation and/or cornering. Repetitive raising and lowering of the openers are more common in no-tillage than in tillage because turning sharp corners with the openers engaged is difficult in untilled ground. Some of the systems used are:

1. Manual return to a guide mark. Where a drill or planter is designed to raise the openers using one or more hydraulic rams on the machine frame, the resetting of those rams to their original positions is achieved by the operator watching a guide mark to indicate where the ram(s) had extended or contracted to previously and stopping the cycle at that point. The potential exists for operators to forget to watch the guide mark and, in any case, such a repetitive manual task adds to operator fatigue. On the other hand, this system allows the downforce on all openers to be altered by the operator without leaving the tractor seat.

If a drill or planter is three-point linkage-mounted to the tractor or has a separately controlled set of transport wheels, the depth adjustment is usually achieved by changing a mechanical linkage, a screw of some description, or the pressure in a second independent hydraulic system. This adjustment remains unaltered during drilling and transportation cycling. Return of the machine to the ground after transportation automatically re-establishes the magnitude of the original downforce, since nothing will have been altered in that respect during transportation. While this reduces operator fatigue, alterations to the downforce often require the operator to leave the tractor.

2. Return to an automated stop or pressure. Where a drill or planter is designed to raise and lower its openers hydraulically, an adjustable hydraulic or mechanical control valve can be positioned on the machine so that a predetermined mechanical movement or oil pressure build-up will trip the valve and halt the hydraulic system at any desired position commensurate with a given downforce. While this increases convenience for the operator, a time delay still results while the tractor hydraulic system moves the ram to its predetermined position, and alterations to the magnitude of the downforce still require the operator to dismount from the tractor.

One tractor manufacturer for many years provided a pressure-modulating system on the internal hydraulic source within their tractors. This system allowed the operator to vary the hydraulic pressure from the tractor seat, useful for pressurizing rams on drills or planters. Repeatability of the system simply relied on the setting of a stop on the tractor's hydraulic controls. The operator returned the lever to this position after actuating the lifting and transport cycles of the hydraulic system.

3. Automated return. A hydraulic 'memory valve' is supplied on some no-tillage drills and planters that utilize the same hydraulic rams for both downforce and lifting. The memory valve increases the repeatability of settings during frequent transportation and drilling cycling by automatically storing the downforce oil pressure in the oil-over-gas nitrogen accumulator(s) when the lifting (transport) cycle is actuated. Upon return of the openers to the ground, the memory valve automatically and instantly returns the original oil pressure to the downforce system without further attention from the operator. This greatly increases the speed of cycling from drilling to transport modes, and vice versa, which is important for field efficiency, operator fatigue and operator accuracy. The operating down-pressure can be changed at any time from the tractor seat.

One of the major problems with all no-tillage drills is that the magnitude of the forces involved for downforce and draught places unusually high stress loadings on drag arms, openers and their supports. This problem is exacerbated when drills or planters are required to operate around corners. The more durable designs have used ball or roller bearings in the drag-arm pivots, where simple bushings would usually have sufficed for the same function with conventional drills in tilled soils.

Unfortunately, some of the previously used simple designs of conventional drills have also been extended to less expensive no-tillage drills. These units often experience early failure of components and loss of accuracy. For example, as pivots of drag arms become prematurely worn, openers are difficult to maintain in vertical or tracking alignment, resulting in inaccurate depth of seeding and uneven row spacing. The frequency of breakages increases and residue handling often suffers. These machine failures cause frustration for the operators and result in a decline of enthusiasm for no-tillage farming.

Wheel and Towing Configurations

A major distinguishing feature of no-tillage soils compared with tilled soils is their long-term ability to sustain wheel traffic without compaction damage and their resistance to surface damage from the scuffing caused by machinery wheels and tracks. Even when compaction does occur, as the populations of soil fauna and bacteria return to sustainable levels in response to decreased tillage disruption and increased organic matter, the natural restorative processes of living soils soon ameliorate most problems.

No-tillage drills and, to a lesser extent, planters are inherently heavier than their tillage counterparts, but it is seldom necessary to increase the footprint area of their wheels, tyres or skids on a proportional basis to their weight because of the increased load-bearing strength of the soils on which they operate. None the less, there is little point in subjecting even untilled soils to footprint pressures from drills that are significantly in excess of the tyres on the tractors that pull them. Tractor tyres usually exert footprint pressures in the range of 50–85 kPa (7–12 psi) and tracks in the 30–50 kPa (4–7 psi) range.

As with conventional drills and planters, there are several optional wheel configurations. Some of these, with their attributes and limitations, are outlined below.

End wheels

End-wheel designs, as the name suggests, have wheels positioned at either end of the drill or planter chassis. Some planters, because of their wide row spacing, have the wheels positioned between the rows some distance from the ends of the machines. This reduces side forces during cornering and allows two or more such machines to be conveniently joined together end to end.

End-wheel designs are suitable for machines up to 6 metres in width. The end wheels provide excellent manoeuvrability and stability on hillsides and are usually less expensive than other options. Most designs use single wheels on each end of the machine, making them unsuitable for end-towing for road transportation without the addition of special transport wheels. Some designs use paired wheels on walking beams, which double the footprint area, reduce bounce and provide an opportunity for convenient conversion to end-towing.

End-wheel drills and planters are not well suited to the joining of several units together side by side. Where joining machines is contemplated, it is necessary to arrange the multiple units in an offset pattern from a common and separate towing frame (as illustrated in Fig. 13.7). On the other hand, no-tillage farming saves so much time that had been previously devoted to tillage before drilling that the need for wide, multiple drills and planters is reduced considerably.

Fore-and-aft wheels

Fore-and-aft wheel configurations involve one or more self-steering wheels on either the front or rear of the machine and at least two fixed wheels on the opposite end. The configuration reduces the lateral distance between the wheel positions, permitting wider machines to be designed than with end wheels. Because there are no wheel structures on the ends of the machines, multiple units can be conveniently joined, as illustrated in Fig. 13.8. Such multiple arrangements need a much less complicated common towing facility than do multiple end-wheel arrangements.

Another arrangement permits two drill units to be used either as a narrow-row drill or as a wide-row planter. The row spacing of each unit is fixed to the desired spacing for the wide-row planter configuration and two such units are arranged end to end to produce a double-width planter (see Fig. 13.9). When narrow-row drilling is required, the two units are arranged in tandem fashion, with the rows of the rear unit splitting the rows of the front unit, thus halving the row spacing.

Of course, for this convenient arrangement to be functional, the seed metering mechanisms must be capable of sufficient accuracy to satisfy the needs of both the planter and the drill. Very few seeders are capable of this degree of flexibility. Either duplicate seeders are used (which is

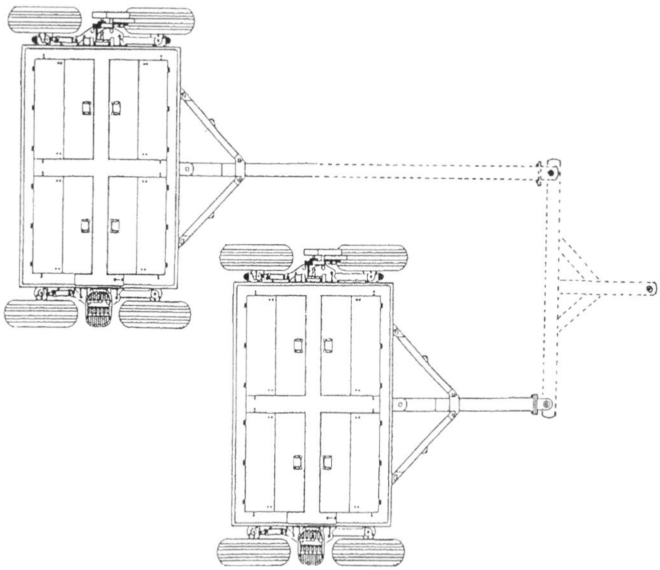

Fig. 13.7. End-wheel drills arranged in an offset multiple arrangement.

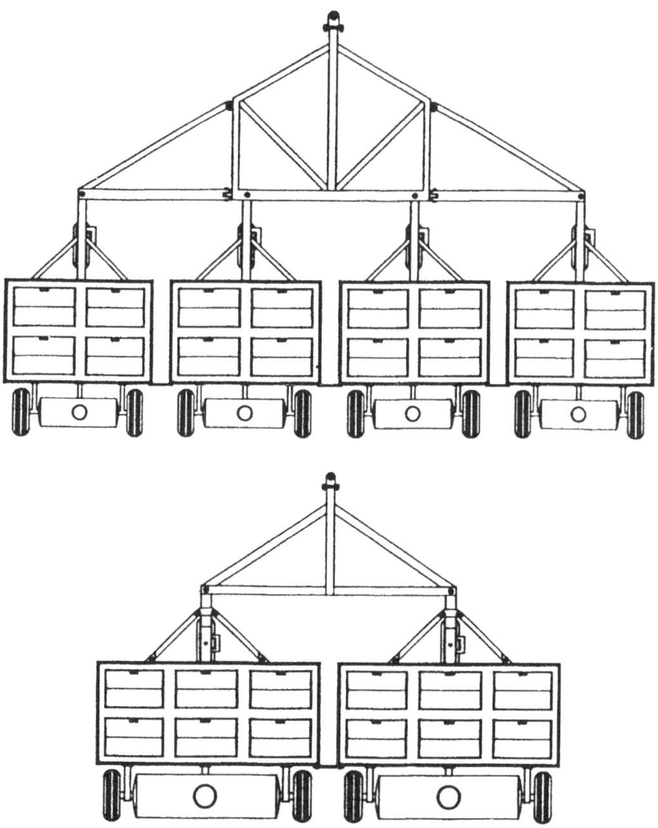

Fig. 13.8. Fore-and-aft wheel drills arranged for multiple unit operation.

Fig. 13.9. Two-wide row units arranged in tandem to produce a drill with half-row spacing.

expensive and mechanically complicated) or one or the other of the two seed metering functions is compromised.

The options for transportation conversions with fore-and-aft wheel configurations are many and varied. An example of a convenient arrangement for a three-unit ganged drill is shown in Fig. 13.10. The two outer drill units fold forwards after the whole machine is raised clear of the ground for transportation. Other options include folding the outer units upwards, but this option is limited to air seeders and planters with lockable lids on their product hoppers to avoid spillage. The product hoppers on air seeders are located on the central drill unit and not involved in the folding process.

Yet another arrangement for transporting two fore-and-aft wheel drills is shown in Fig. 13.11.

Matching Tractors to Drills and Planters

In conventional tillage, tractors are usually selected to match the heaviest power-demanding implement(s) used, from primary tillage (usually ploughing) to drilling. Since drills and planters in conventional tillage are among the least power-demanding implements, tractors are seldom selected to match drills and planters, or vice versa. Indeed, often a smaller available tractor than the main tillage tractor(s) is used for drilling and/or planting.

In no-tillage farming, the sprayer is the only light power-demanding implement in the system. Drills and planters are the heaviest power-demanding implements, and this power requirement may exceed the power required by any one of the tillage implements it replaces. This is not to say that no-tillage is energy-inefficient. On the contrary, this single input of energy is several times more energy-efficient in terms of total litres of tractor fuel used per sown hectare than the sum of all of the multiple smaller inputs of energy during tillage.

With planters, the maximum number of widely spaced rows to be sown by any one machine seldom exceeds 12. The power requirement for such machines is therefore less likely to be a limiting factor, even under no-tillage, than with no-tillage drills, which may have up to 50 such openers.

Fig. 13.10. A folding arrangement for multiple drills.

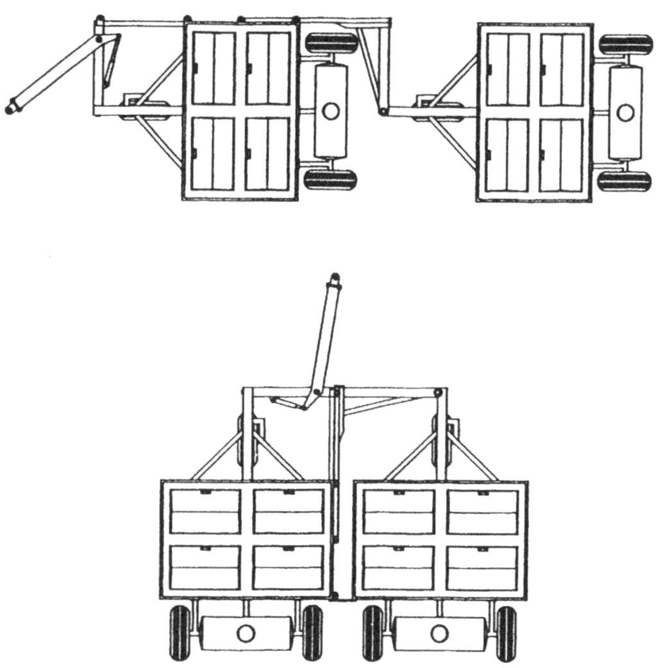

Fig. 13.11. A towing arrangement for two fore-and-aft wheel drills.

First-time no-tillage farmers must often change their evaluations to correctly match tractors with drills and planters. Difficulties arise in several ways:

1. Farmers are not used to thinking of drills in terms of their power requirements.
2. There is little information available to inform farmers about the specific power

and/or draught requirements of different drills and/or openers.

3. Because no-tillage drills are often considerably heavier than their tillage counterparts, some of the power requirements will be needed to move the machine weight, especially on hilly land.

4. Since no-tillage drills and planters break untilled and often hard ground, they are more sensitive to speed than tillage drills as far as power demand is concerned.

5. On the other hand, because no-tillage is so much more time-efficient than tillage, high drilling/planting speeds may not be important.

6. Often, in no-tillage, the traction of a tractor will be more important than its available engine power. Thus, four-wheel-drive and tracked tractors are likely to become more useful.

7. Because turning corners while drilling with no-tillage drills is more difficult than with tillage drills, more fields are drilled in strips ('lands'). This demands sharp turning on headlands or looped turns on corners, requiring a tight turning-circle capability from the tractor and drill.

8. The annual tractor use for drilling/planting is likely to be reduced substantially under no-tillage compared with tillage. This means that total annual tractor costs are lower, tractors last longer in terms of time and replacement scheduling, but the actual hourly cost may be increased.

9. The necessity to continuously monitor drill/planter functions from the tractor seat is increased, because under no-tillage a farmer has but one chance to get everything correct. Tractors therefore need to be electronically as well as mechanically compatible with their drills and planters.

10. The soil in wheel tracks under no-tillage is often loosened because of the high demand for traction, whereas under tillage the result is almost invariably compaction in the wheel tracks. Tractors working near the traction limit in no-tillage will cause more soil loosening and therefore greater differences of opener performances between those within and outside the wheel track areas.

It is difficult to generalize power requirements of no-tillage drills because they have a large range of weight and draught. Ignoring the weight of the drill, some generalizations can be made about the power requirements of individual no-tillage openers from Table 13.1. While draught requirements for only two openers (triple disc and winged) are shown, these two designs are near either end of the range of draught requirements for no-tillage openers. Thus, their requirements may reflect a range of power requirements for no-tillage openers in general.

The power required to pull an opener through the soil is given by the expression:

$$\text{power (kW)} = \frac{\text{pull (newtons)} \times \text{speed (km/h)}}{3600}$$

or

$$\text{power (hp)} = \frac{\text{pull (pounds)} \times \text{speed (miles/h)}}{375}$$

It can be seen in Table 13.1 that at a speed of 5 km/h (3 mph) a single triple disc opener would require up to 2.3 kW (3 hp) and a single simple winged opener up to 2.9 kW (3.8 hp). At 10 km/h (6 mph) the respective power requirements would be 4.6 kW (6 hp) and 5.8 kW (7.6 hp).

In general, the power requirements of no-tillage drills and planters might range between 2 and 6 kW (2.5 and 8 hp) per opener, depending on the drilling speed, the ground conditions, the soil type, the density and state of decay of root material in the soil, the contour of the field, the method of working the field, the design of the opener and the weight of the machine. Allowing for a tractive efficiency of 65% by the tractor, this would require a tractor engine size range from 3 to 9 kW (4 to 12 hp) per opener, which closely matches field experience.

Product Storage and Metering

For handling products such as seed, fertilizer and insecticides, the most distinguishing feature of no-tillage drills in comparison with their tillage counterparts arises from the need for openers to be spaced widely apart to clear surface residues. With planters, the openers are spaced widely apart, usually in a single line, anyway. So no major distinction is made in this regard between planters for tillage and for no-tillage.

With drills, the wider-than-normal opener spacing is usually achieved by increasing the longitudinal staggering of alternate openers, since the row spacing between openers cannot be altered without affecting the agronomy of the crop. This increase in longitudinal spacing results in long seed delivery tubes and shallow drop angles between the hoppers and openers for these tubes if supplied by a single hopper. Such shallow angles interrupt normal gravity flow, especially on hilly land. The problem is overcome in one of three ways:

1. Raising the product hoppers to greater heights above the openers so as to increase the angles on the delivery tubes (Fig. 13.12).
2. Doubling the number of hoppers so that each hopper is positioned over the openers at normal height and delivery tube angles.
3. Utilizing air delivery of product to the openers from a central hopper (Fig. 8.14).

There are arguments for and against each option. Doubling the number of hoppers, for example, adds to the capital cost of the drill but increases the amount of product that can be carried and therefore reduces the number of times the machine needs to be out of service for filling, as well as temporarily adding to the weight of the machine, which may help with downforce. Air seeders are inexpensive but larger designs carry the weight of the product on a separate axle where neither it nor the weight of the hoppers themselves contributes to the overall weight of the machine to assist downforce.

High hoppers are inexpensive but are difficult to fill and contribute to drill instability on hillsides. On very steep hills, at least one drill that carried liquid fertilizer tanks provided a facility to slide the tank to the uphill side of the drill to assist stability (Fig. 13.13). There are no known designs that shift dry hoppers on the move.

Because the surface residues common in no-tillage provide a habitat for pests (and their predators), it is often necessary to apply insecticide(s) with the seed at drilling. Thus, dry granule hoppers and/or liquid insecticide facilities are common on

Fig. 13.12. A no-tillage drill with elevated product hoppers.

Fig. 13.13. A no-tillage drill with sliding liquid fertilizer tank.

some no-tillage drills and planters. Some planter manufacturers have cooperated with chemical manufacturers to provide closed transfer systems for insecticides. This provides for safer handling of chemicals, although operators need to be cautious of pesticide residues on drill and planter components during maintenance.

The concept of drilling and spraying simultaneously by mounting a spray boom on the drill or planter was investigated in New Zealand. While such an achievement would have made no-tillage a truly one-pass operation, the idea was judged not practical for several reasons:

1. It was possible to drill on days on which it was not wise, or possible, to spray because of wind or rain that might otherwise compromise the efficacy of weed and pest control formulations. By restricting drilling opportunities to those times when spraying was possible, some of the time advantage of no-tillage would have been lost.
2. It introduced yet another function to be observed by the operator and/or monitored, increasing the potential for error.
3. Some openers displace, or indeed throw, soil, causing dust, which inactivates the most commonly used herbicides in no-tillage (glyphosate and paraquat). Spraying is better performed with a separate operation by a specialist prior to drilling.

Although blanket application of herbicides at the time of drilling appears to be impractical, banded application on each row has been used successfully (see Chapter 12).

Summary of No-tillage Drill and Planter Design – Large-scale Machines

1. Designs of no-tillage drills need to be more sophisticated than those of tillage drills.
2. No-tillage drills are invariably heavier than tillage drills and are more stressed during operation.
3. Wear and general maintenance are more important and expensive on no-tillage drills and planters than on tillage drills and planters.
4. The tractor engine power required to operate no-tillage drills and planters ranges from 3 to 9 kilowatts (4 to 12 horsepower) per opener.

5. The power requirements for no-tillage drills and planters are more sensitive to operating speed than those for tillage drills and planters.
6. Larger tractors are generally required for no-tillage drilling.
7. Because tractors are operated fewer hours per year than tillage tractors, their hourly operating costs are higher than the latter but their total annual costs are reduced.
8. The total energy expended per sown hectare and the annual operating cost of all equipment are much lower in no-tillage than for full tillage.
9. No-tillage drills are generally narrower than tillage drills because of the increased power requirement. No-tillage planters may be the same width as tillage planters because of fewer openers.
10. Although it is not as necessary to travel as fast during no-tillage drilling or planting as in tillage because of the time efficiency of the system as a whole, some no-tillage drills and planters are actually capable of higher speeds than their tillage counterparts. On the other hand, other no-tillage designs require low speeds.
11. Time analyses to cover a field with a relatively narrow no-tillage drill compared with a wider tillage drill often fail to account for the multiple tillage passes made before the tillage drill begins work.
12. Downforce systems on no-tillage drills and planters need to be more sophisticated, exert greater force and have a greater range of travel than for tillage machines.
13. The geometry of no-till opener drag-arm attachments must compensate for the increased drag forces.
14. Parallelogram drag arms with either gas or oil-over-gas hydraulic pressurized downforce systems provide the most consistent downforces and seeding depths.
15. Drill and planter frames should be suspended on wheel arrangements that minimize bounce from uneven ground.
16. Turning corners while drilling or planting is more difficult in no-tillage than in tillage because of the firmer soils.
17. The firmer ground in no-tillage is better able to withstand scuffing from the wheels when turning corners than with tilled soils.
18. Automated systems that return the opener downforces quickly to preselected values after raising the openers for transport are desirable in no-tillage because of the need to raise the openers more frequently during operation.
19. End-wheel drill and planter configurations are generally the cheapest option but have a maximum width of approximately 6 metres (20 feet).
20. Fore-and-aft wheel configurations allow greater drilling widths and simpler side-by-side joining of two or more drills or planters.
21. Delivery of product from hoppers to no-tillage openers is somewhat more demanding than for tillage drills because of the need for wide spacing between adjacent no-tillage openers to clear surface residues.
22. Because both tillage and no-tillage openers on planters are widely spaced, there are fewer special requirements for product delivery on no-tillage planters compared with drills.

14 No-tillage Drill and Planter Design – Small-scale Machines

Fatima Ribeiro, Scott E. Justice, Peter R. Hobbs and C. John Baker

Small-scale no-tillage farming is not only practical but may be the most important improvement to crop production and resource protection for developing nations to be advanced this century.

Characteristics

Small-scale no-tillage is usually characterized by small field sizes and limited availability of energy, often also accompanied by limited financial resources. Operation of large-scale tractor-drawn implements is neither practical nor possible for many farmers on small properties. For these reasons, most small-scale farmers use either hand-operating jabbing devices or drills and planters with one or two rows. Some triple-row planters are also available but are reasonably rare.

The limited number of rows influences several functions, including opener design. Some of these influences are beneficial. Others are not. For example, many of the more advanced opener designs discussed elsewhere in this book require up to 12 horsepower per opener, which is often beyond the resources of small farmers. Also, non-symmetrical openers, such as angled discs, are seldom regarded as an option on single-row machines because the side forces are too difficult to counteract while keeping the machine heading in a straight line.

But small-scale no-tillage is benefited by the operator attention to each square metre being planted, and weeds and residues are often manipulated by hand or collected for heating fuel or animal bedding.

Another benefit is that most small-scale planters sow fertilizer and seed simultaneously in separate slots. In this way they may be considerably more sophisticated than many of their larger counterparts, some of which do not sow fertilizer at all under no-tillage because of the mechanical complexity of achieving such a desirable function with multiple rows spaced closely together.

Thus, while small-scale no-tillage might be disadvantaged in some respects by the necessary simplicity of drills, planters and available power, it may also benefit in other respects for the same reasons.

Range of Equipment

There is a wide range of small-scale no-tillage seeding equipment available, each suited to different sources of power and field conditions. The range includes hand jabbing, animal-drawn planters, power tillers and planters for limited-powered tractors. Despite the differences in power requirements, the designers of most small machines recognize the need to be able to handle residues, open an appropriate slot, meter seed and perhaps fertilizer, distribute this to the opener(s), place it in the soil in an acceptable pattern, and cover and pack the seed and the fertilizer.

Hand-jab planters (dibblers)

Hand-jab planters are popular amongst small-scale farmers. Some form the primary means of sowing seeds under no-tillage. Others are kept in reserve for filling in spaces in crops otherwise sown with openers in rows. Since the residue-handling ability of small drills and planters is often limited, spaces occur if and when residue handling suffers along the row.

Hand jabbers may have either separate hoppers for seed and fertilizer or one hopper for seed only. Figure 14.1 illustrates a typical double-hopper jab planter.

A common seed metering device used on hand jabbers is a rectangular plate placed inside the hopper. When the handles are pulled apart, the seeds drop into the holes, which are delivered to the outlet and the discharge tube. Plates with different hole sizes are available according to the seed size. Seeding rates can be adjusted according to the number of holes in the seed plate that are exposed in the outlet.

Part of the attraction of hand-jab planters is that they do not require access to animal or tractor power and they are low-cost, light and easy to operate, although some skill is required (Ribeiro, 2004). For these reasons they are often used by women, which increases the available labour pool for small farmers, although no-tillage itself reduces labour demands significantly anyway.

By planting seeds in pockets, there is minimal soil disturbance so weed seed germination is minimized, resulting in easy hand hoeing between plants. The small size of the devices makes them suitable for operation on hilly, stony and stumpy areas and for intercropping (e.g. sowing mucuna between maize rows) and for planting in fallow areas.

Fig. 14.1. A hand-jab planter with seed and fertilizer hoppers.

Their use is most suited to light soils since penetration is sometimes too difficult in harder soils in the absence of some form of tillage. Some clay soils may also stick to the blades when working in wet conditions and seed coverage may be affected by the V-shaped pockets and minimal disturbance (Ribeiro, 2004). This limitation is common to that experienced with V-shaped continuous slots and is not restricted to discrete pockets. But during the transitional phase from conventional tillage to no-tillage it may be difficult to use a hand-jab planter, in which case a ripper may be used to loosen a narrow strip where the hand-jab planter will place the seeds.

Many hand-jab planters for no-tillage are adaptations of similar devices designed for use in tilled soils. The main modification has been to provide longer and narrower points to improve penetration. Such improvements require less downward force from the operator and help to cut residues and penetrate the soil, resulting in less-open slots. They have resulted in 28% and 23.6% increases in emergence of maize and cowpeas seedlings, respectively, compared with shorter points operating in heavy residues (Almeida, 1993).

Row-type planters (animal-drawn and tractor-mounted)

The principles of operation of animal-drawn and tractor-mounted small no-tillage planters are the same as for larger machines. Some of these features are discussed below and comparisons drawn between small and large machines in terms of the conditions under which they each operate.

Downforce

With small machines, an opportunity exists to use weights as the method of downforce. Springs are also used but hydraulic downforce systems are very rare. But weights have the same advantages as hydraulic systems at a much lower cost. In its simplest and cheapest form, weight can be applied by an operator standing on a platform on the machine. Figure 14.2 shows such a single-row machine directly mounted on a small tractor. The advantage is that the weight is easily applied and removed by simply stepping on and off the operator's platform.

Since weights apply a consistent downforce regardless of the vertical position of the opener, they act in a similar manner to oil-over-gas hydraulic systems applied to individual rams on each opener, which are a feature of some of the most advanced larger no-tillage drills.

Therefore, some small-scale no-tillage drills and planters may provide a more sophisticated downforce system than some of the less-advanced larger machines. The electronic modulation of downforce in response to ground hardness is not possible on the smaller machines. But, then again, nor is the direct application of weights a practical option for larger machines. Operators would need to be adding and removing multiple weights every time the downforce

Fig. 14.2. A tractor-mounted single-row drill that relies on the weight of an operator for penetration.

was changed. Doing so might be acceptable on a single-row machine but would soon fall out of favour on a multi-row machine.

Figure 14.3 shows the main components of typical small-scale no-tillage planters. The disc (1) cuts straw (although the effectiveness of cutting straw in this manner often leaves much to be desired – see Chapter 10). Metering devices are positioned at the bases of the seed (2) and fertilizer (3) hoppers. The openers (4 and 5) open slots for placement of fertilizer and seed, respectively. Usually the fertilizer opener (4) operates deeper or off-line compared with the seed opener (5), in the same manner as bigger machines. The packing wheel (6) controls the depth of seeding and firms the soil over the slot. The effectiveness of packer wheels operating on the soil over the slot, compared with operating in the base of the slot before covering, is discussed in Chapter 6. In general, the value of packer wheels operating in the manner shown in Fig. 14.3 is more one of covering (which is important enough) than of improving seed-to-soil contact.

Discs

All of the principles of discs and residue handling, discussed in Chapter 10, apply equally to small-scale machines as they do to large-scale machines, except that with single-row small-scale machines there is greater clearance around the opener for random residues to fall away without blocking the machine.

Most small-scale no-tillage planters have discs, the effectiveness of which are dependent upon the disc diameter and design (plain, notched, wavy, flat or dished), soil conditions, residue conditions and adjustments provided on the planter. Ineffective residue cutting results in clogging of straw on the seed components, which in turn results in problems for seed and fertilizer placement and coverage, and even seed and/or fertilizer metering.

Uneven straw results in hairpinning by discs and wrapping of residues on tined openers, although Casão and Yamaoka (1990) claimed that the severity of blockages could be reduced (though seldom eliminated) with increasing distance between the disc and any stationary tines that follow (they recommended a minimum distance of 25 mm).

On the other hand, some of the more successful combinations of tines and discs have the discs in close association with the tine. One example is shown in Fig. 14.4 (centre tine), in which a groove is created in the leading edge of the tine especially for the disc to operate within. Figure 4.27 shows the disc version of a winged opener in which two tines actually rub against the flat face of a disc.

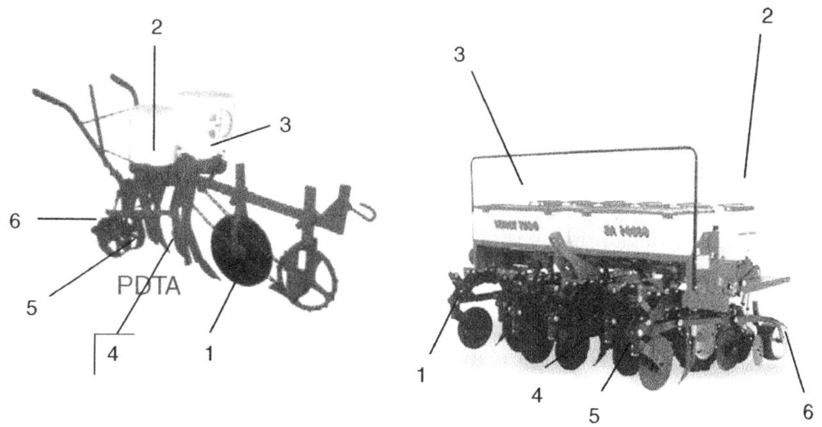

Fig. 14.3. The main components of typical small-scale animal-drawn and tractor-mounted no-tillage planters.

Openers

The functions of openers for small-scale no-tillage are no different than their functions for larger-scale machines and are discussed in detail in Chapters 4, 5, 6 and 7. On small-scale planters with tined openers, there should be independent adjustment of the fertilizer opener so that fertilizer can be placed deeper than the seed (Van Raij *et al.*, 1985). Although placing fertilizer beneath the seed in no-tillage does not always result in the best crop yield (see Chapter 9), with small-scale drills and planters it is a more realistic option than placing fertilizer alongside the seed because the latter option requires the fertilizer opener to be operating in new ground, which requires more energy than when both openers (seed and fertilizer) operate at different depths in a common slot. In any case, placed fertilizer within the seed zone is far superior to surface broadcasting causing slow crop access and increased weed growth.

As with larger machines, there are advantages for slots with minimal disturbance (see Chapters 5, 10 and 13). While the choice of opener type might depend on soil resistance to penetration and the amount and resistance to cutting of residues, it is no more feasible for small-scale no-tillage farmers to possess more than one no-tillage machine in order to cope with varying conditions than is the case for large-scale farmers.

Therefore, to be universally useful for practising farmers (large or small), it is inevitable that the choice of preferred opener types will, over time, gravitate towards those that function best in the widest possible range of conditions. Tillage has as one objective to reduce the physical variability between different soils so that drills do not have to cope with widely varying conditions. But, when the tillage process is eliminated altogether, emphasis then shifts to the capability of no-tillage openers to cope unaided with this variability. By definition, this demands increasing sophistication from the designers of no-tillage openers, regardless of their scale of operation.

Double disc openers (V-shaped slots with Class I cover) are commonly used on small-scale drills and planters. The slots are narrow at the surface and may be compacted at their bases and sides, but are less power-demanding than tine-disc openers that have less compacting tendencies. With unequal-diameter double disc openers, because the smaller disc rotates faster than the larger disc a degree of cutting, or 'guillotine', effect is created (Fig. 4.3 – Chapter 4).

A range of tined openers is shown in Fig. 14.4. Generally, tines require less downforce than double disc openers, which contributes to maintaining a uniform seeding depth if a suitable depth-control mechanism is included. Tines are preferred in hard soils, although their drag force may become excessive for the power available. And tines are more susceptible to blockage with residues and are unsuitable in stony areas.

None the less, most of the planters used in small-scale agriculture have tines because of their better penetration of hard ground and ease of manufacture. In situations where

Fig. 14.4. A range of tined openers used with animal-drawn no-tillage planters. The centre opener has a groove cut into its leading edge, in which the leading disc rotates.

soil crusting is a problem (such as where cattle have trampled the soil when wet), only tractor-mounted planters with tined openers will break the compaction in the soil surface, although this is often only 100 mm deep.

Seed metering devices

There continues to be debate amongst researchers about the importance of seed spacing along the row with row crops such as maize (Sangoi, 1990; Rizzardi *et al.*, 1994). More recent evidence has shown that uniform plant emergence along the row may be more important than plant spacing to reduce plant competition of smaller plants by larger plants. But the fact remains that, if 'perfect spacing' has become the accepted norm in conventionally tilled seedbeds, no-tillage exponents need to match this norm in untilled seedbeds in order to avoid introducing an unnecessary negative factor against no-tillage.

Seed metering devices are responsible for governing seed rate (number of seeds/m) and seed spacing (consistency of spacing between seeds in the row); thus their accuracy must be assured.

Most crops sown by small farmers are in wide rows. Singulation of seeds is therefore important. So emphasis is placed on seeding mechanisms and power requirements as priority design criteria. This contrasts with larger no-tillage planters where slot micro-environment, residue management and fertilizer banding assume at least equal importance to seed spacing and energy requirements.

No-tillage farming in Brazil provides an interesting comparison and contrast of small-scale machines and tractor-drawn machines. Both systems are practised widely in a country that spans many climatic and socio-economic zones, often in relatively close proximity to one another.

Seed metering devices used on animal-drawn no-tillage planters in Brazil all feature the same gravity seed plates that are used on local tractor-mounted planters, namely plastic or cast-iron horizontal plates. Figure 14.5 illustrates a horizontal plate-type metering device along with several alternative plates. Some manufacturers provide seed plates suited to small seeds (e.g. canola, hairy vetch, forage radish) as well as maize and other larger seeds.

The use of such devices has been driven by their relatively low cost, since most singulating seeders used in countries that do not have small-scale agriculture are now of the vacuum, air pressure or 'finger-picker' type, which involves seeds being sucked, blown or clamped against vertical plates rather than falling under gravity into holes or notches in horizontal plates. Vertical plate seeding mechanisms are faster and less sensitive to seed shape and size than horizontal plate-type seeders, but are also more expensive. Of course, vacuum and air singulators also require a powered air fan as the basis of operation and this would be

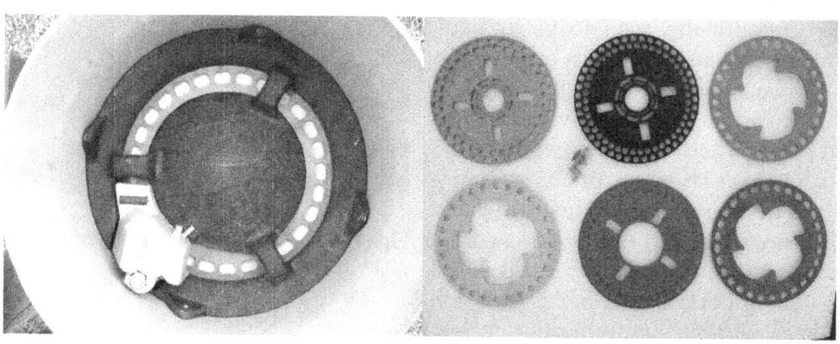

Fig. 14.5. A horizontal plate metering device (left) used in precision planters, with an array of optional seed plates (right).

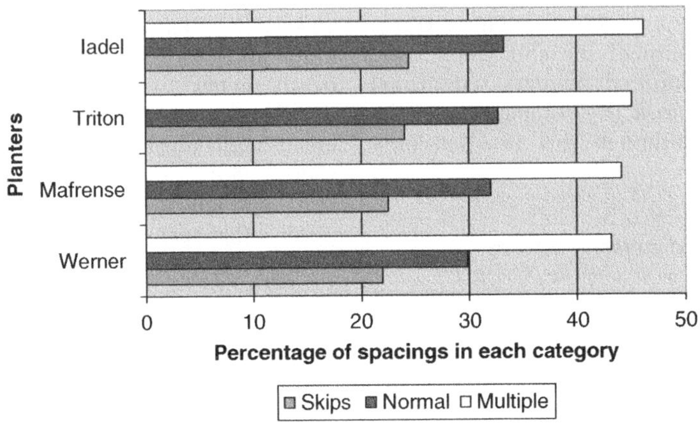

Fig. 14.6. Percentage of normal spacings, skips and multiple seeds provided by four models of animal-drawn no-tillage (NT) planters (Ribeiro *et al.*, 1998). The criteria for classification of spacing is based on Kurachi *et al.* (1993). Each crop has an ideal spacing (X_{ref}), which depends upon the recommended number of plants/m. For example, if for maize the recommendation is 7 seeds/m, then X_{ref} is 1.00/6 = 0.17 m. In this manner the following classes are established: normal ($X_{ref} < X_i < 1.5\ X_{ref}$); doubles ($X_i > 1.5\ X_{ref}$) and skips ($X_i < 0.5\ X_{ref}$).

difficult to facilitate on an animal-drawn machine without resorting to a stationary engine.

Horizontal plate singulators are a very old, well-proven and refined system that pre-dated the vertical plate systems now in common use on larger planters. It is no surprise, therefore, that, when Ribeiro (2004) evaluated the uniformity of distribution of maize seed along the row with four models of plate planters in Brazil, she found no significant differences between models in the proportion of normal spacings, skips and doubles. The results are summarized in Fig. 14.6.

To be most effective, horizontal plate singulators require the seed to be graded into uniform sizes and the holes or cups in the plates to be matched to the chosen seed size. This requires having several plate sizes and some experimentation when seed lines or batches are changed. But, with limited numbers of rows and small quantities of seed, this is not a difficult undertaking compared with multi-row machines. But it does highlight the importance of being able to change plates without emptying the entire seed hopper. Figure 14.7 illustrates a closed hopper system that allows the plate to be changed without spillage of seed.

Fig. 14.7. A closed hopper system for easy seed plate change.

Fertilizer metering devices

The types of fertilizer metering devices found on small-scale no-tillage machines include rotating bottom, auger type, edge cell and star wheels (Figure 14.8). The discharge rate for star-wheel and rotating-bottom types is controlled by adjustable outlets, while auger and edge-cell types are controlled by changing their speed of rotation relative to the ground speed (Ribeiro *et al.*, 1998).

Packing wheels

While seed row packing wheels vary in design, most are of either steel or plastic construction. V-shaped wheels are used where soil disturbed by tined openers needs to be collected and thrown into the open slots. Good coverage/compaction depends on the depth of seed placement, the type of seed compaction wheel and soil moisture. Open-centred wheels are better for soils with a tendency towards crusting as they press the soil laterally towards the seed.

Power requirements and ease of operation

Small-planter operation requires more intimate operator involvement than for larger machines. Therefore ease of operation is important. For example, most small planters require the operator to hold a pair of handles and steer the machine, as well as controlling the animals that may be pulling them. With small tractor-drawn machines, a second operator usually controls the tractor. In either case, energy requirements are important. But, since the openers used on most small planters are similar to those used on larger machines, all of the forces and principles of soil reaction apply equally to both classes of machine.

Of the seven machines reviewed by Ribeiro *et al.* (1998), four featured tined seed openers and three featured double disc openers. Ralisch *et al.* (1998) evaluated the draught and energy requirements of a small planter with tined seed and fertilizer openers in an untilled soil of quite low bulk density, 1.07 g/cm^3, operating at 100 mm depth. They recorded a draught force of 834 N, which is less than half the values recorded by Baker (1976a) for a single simple winged opener (see Chapter 13).

Draught forces vary widely with soil strength, which is itself influenced by soil moisture content, soil type, SOM and the time under no-tillage. So it is difficult to compare opener (or, indeed, drill) types in different conditions. But, at 2.4 km/h, the machine tested by Ralisch *et al.* (1998) would require 1.4 kW of draught power or approximately 3.6 kW (5 hp) of tractor engine power (at a tractive efficiency of 0.65). This compares with larger drills, which commonly require 4–9 kW (5–12 hp) of engine power per opener to operate at up to 16 km/h. Such high forward speeds are unobtainable by small machines, even if sufficient power is available, because of the difficulty in controlling them at high speed, especially if the operator walks behind the machine. Therefore the lower power requirement for small machines probably reflects the lower operating speeds more than other variables.

Fig. 14.8. Two examples of fertilizer metering devices used on small-scale no-tillage planters. Left: edge cell (or fluted roller); right: star wheel.

According to Siqueira and Casão (2004), differences in power requirements are primarily due to the design of the openers, the weight of the planter and the number, and the contact surface area of the residue-cutting and groove-opening components. The main characteristic that makes such machines suitable for small tractors or animals is the small number of rows: two and three rows for maize and soybean planters and six to seven rows for wheat and rice drills.

Some of the factors that contribute to the physical effort by the operator in controlling the machine are the weight of the planter, the height of the handle(s), manoeuvrability, stability and ability to operate on sloping ground. The height of the handle(s) becomes particularly important during headland manoeuvres and in most cases is adjustable. Multiple-row models generally require less manual effort from the operator than single-row models, because seats or standing platforms are provided.

Models with two rear support wheels provide good stability when working on flat land but may be constrained on hillsides. Models with only one wheel are more adapted to stony and stumpy areas because it is easier to steer such machines around obstacles. For those models that evolved from 'fuçador' ploughs, improved stability occurs when fixed-shaft systems are used rather than chains. The 'fuçador' plough consists of a wooden drawbar, which is fastened to the yoke of the draught animal(s), on which is mounted a leg and a shovel-like plough body (Schimitz *et al.*, 1991). For no-tillage, the mouldboard plough body is replaced with no-tillage openers. The device is used in the stony and hilly areas of south Brazil.

Adjustment and maintenance

All models offer adjustments of both seed and fertilizer sowing rates. But some models do not offer many adjustments either for seed and fertilizer sowing depth or for residue handling. On the other hand, the most sophisticated openers do not require adjustments to handle a wide range of residue types, but these are seldom used on small drills or planters. In general, tined openers have the poorest residue handling characteristics (see Chapter 10) and disc openers the best. But certain disc openers (e.g. double disc) have a tendency to hairpin pliable straw into the slot, where it interferes with seed germination in both wet and dry soils. These disadvantages apply equally to small planters as to larger equipment.

For this reason, several small planters with tined openers provide adjustments that affect their residue-cutting ability. The two main adjustments are the hitching point and the front ground wheel. Adjustments made to the disc will also affect the depth of the fertilizer slot. For the same depth of the fertilizer, different depths for seeds are possible through adjustments of the rear ground wheel.

In the simplest models, seed rates are adjusted by changing to different seed plates, while multiple-row models often provide sets of gears to change the plate speed. Other models that do not sow widely spaced rows provide geared adjustment of the speed of bulk seeders.

Animal-drawn planters

Figure 14.9 shows a range of no-tillage drills developed in Brazil. The models shown in the two top photographs are more sophisticated, have a greater range of adjustments and are likely to produce better results than the models shown in the two middle photographs, which have evolved from 'fuçador' ploughs. They are lighter, less expensive and more adaptable to hilly and stony areas. The model shown in the bottom photograph features disc openers and platforms for an operator.

Planters adapted from power tillers

Power tillers that are normally used for conventional tillage are sometimes used for strip tillage by eliminating some of the powered blades to till narrow strips (20 to 200 mm wide), leaving the ground between

Fig. 14.9. A range of small-scale no-tillage planters developed in Brazil.

the rows (up to 500 mm wide) untilled. Chapter 4 addresses the issues of how larger versions of such machines have been adapted to follow the ground surface and Fig. 4.22 shows an example of one such machine producing narrow strips.

Tractor-drawn planters

Small farmers also use animal-drawn or small tractor planters requiring up to 50 hp. The machines have the same straw-cutting (smooth disc) and slot-forming (tine or double disc) openers as the single-row machines and most are capable of applying fertilizer at seeding time.

Some models provide bulk seed and/or fertilizer hoppers in a similar manner to larger machines (e.g. Figs 14.10 and 14.11) while other models are set up as multi-row precision seeders (e.g. Fig. 14.12).

No-tillage farming in Asia

Zero-tillage (or no-tillage) has been adopted on about 10–15% (2 million out of 13.5 millions hectares) of the wheat planted after rice in the rice–wheat cropping system in India and Pakistan. Spring wheat planted in

Fig. 14.10. A small tractor-drawn no-tillage drill.

Fig. 14.11. Two small planters with bulk fertilizer hoppers and precision seeders.

the winter season and, increasingly, other winter crops, such as lentils, are being zero-tilled. Yet the gains in soil health from the winter season are countered by puddling of summer rice. In addition, the vast majority of the zero-tillage occurs in fields where the rice residue either is removed as fodder or fuel or is burned, because the current low-cost zero-tillage drills have no residue-handling capacity. In many cases, only anchored straw remains. This leads to a hybrid system where yields cannot and will not be maintained due to soil degradation.

Long-term experiments in Mexico have shown that zero-tillage without residue retention in intensive maize–wheat systems results in more rapid decline of yields than where a full tillage system is retained in which residues are buried. But the best treatment has been no-tillage with residue retention (Govaerts *et al.*, 2004). This points out the need for 'rational residue retention' in the humid tropics and subtropics with heavy monsoons and sometimes triple-crop annual intensity (K. Sayre, 2004, personal communication).

There is currently research being initiated and undertaken in some parts of South Asia on direct-seeded or zero-tilled rice (RWC website). There is little or no

Fig. 14.12. A small planter adapted from animal operation for tractor mounting.

prior research on how to plant zero-tilled rice under monsoon conditions. The major problems facing scientists and farmers are: (i) planting time decisions influenced by erratic onset of pre-monsoon and regular monsoon rain and little or no assured irrigation schedule that can otherwise keep machinery from entering fields when they are too wet; (ii) the enormous weed management problems brought about by the loss of puddle conditions in sandy soils that allow fast infiltration and therefore reduce the ability to control weeds by impounded water; and (iii) the lack of drainage, especially in the lowlands, which can submerge and kill recently emerged seedlings. Current experiments include zero-tillage of transplanted rice, newly available herbicides, rice varieties that can withstand submergence and varieties that do well in alternating flooded and dry conditions.

Table 14.1 summarizes the special problems for zero-tilled rice.

Research into residue retention is progressing, but the normal Western technologies, such as double disc openers, are probably too expensive, heavy and need excessive power. Indigenous or locally made systems, such as openers, with inverted-T, double disc and star-wheel injector planters are moving forward. But research suggests that much cheaper strip-tillage systems might provide the answer to low-cost handling of residues, especially for wealthier farmers. For poorer farmers, residues are highly valued for fuel and fodder and will probably remain so for several decades.

Two-wheeled or four-wheeled tractors?

It is a problem to learn how to apply conservation agriculture methodologies in the intensely poverty-stricken areas of South Asia. Although zero-tillage drills are becoming more available, there is a dearth of four-wheel tractors. As a result of poverty, many holdings are small and scattered. Intense monsoon rains provide large challenges to researchers, conservation agriculture proponents and machinery designers. Whichever system(s) become dominant, it is likely that the majority of small and poor farmers will not own their own equipment but will rent from service providers.

There have been efforts in recent years to bring conservation agriculture to two-wheeled tractor farmers. Although the area

Table 14.1. Problems and possible solutions for zero-tilled rice.

Problems	Possible solutions
1. Majority of rice is rain-fed. Major problems are erratic monsoon and therefore problems of entering fields for seeding operations.	1. Planting needs to be done as quickly as possible when the proper soil moisture is reached. Once the field is too wet serious compaction will occur. 2. Smaller, lighter machinery (two- and four-wheel tractors) may help. 3. Farmers may want to have the option of transplanting by hand or machine into zero-till fields if direct seeding is impossible. 4. Move to early dry-season irrigated rice.
2. Lack of drainage and flooding kills off emerging seedlings after a heavy downpour of monsoon rain.	1. Permanent beds and introduction of some drainage capability. 2. Flood-tolerant rice varieties are also possible. 3. Transplanted zero-tilled rice.
3. Problems of weed control when soils are not kept flooded (more serious on research stations than in farmer fields).	1. Integrated weed management will be the key, using competitive varieties, mulching, preventing seed set of weeds, rotation and various herbicide strategies. Untilled seedbeds where the first flush of weeds are allowed to germinate and then controlled with herbicide is another strategy. In this system, avoiding ploughing will avoid a new flush of weeds germinating. 2. Planting of a cover crop after wheat and killing the cover crop and weeds with herbicide before zero-tilling rice.

of adoption is still small, engineers and researchers feel they are finding attachments to fit into this complicated socio-ecological system.

Four-wheeled tractors

India is the largest tractor manufacturer in the world in terms of numbers. Still today, only 50% of tillage is mechanized in India (perhaps 90% in the rice–wheat areas) and less than 20% in Nepal, but greater than 70% in Bangladesh. The surprising gap between Bangladesh and the rest of South Asia is discussed later. Further, the Indian government laws prohibit tractor manufacturers from manufacturing implements such as seed drills in order to promote local small manufacturing.

TOOLBARS AND TOOLS. Many machine toolbars in India and Pakistan are based on early 'rabi' (winter wheat, lentil) seed drills that were developed in the 1970s and 1980s. The manufactures of conservation agriculture machinery have for the most part simply strengthened the frames, bars and shanks (Hobbs and Gupta, 2004). The toolbars are flat (i.e. not diamond) and generally made from two pieces of 50 mm angle steel welded together to form a square toolbar. Two or three bars are positioned at fixed distances. There are various systems for attaching the shanks to the toolbars. Farmers are learning that an adjustable shank length provides more adaptability but has a tendency to swing to one side or another if not properly tightened or if of inferior quality.

ZERO-TILLAGE DRILLS. The current level of enthusiasm for conservation agriculture research and development in South Asia was sparked by a CIMMYT (International

Centre for the Improvement of Maize and Wheat, Mexico) programme that imported simple inverted-T drills from New Zealand (Baker, 1976a, b, Fig. 14.13) into Pakistan in the early 1980s for wheat. Over a period of time, various national and international programmes in Pakistan and India reduced the size and cost of the initial machines and 'indigenized' them. Specifically, the popular locally made 'rabi' or winter wheat drills were strengthened and locally made inverted-T openers attached (Hobbs and Gupta, 2004).

Toolbar platforms and tools for zero-tillage have become as uncomplicated and light as possible (Fig. 14.14). Nearly any medium-sized workshop is able to produce them. The first system to fail on locally made tractors is the draught control system and the second is the hydraulic lift. Many farmers who purchase zero-tillage machines therefore find that their three-point-hitch hydraulic lifts soon need overhauling. So most zero-tillage drills come with various types of depth-control wheels. In Pakistan, pneumatic tyres are often used, but the cheaper Indian and Pakistani models have metal wheels.

STRIP TILLAGE DRILLS. Much less popular than either zero-tillage or bed planting are strip tillage drills for four-wheeled tractors (Fig. 14.15). These drills were developed by Indian scientists and engineers at Punjab Agricultural University, Ludhiana, in the late 1980s. Typically they comprise a simple 2.2 metre PTO driven 'rotavator' with four blades or six blades per strip and they come in nine to 11 row models. Such machines cost 50% more than zero-tillage drills. Fuel consumption is greater than zero-tillage but much less than conventional tillage. Farmers remark that strip tillage helps in fields where residue levels are too high for the simple inverted-T zero-tillage shanks. Yields are comparable to those of zero-tillage (Hobbs and Gupta, 2003). Pakistan research on rotating discs, smooth and serrated, reported that the disc wear was high.

STAR-WHEEL (PUNCH) PLANTERS. In an attempt to solve the problem of planting into

Fig. 14.13. Inverted-T openers mounted on rigid shanks attached to a square hollow toolbar. (Note that details of typical inverted-T openers can be seen in Figs 4.22, 4.23 and 4.24 – Chapter 4.)

Fig. 14.14. A typical zero-tillage drill on a typical Indian tractor.

Fig. 14.15. Strip tillage drill from India.

heavy residue, star-wheel or rolling punch planters (originally developed in Zimbabwe) have been added to existing zero-tillage frames (Fig. 14.16). Modifications have been made to assist with synchronization of seed delivery and to prevent seed from falling outside the punch (RWC website). Perhaps the biggest problem facing

Fig. 14.16. A multi-row rolling punch planter.

this system in South Asia is its relatively high cost.

BED PLANTERS (RIDGE AND FURROW PLANTING). Bed systems for wheat were originally developed by Mexico's Yaqui Valley farmers to compensate for dwindling water supplies. Irrigation water is saved by applying it through the furrows between the beds, which greatly enhances water conservation and drainage. Bed-planted wheat also allows access to the field after planting for chemical applications and mechanical weeding. More than 90% of Yaqui Valley farmers have now adopted the practice (Aquino, 1998), but they still completely knock down the beds and reshape them for the next crop.

Work began on bed-planted wheat in South Asia in the mid-1990s and current adoption is increasing (Hobbs and Gupta, 2004). The goal is to eventually have permanent beds, especially on the dry sandy soils, where groundwater supplies are fast receding, or on clayey soils, where wheat is prone to waterlogging. Some variations exist for adapting to the erratic monsoon problems and low-yielding direct-seeded rice by transplanting rice by hand on to beds using inverted-T openers to open the slots for transplanting. There might be good prospects for bed-planting of rice–vegetable rotations in India or cotton–wheat rotations in Pakistan.

Work is still needed to successfully grow dry-seeded rice on beds, including selecting sowing dates, weed management, soil types and climatic and socio-economic situations under which permanent beds will be of benefit. There are still questions to be answered about the shift from anaerobic to aerobic fluctuating conditions for rice. And there are questions about the most appropriate machinery to be used, since the more complex monsoon systems of Asia might require more adaptation of designs first created in the Yaqui Valley (Mexico) ecosystem (Sayre and Hobbs, 2004).

The majority of current commercial bed-planter designs are derivatives of zero-tillage drills, using the same frames and fluted roller seed meters, but with simple adjustable-width furrower shovels added. Much work has been undertaken on the agronomy of wheat and rice and two rows sown on 72.5 cm beds has become the standard in rice–wheat rotations, although most planters can be adjusted to three rows and varying bed spacings. Some designs offer zero-tillage bed-planter combination machines that have extra inverted-T openers, shovels and shapers. But these designs seem to be inadequate for permanent beds and increased residue levels, and work has started on adding double disc openers and star-wheel punch planters.

'HAPPY SEEDER'. The 'happy seeder' (Fig. 14.17) was designed to handle high rates of residue and seed either on beds or on the flat. The drill is a combination of two machines, a forage harvester and a zero-tillage drill using inverted-T winged openers (RWC website). The forage harvester cuts, chops and lifts the straw, providing the drill with a clean surface for zero-tillage drilling. The chopped material is blown directly behind the drill and floats down as mulch. Field trials in India have confirmed the usefulness of the approach. But problems with germination and skips have persisted and resulted in the need for adjustment for the cutting height as well as strip tilling in front of each inverted-T opener. Adaptations in Pakistan have resulted in optional separation of the two halves of the machine.

Fig. 14.17. An example of a 'happy seeder'.

Two-wheeled tractors

Relative poverty results in landholdings becoming smaller and more fragmented. A successful small farmer might own 5 hectares while a financially poor small farmer will own less than a hectare with an average of five fragmented parcels. The number of four-wheeled tractors declines to virtually zero for poor farmers, as does other modern machinery. The eastern India and Bangladesh areas (Fig. 14.18) have arguably the most fertile land in all of South Asia; yet poverty and very high population density offer conservation agriculture researchers a particular difficult and restrictive socio-economic situation.

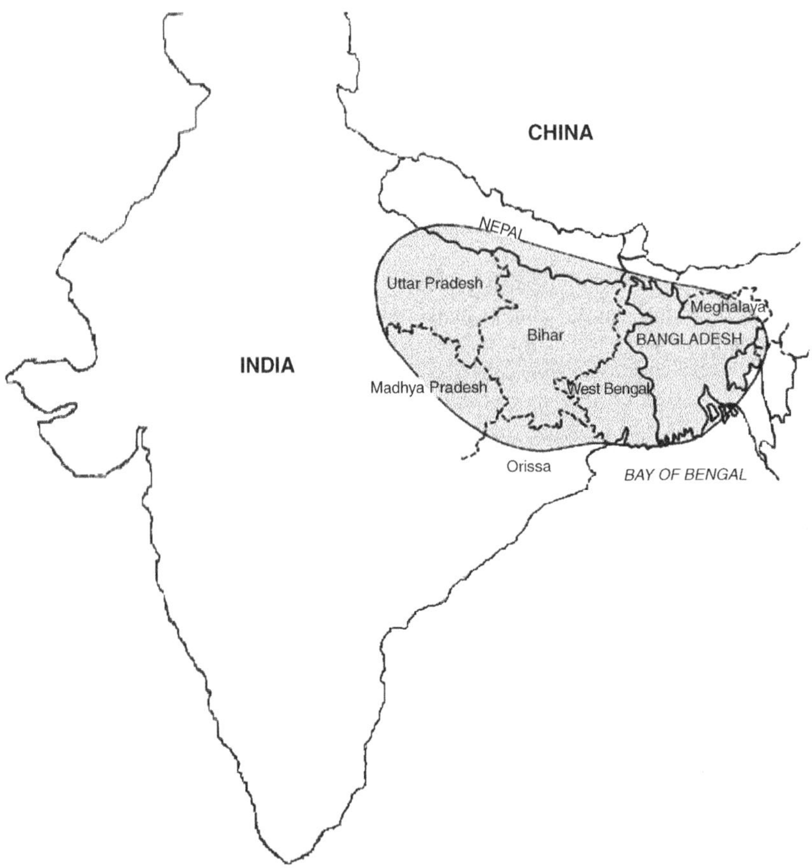

Fig. 14.18. The South Asian 'poverty square', where 500 million farm-supported families each live on less than 1 hectare of land per farm.

If conservation agriculture is to be introduced and adopted by farmers of this region, the equipment must be adapted to either bullock or two-wheel tractor power sources. These power sources must also be made widely available, as there are currently large areas where even the simplest power sources are not available. Two-wheeled tractors have been seen as appropriate and socially equitable (Justice and Biggs, 2004a), since the cost of keeping a pair of bullocks for land preparation and some transport are becoming prohibitively expensive. Many farmers seek alternatives to animal-drawn options, but developers here and perhaps in other underdeveloped regions face many extra hurdles:

1. The inherent conservative nature of all farmers, but particularly those who are resource-poor and can ill afford to take cropping risks.
2. A substandard infrastructure, including local manufacturers and extension systems, together with low literacy, slows interest in or adoption of any technology.
3. All farmers focus on low-cost machinery investment and forgo quality for price.
4. The limited research and development on conservation agriculture attachments for two-wheeled tractors compared with four-wheeled models.
5. Emphasis on four-wheeled tractors and indigenous production has limited the availability and competitiveness of two-wheeled models.

THE ROLE OF TRANSITIONAL TECHNOLOGIES. Despite these hurdles, sales of two-wheeled tractors and the common 'rotovator' have increased in the last decade, especially in Bangladesh, where it is estimated that more than 400,000 Chinese-made two-wheeled tractors undertake more than 70% of land preparation by Bangladeshi farmers. This dramatic increase was brought about by changes in government policy and development of a vibrant market for tractors following a severe cyclone disaster and floods in 1987 that decimated the animal population. A similar picture is emerging in Nepal and to some extent in India. Special projects in Nepal have made farmers more aware of the benefits of owning such power sources to generate income or to provide contractor services for non-owners of tractors (Justice and Biggs, 2004b). The availability of such power sources now allows conservation agriculture methods and techniques to be made available to farmers in these regions.

Besides providing power for conservation agriculture, these tractors undertake a multitude of other activities, such as reaping, pumping, seeding and tillage. The tractor, or its engine, is also used as a power source for threshers, winnowing fans, milling and transport for people and goods, both on land (pulling 2 t trailers) and on water (thousands of country boats in Bangladesh). They also reduce the drudgery of puddling rice paddies when cage wheels are fitted. All these functions speed up farm operations (timely land preparation, sowing and harvesting), improve yields and increase cropping intensity and efficiency of crop production. These results are all vital for an area where population densities exceed 1000 people per square arable kilometre.

Land preparation costs for both winter crops and summer puddling of rice are one-third less per unit of land with two-wheeled tractors than with four-wheeled tractors (Sah *et al.*, 2004). The time spent by four-wheeled tractors in turning and backing is also eliminated with two-wheeled tractors, especially in small fields. The challenge has been to extend these advantages to conservation agriculture. First, a toolbar concept has been used in zero-till and bed planters; and, secondly, a reduced-till/shallow-till seed drill has been modified to strip-till and form beds in one operation.

TOOLBARS. As with four-wheeled tractors, toolbar designs for two-wheeled tractors are largely based on modifications of the familiar 'rabi' flat-bar seed drills. The mounting plate for the toolbar is bolted to the rear of the transmission of two-wheeled tractors. Such a rigid mounting system in uneven fields is a problem compared with more flexible three-point-hitch systems. None the less, it has proved to be a robust platform for conservation agriculture implements. Generally, two bars are used to attach tools and implements.

TOOLBAR ZERO-TILLAGE DRILLS. Most two-wheeled tractors are capable of pulling up to four-row zero-tillage seeders. Designers have simply adapted the designs of the four-wheeled tractor zero-tillage drills to the reduced row numbers, using full-sized inverted-T openers and the same shanks but having downsized the seed and fertilizer hoppers (Fig. 14.19). The effective field capacity of such machines is typically 0.20 ha/h for simultaneous seeding and fertilizer application. Planting cost for wheat and maize has been reduced by some 50% compared with conventional tillage methods.

TOOLBAR BED PLANTERS. Bed planters that simultaneously till the soil and form the bed are not considered, regardless of whether or not they also sow seed and fertilizer, although such a practice may eventually lead to a full no-tillage programme involving permanent beds.

Bed width is limited mainly by limitations on wheel spacing of two-wheeled tractors. The standard rice–wheat bed is 65–70 cm wide. Problems occur when first forming beds if the land is not previously prepared. The shovels grab at clods, pulling the machine off course, which may cause handling problems if one wheel travels into a furrow and tilts the bed former. Clods are less of a problem under permanent bed conditions where light reshaping of the bed is performed and the wheels track nicely in the furrows and greatly reduce fatigue of the operator.

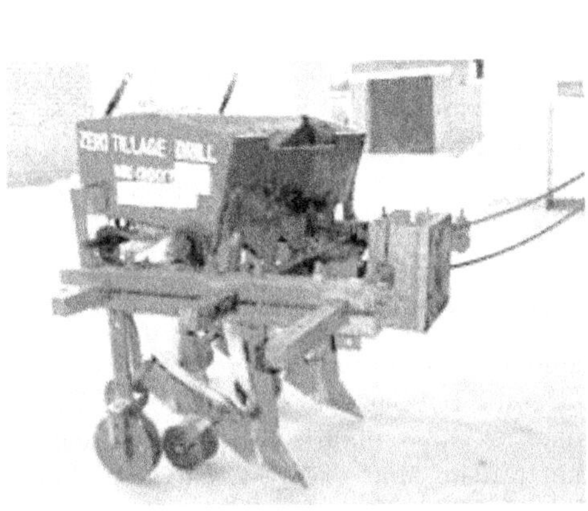

Fig. 14.19. A zero-tillage flat-type toolbar showing the mounting plate for a two-wheeled tractor.

REDUCED-TILLAGE SEED DRILL. A Chinese-designed reduced-tillage/single-pass seed drill was introduced into Nepal in 1989 and Bangladesh in 1996 by CIMMYT. It has been the only conservation technology available from China for two-wheeled tractors in those regions and has undergone much research by Pradhan *et al.* (1997), Meisner *et al.* (2003) and Sah *et al.* (2004), who demonstrated consistently high yields for the following reasons:

1. It was able to drill wheat, lentils and other winter crops into very wet soils (up to 30% moisture content) immediately following the rice harvest, avoiding late planting.
2. It provided a very fine soil tilth, which ensures germination.
3. It placed seeds at a uniform depth.
4. It reduced weed problems associated with the previous rice crop.

Although the machine cannot be considered a true no-tillage drill when in its full-tillage mode (Fig. 14.20), it represents an excellent transitional (and flexible) technology from multiple ploughing to zero- or strip-tillage (Fig. 14.21). The drill's three main components are:

1. A 48-blade, 120 cm wide high-speed shallow tillage (maximum 10 cm deep) 'rotovator'.
2. A six-row fluted roller seed meter (11 and 17 flutes available) and seed bin.
3. A 120 cm roller for planking, compaction and depth control.

STRIP TILLAGE. Research on strip tillage is more recent (Justice *et al.*, 2004), but results have been promising using the Chinese-designed machine. Field efficiency improves by 15–20% with less fuel and time consumption. The soil area disturbed can be adjusted from 15 cm to as little as 2–3 cm (with straightened blades). For narrow

Fig. 14.20. A two-wheeled reduced-tillage machine in full-width tillage mode.

Fig. 14.21. A two-wheeled reduced-tillage machine in strip-tillage mode.

stripping, additional blade holders are welded to the axle to compensate for the absence of a normal spiral pattern and to reduce vibrations. Work in Mexico, Bangladesh and Nepal has shown that this system's high-speed 'rotovator' blades (which rotate at greater than 400 rpm) are able to cut and seed into loose straw and may present an inexpensive machinery solution for the residue retention problems throughout this region for two- and four-wheeled tractors. Figure 14.21 shows a self-propelled two-wheeled strip tillage machine creating 50% disturbance and sowing wheat in 100 mm spaced rows.

PERMANENT CONSERVATION AGRICULTURE BEDS. The flexibility of the Chinese-designed drill has recently been extended to making new beds and seeding in permanent beds with very few modifications. When it is necessary to reshape permanent beds, the toolbar system with shovels can be used, or only a few rotary blades in the furrow might move soil back on to the undisturbed bed.

STRIP TILLAGE ON PERMANENT BEDS. If the beds do not require reshaping, the same machine simply strip-tills on the existing beds. In Mexico and Bangladesh, modifications to conventional strip tillage machines have been carried out by CIMMYT as follows:

1. Two depth-control wheels are positioned in the furrows in place of the roller.
2. The furrow openers are extended down about 7 cm.
3. The standard 'C'-type blades are straightened to cut through residue and reduce the amount of soil movement.
4. Extra blades are added to reduce vibration (circled in Fig. 14.22).

Figure 14.22 shows a modified strip tillage machine/seed drill, in this case used for drilling mung bean after wheat on permanent beds. The straightened 'C'-type blades (inset) are able to cut the residue, leaving it on the surface of the bed with minimal soil disturbance or raking, which is otherwise found with fixed inverted-T openers.

There has been much debate about the most desirable height for beds of this type. Most bed planters can only make beds up to 10–12 cm high. Early attempts to create higher beds are now recognized as wasting energy and are often agronomically undesirable as they dry out more quickly. It is now generally accepted that beds need only to be as high as is necessary to allow water to move from one end of the field to the other for irrigation or to drain the field. Because many fields are small (average less than 0.2 ha), lower beds are sufficient.

Strip tillage systems based on two-wheeled tractors also involve comparatively lightweight machines that allow seeding into wetter soils compared with four-wheeled tractors and their associated bed planters. This is important in conservation agriculture systems in South Asia with both flat and low-bedded applications.

On the negative side, two-wheeled strip tillage on permanent beds does not allow access back into the field after the crops are established. It would be desirable to facilitate banded top dressing, inter-row cultivation and spraying as with four-wheeled tractor models.

Results of recent tests with wheat establishment in Bangladesh (Rawson, 2004) found full tillage and strip tillage to be initially superior to bed planting and

Fig. 14.22. A strip tillage drill operating in heavy residues on a permanent bed, sowing mungbean.

zero-tillage, but also noted that results improved after operators had learned to plant at the correct soil moisture content, especially with no-tillage. As a result, it is now believed that bed-planting and no-tillage with two-wheeled tractors may be the future of conservation agriculture in that region.

Summary of No-tillage Drill and Planter Design – Small-scale Machines

1. Most small-scale farmers use either hand-operated jabbing devices or drills and planters with one or two rows pulled by animal or small tractor power.
2. Small-scale no-tillage farming benefits from increased operator attention to seeding and weeding details.
3. Many designs of hand or animal planters have evolved from simple ancient designs.
4. Small-scale opener designs have many of the same requirements and designs used on larger-scale farming presented in previous chapters.
5. Some small-scale opener designs are restricted by power, downforce and symmetry requirements.
6. Providing separate fertilizer and seed placement at seeding time is important to enhance early crop availability and reduced weed growth.
7. Seed and fertilizer metering devices most commonly resemble adaptations of those used in larger machines.
8. Hoe openers are more common in small-scale farming due to increased penetration capability compared with disc openers.
9. Residue handling is often easier with small-scale machines as a result of fewer rows and openers.
10. No-tillage in Asia presents special problems associated with rice–wheat rotations and monsoonal rains.
11. Extreme poverty is a further problem in areas of Asia, which limits the sophistication of no-tillage equipment and consulting services to service farmers.
12. Widespread use of simple winged (inverted-T) openers has opened opportunities for no-tillage in Asia.
13. Bed planting and/or strip tillage is seen as an interim step towards full no-tillage in Asia.
14. 'Happy seeders', which combine forage harvesters and seeders, allow residues to be placed over the seed during no-tillage, simulating some of the advantages of larger-scale no-tillage machines.

15 Managing a No-tillage Seeding System

W. (Bill) R. Ritchie and C. John Baker

The overall success of a no-tillage seeding system will be no greater than the least successful component of that system.

Most of this book relates to the physical, biological, chemical and economic risks associated with equipment. But even the best equipment available will not provide optimum results if other input factors are not of equal or similar standard. Consequently, we must seriously consider the other factors required to put together a successful no-tillage seeding system that will fully minimize the risks. We obviously cannot provide a 'recipe' for fail-safe no-tillage seeding in every condition. Each successful package must be tailored to suit an individual farm, field or field component.

This chapter briefly highlights the range of factors that can influence the outcome from no-tillage crop or pasture seeding when undertaking a no-tillage system. A more detailed outline of the way such factors interact and how they determine the success or otherwise of a no-tillage system as a whole is given in *Successful No-tillage in Crop and Pasture Establishment* (Ritchie *et al.*, 2000).

Site Selection and Preparation

There is often little choice as to which field or fields will be no-tilled. In other cases, however, farmers may be in a position to be more selective about fields, especially if they are just beginning to convert from tillage to no-tillage. If this is the case, it is important to review the criteria that should be considered.

Many who convert to no-tillage farming do so on areas with a history of intensive tillage that has resulted in poor soil structure, low SOM, low soil microbial activity, low earthworm numbers and possibly high soil compaction. Such conditions are not conducive to high yields from crops under any crop-establishment system. Although no-tillage would be expected to repair the damage over time, the technique may be disadvantaged in the short term. No-tillage may not be an overnight cure for such conditions, even though it is certainly a long-term cure.

If correctly managed, no-tillage can provide a sustainable method of crop production while at the same time allowing the natural processes of soil formation to continue. These processes take time, perhaps years and decades. Until a certain degree of repair has occurred, yields may even be reduced, especially if the farmer does not apply the best-known inputs into the system. But in other cases, where farmers have used high levels of inputs, including banded fertilizer, there are numerous field examples where crop yields have not suffered, even in the first year; and most have steadily improved thereafter, often to new levels never before experienced in that field.

Best results for converting to no-tillage will come where a farmer has the option to select fields that have high potential returns from the outset. On an integrated pasture (sod) and crop farm, it may be most appropriate to begin a no-tillage crop rotation in a field that has been in pasture or lucerne for some time and contains soil in better condition than fields that have been cropped for many years.

On farms that have been entirely tilled in the past, fields that have been least affected by the destructive aspects of tillage should be chosen. It is unrealistic to expect to objectively assess the potential of a system such as no-tillage unless it has been given a realistic opportunity to show its true potential.

Effective soil drainage will have a major influence on soil condition. While no-tillage will improve the natural drainage capabilities of a soil over time, some artificial drainage may also be required. Well-drained soils or fields will provide the best results.

The importance of no-tillage openers being able to faithfully follow ground surface undulations has been outlined in Chapter 8. But, whatever the merits of any given technology in this respect, it will perform more effectively and will allow higher operating speeds to be used if the field is smooth. When tilling a field prior to converting to no-tillage, extra effort should be put into smoothing the final surface, a good investment for later no-tillage farming.

It is worth noting, however, that, over time, earthworm casting is capable of completely levelling ruts as deep as 75–150 mm (3–6 inches). But, of course, increases in earthworm numbers are a medium-term result of no-tillage rather than a short-term effect.

Seeding with no-tillage drills or planters will also be enhanced if fields are shaped so as to provide relatively straight lands. The firmer nature of untilled soils limits the ability of many no-tillage machines to turn sharp corners. Pre-planning during subdivision can assist in this respect.

Weed Competition

Considerable discussion has centred on weed competition in relation to openers. It is important to remember that most of the operations during conventional tillage are designed to control competition with the crop arising from weeds (unwanted plant species). Consequently, the importance of the spraying operation(s) in no-tillage cannot be overstressed. Good management will include careful identification of the weed species, followed by careful selection of the most appropriate herbicides or other weed control strategy, such as mulching. Adequate planning is important to ensure that any residual herbicides used will be compatible with the immediate and other future crops, as well as desirable soil fauna such as earthworms. Some herbicides and pesticides, for example, are toxic to earthworms.

Having chosen the herbicide(s), additional management input is required to ensure that the specific chemical is applied at the correct rate of the active ingredient, with the correct rate of the carrier (usually water) and any other allied chemical (e.g. surfactant). Appropriate weather conditions during and for a specified period after spraying may be necessary. The particular stage and vigour of growth of the plants or size of leaf material may influence the activity of the herbicide. With some herbicides, there may be a minimum time period between spraying and drilling. In most cases, it is more critical to ensure that the timing of herbicide application is optimized with regard to that particular formulation and the stage of growth of the weeds unless there is residual activity from the herbicide in the soil or danger of the 'green bridge' effect (Chapter 3).

One principle that has repeatedly occurred has been the shift in troublesome weed species with continued years of no-tillage. Each weed species has an optimum pattern of tillage, crop competition and moisture to establish. Almost all long-term no-tillage studies with weed observations have noted this distinct shift of both species and intensity. But the same and other longer-term studies show a significantly reduced total weed incidence with continued no-tillage systems that have used appropriate control and crop rotation strategies.

Pest and Disease Control

Most of the same management principles that apply to the control of weeds also apply to the control of pests and diseases. Accurate identification is essential to ensure appropriate and cost-effective control. Most importantly, it is necessary to recognize that some pests and diseases behave differently under no-tillage compared with tillage. It can often be quite misleading to assume that the control measures appropriate to tilled soils can be applied without modification to untilled soils. These principles apply to both pre- and post-drilling/planting management.

Chemical control measures may also be complemented by other management techniques, such as crop rotation, which is an essential tool in the development of sustainability. Not only is rotation effective to control pests and diseases, but it can also enhance weed control by allowing a wider range of herbicides to be used and/or enhancing the activity of particular herbicide treatments, modifying soil fertility and helping to raise SOM levels. Care must be exercised, because the chemical eradication of one unwanted pest species may be detrimental to other wanted species, especially earthworms.

Managing Soil Fertility

The development of no-tillage drilling and planting technologies that provide separate banding of fertilizer at the time of drilling/planting has opened the door to new opportunities for fertility management under no-tillage. However, all of the old principles apply.

The key to cost-effective fertilizer use is accurate assessment of fertilizer levels and crop requirements. Soil and plant tissue analyses are useful tools in this process, as is accurate interpretation of the results. These results should then provide the basis for the selection of the most cost-effective fertilizer options, some of which might be restricted by machine limitations while others will not.

Considerably more site-specific research may be needed under no-tillage to determine the most appropriate fertilizer regime for any given combination of crop, soil type and climate under no-tillage. Fertilizer responses under no-tillage can differ from those under tillage in the same soil type. So the extension of experiences and research results under tillage may not necessarily be appropriate when applied to no-tillage systems. But plant requirements are generally not changed. No-tillage seeding with banded fertilizers offers an opportunity for increased application efficiency, but the total quantities of nutrients required, with the exception of nitrogen, may not be altered greatly.

Seeding Rates and Seed Quality

There is often considerable discussion about optimum seeding rates for no-tillage. Some have argued that seeding rates should be increased, presumably to counter some expected reduction in seed germination and/or seedling emergence. This practice has become known as using ‘insurance’ seeding rates. But doing so, even with no-tillage openers that have low emergence, can be counterproductive if ideal conditions are experienced that result in plant populations exceeding the optimum. And high seeding rates involve unnecessary extra seed cost.

There are few, if any, reasons for seedling establishment from no-tillage to be any lower than from conventional tillage if appropriate equipment is used. In fact, with advanced equipment and an appropriate system, no-tillage has the potential for higher establishment percentages than tillage.

In any case, it is not how much seed that is sown that is important. Established seedlings are the final measure. Therefore, seeding rates should be based on an assessment of the degree of risk associated with any given situation, leading to a prediction of the likely effective seedling emergence (Ritchie *et al.*, 1994, 2000). The first factor to incorporate is the germination potential of the seed, which is specified on the seed certification data. Seeding rate can then be calculated using the following formula:

$$SR = \frac{TSW \times TPP}{EFE}$$

where: SR = seeding rate (kilograms per hectare); TSW = thousand seed weight (grams); TPP = target plant population (plants per square metre); EFE = effective field emergence (per cent).

The important principle is cost-effectiveness to produce the proper plant density. To be confident of achieving a target plant population, a farmer must use seed of good quality in conjunction with seeding equipment that provides reliable seedling establishment under a wide range of conditions.

Another important factor is accurate calibration of both seed and fertilizer output from the drill or planter. Because different lines of the same seed species can vary quite markedly in their seed weights and sizes according to the vigour of the crop and weather conditions and even the geographical location at the time of harvest of the particular line of seed, it is important to calibrate the metering mechanism when changing seed lines or varieties. A check on calibration should be kept during drilling/planting by matching seed and fertilizer used to the area covered if monitors are not available. Some seeders actually change their metering rates with changing ambient temperatures. The warming of the day from morning to afternoon may bring about an appreciable change in seeding rate with such seeders.

Farmer experience in Western Australia with the disc version of winged no-tillage openers showed that seeding rates for an equivalent canola stand could successfully be reduced from 9 kg/ha under tillage to 4–5 kg/ha with no-tillage using an advanced machine design (J. Stone, 1993, personal communication). The resulting saving in seed cost alone was equivalent to the additional machine cost. Prior to reducing the seeding rate, the experience of this operator from sowing at the higher rate with this no-tillage drill had been an overpopulated crop, which remained largely vegetative with poor crop yield.

Operator Skills

No-tillage is a relatively new technique to tillage farmers. When undertaking conventional tillage, farmers can draw on a long history of tillage experience from most soil types of the world, even if that experience was not personal. However, only a limited experience-base exists with no-tillage. Further, that limited experience-base has already shown that the two techniques are quite distinct and that new skills must be learned.

The 'one-pass' nature of no-tillage leaves little latitude for error. On the other hand, the range of implements and functions involved is much smaller. Therefore, a detailed knowledge of the key machines (sprayers and seeders) can be more easily gained.

Since soil physical conditions are more likely to vary under no-tillage from field to field, or even within a field, there is a much greater need for the operator to understand the principles involved under the conditions and to be able to adjust the machine accordingly. Of course, no-tillage drills and planters vary widely in their respective abilities to ignore soil variations by automatically adjusting to them, but all will require a reasonable level of operator skill to achieve optimum performance.

It is likely that in the future we shall see an increase in the use of electronic monitoring and control of no-tillage drill and planter functions to enhance performance and reduce dependence on operator skills. It is also likely that the operation of no-tillage drills and planters will become a more specialized task, with an increased emphasis on operator training.

Post-seeding Management

A key catchphrase that has been coined for the modern age of intensive agriculture is 'knee-action farming'. The principle conveyed by this term is the importance of monitoring crop performance carefully and regularly at close quarters throughout the growth cycle. In many situations, this monitoring involves kneeling down to inspect the crop, rather than inspecting it from a distance in a standing position, and often with a magnifying glass in hand.

The 'knee-action farming' principle is not exclusive to no-tillage systems but is crucial to achieving consistently good cropping results, and is especially important to no-tillage because so many of the rules of crop husbandry differ from those common under tillage. No-tillage as a technique has suffered in the past from a lack of analysis of the reasons for poor results. Too often, farmers and researchers have been prepared to condemn no-tillage as a system on the basis of a poor result without determining the specific reason for the failure. This often contrasts with an acceptance of failure in a conventional tillage system on the basis of poor weather, an 'act of God' or just plain bad luck.

At times, there seems to have been a lack of realization that tillage crop failures due to severe wind or water erosion are not caused by unfortunate timing but an inherent failure of the tillage system to protect the crop from such a risk in the first place. No-tillage reduces some of those risks, but may introduce other risks of a different nature. For example, pest control becomes more important in some no-tillage situations because there is no physical destruction of their environment by the tillage process. All of this means that a farmer must maintain vigilance over the crop to promptly react to crop management problems that might arise. It is a necessary advantage to have the skills to identify specific problems and how to solve them or know where to go for assistance. Regular, close observation is an important tool for 'knee-action' farming.

Planning – the Ultimate Management Tool

No-tillage is potentially a very flexible system. It provides farmers with the opportunity to respond at short notice to changes in soil or climatic conditions or market indicators. It is also a system, however, that benefits from effective long-term planning and regular reviews of the plan. The success of a crop may well depend on the implementation of a plan from several previous months. For example, crop rotation will influence weed management, pest and disease management, fertility levels and residue levels. Forward planning may well provide key opportunities to take advantage of these changing circumstances and markets.

Residue management for no-tillage systems is a case in point (see Chapter 10). Obviously, decisions at harvest of the previous crop will significantly influence the next phase of the farming rotation, which might occur several months hence. These connecting events apply to chemical use, equipment selection, fertilizer programmes, crop rotation and harvesting patterns, all of which emphasizes the role of forward planning as a management tool.

Another example is the application of lime to raise soil pH, which with no-tillage should take place at least 6 months in advance of drilling because without tillage there is limited opportunity to mix this low-solubility fertilizer with the soil.

Most other general aspects of managing a crop production programme apply, such as rigorous and regular maintenance of drilling, planting and allied equipment and maintaining regular contact with suppliers and contractors to ensure that all components of the programme come together when required. Accurate record keeping is an integral part of any effective management programme.

Table 15.1 outlines the timing of many of the key in-field management decisions that need to be made in New Zealand if a no-tillage programme is to succeed. It is not intended as a recipe, but only to highlight the important issues. Since many of the

Table 15.1. An example of management steps for a no-tillage programme in New Zealand.

When	What to do	Implications
Any time before drilling	Ensure that drainage is OK	No-tillage will not rectify poorly drained soils
Any time before drilling	Determine how much risk you are prepared to take	Risk will be influenced by your choice of: herbicide (effectiveness is a function of conditions – poor conditions need better formulations); slug bait (heavy infestations and wet conditions need better formulations); pesticide (ensure you have identified the target pest and have chosen the correct treatment); drill (difficult conditions and small seeds need better technology); seed (difficult conditions will place more pressure on seed quality)
Any time before drilling	Check for pests that are not specific to no-tillage	Some pests may need treating before or at the time of drilling. Consider using insecticide-treated seed
Sometime before drilling	Subsoil to alleviate compaction if it exists. Best done when soil is dry	Use a subsoiler that does not disrupt the surface sufficiently to require tillage to smooth it out again. Slant-legged or shallow subsoilers are best in this regard
When heavy stock is removed from field	Smooth out hoof marks greater than 75 mm deep	Most drills will smooth out 75 mm deep hoof marks as they drill (some do it better than others). With deeper hoof marks use a 'Ground Hog', shallow subsoiler or leveller to knock only the surface humps off when the soil is somewhat crumbly on top
6 months before drilling	Apply lime if soil pH is low	Lime takes longer to act when there is no cultivation to incorporate it. Do not apply lime close to spraying time. Lime on plant leaves may affect the glyphosate and is slow to dissolve and wash into the soil
3 months before drilling	Take fertility samples	It takes time to get the results, analyse fertilizer options and take action. In long-term no-tillage 75 mm sampling may be more appropriate than 150 mm sampling
3 weeks before drilling	Aim to spray with glyphosate plus chlorpyrifos if springtails, aphid or Argentine stem weevil are a risk	Where farmers do not want to use the higher rates of chlorpyrifos, control of Argentine stem weevil may be obtained by waiting 3 weeks between spraying and drilling. However, you need to be aware that a low rate of chlorpyrifos may still be necessary to control springtails or aphids

Continued

Table 15.1. *Continued*

When	What to do	Implications
At least 2 weeks before drilling	Remove stock from the field (if it is in pasture that has not already been sprayed)	To be most effective, glyphosate should be sprayed on to as much clean, freshly growing leaf as possible. This also produces a heavy mulch, which will help control weeds and retain moisture, so long as the drill can handle the heavy mulch. If necessary, pastures can be grazed after spraying, provided that chlorpyrifos has not been used. Do not graze just before spraying, as leaf area will be reduced. Besides, fresh animal manure will reduce weed control and adversely affect some drill openers. The time needed to 'freshen up' a pasture will vary with growing conditions at the time
10 days before drilling	Check for the presence of slugs	Scatter short lengths of smooth timber about each field and leave for 2 or 3 days. One or two slugs on the underside of a 300 mm length of 150 × 20 timber indicates sufficient numbers to treat for
1 week before drilling	Pre-bait for slugs	This is only necessary for severe infestations. Moderate or low infestations can be effectively treated by applying baits at the time of drilling. With heavy infestations, apply half the bait 1 week before drilling and the other half at (or immediately after) drilling. Some drills can apply slug bait as they drill, either surface broadcast or 'down the spout'
1–10 days before drilling	Spray glyphosate (to control competition), together with chlorpyrifos (to control pests)	Tank-mix chlorpyrifos with the glyphosate where necessary to control pests. The longer the gap between spraying and drilling, the more crumbly the soil will become as roots decompose. But also be aware that soil dries more slowly after spraying because the plants are dead. In the event of rain after spraying, the soil may stay wet for longer. When cutting pasture for silage, wait 3–4 days after spraying before harvesting
1–3 days before drilling	Look at the moisture content of the soil	With most drills no-tillage works best when the soil is a little on the dry side. Being patient and waiting a few extra days often gives a better result
At the time of drilling	Preferably apply all of the crop's fertilizer requirements 'down the spout'. Crops like winter wheat and maize may also need further fertilizer after emergence	Only apply fertilizer 'down the spout' if the drill is sophisticated enough to band it separately from the seed (not mixed with the seed). Crop yield responses to placed fertilizer under no-tillage can be spectacular and there are generous limits to what and how much can be applied. But only a few advanced no-tillage drills do this. Where such drills are not available, avoid putting any fertilizer 'down the spout' at all, or be very careful to select non-burn-type fertilizers. Broadcasting is then the main option, although some people go

Table 15.1.

		to the trouble of drilling fertilizer alone first and then drilling seed at a shallower depth as a second operation
At the time of drilling	Ensure all seed is sown at the target depth and covered	This is sometimes easier said than done unless you have sophisticated no-tillage openers. Where openers are not so sophisticated, a level of risk must be accepted since germination and emergence will then be highly dependent on good weather, smooth fields and low residue levels
At the time of drilling	Apply slug bait	This is most important with spring drilling but may also be important in autumn. Moderate to light infestations of slugs can usually be controlled by applying slug bait either with the drill or as soon as drilling has finished. Get specific information from the experts on the effectiveness of different baits
In the first 3 weeks after drilling	Open slots and check for slug damage	There is often a small window of opportunity to apply slug baits after drilling if you have not already done so and you find slugs feeding in the slots
In the first 3 weeks after drilling	Open slots and check for twisted seedlings	Contrary to popular belief, twisted seedlings do not indicate fertilizer burn. They indicate low-vigour seeds. Do not be reluctant to have a sample of seed tested for vigour (not to be confused with germination) at a seed-testing laboratory. Almost every case of twisted seedlings we have seen has been caused by low-vigour seeds, which you would need to talk to your seed merchant about
In the first 3 weeks after drilling	Check for damage by Argentine stem weevil, springtails or aphids	All should have been controlled by tank-mixing the appropriate amount of chlorpyrifos with the glyphosate. But, if that did not occur, be extra vigilant because these are the main pests of no-tillage and can decimate an entire crop or pasture
In the first 3 weeks after drilling	Check for other pests not controlled with chlorpyrifos	Most normal pests of crop and pasture could also be troublesome under no-tillage. Be at least as vigilant as you would be with a tilled crop
4–6 weeks after drilling pasture	Check new grass plants for resistance to pulling (by hand)	When new grass plants are not easily pulled from the ground, they should be ready to be grazed lightly. Use light stock in large mobs for a short period, rather than set-stocking smaller mobs for long periods
After 6 weeks	Treat crops or pastures normally	That does not mean relax. It means that any problems that do arise will be no worse than under tillage. In fact, new no-tilled pastures, because of the firmness of the soil, can often be treated similarly to already established pastures. Utilization of turnips and swedes will improve because a greater proportion of the bulbs will be above the ground
At harvest	Spread crop residues evenly	Do not burn crop residues except where the drill to be used next will not handle them. Baling is

Continued

Table 15.1. *Continued*

When	What to do	Implications
		acceptable but will slow down the build-up of SOM. With some drills, chopping of residues will be necessary. Others can handle any residue in any form. Still others cannot handle any residues at all. Operators need to know what drill will be used for the next no-tilled crop before making decisions about what to do with the residues from the present crop
After 1–5 years of no-tillage	Examine your soil and bank balance	Both will probably be improved. Soil structure, health, porosity, organic matter and earthworm activity will be noticeably improved. Provided you have managed the system correctly and used the appropriate levels of inputs for your chosen level of risk acceptance, gross margins should increase progressively

Table 15.2. Comparative ownership and operating costs (US$/ha) of no-tillage drills.

Area drilled (ha/year)	Simple low-cost drills	Conventional no-tillage drills	Sophisticated heavy-duty drills
50	69	107	182
100	45	62	95
200	32	39	53
400	26	29	30
600	24	26	23
800	23	23	20
1000	23	21	18

Simple low-cost no-tillage drills = US$15,000.
Conventional no-tillage drills = US$30,000.
Sophisticated advanced no-tillage drills = US$65,000.
Other important assumptions: 24% marginal tax rate; inflation = 4%; interest rate = 11%; depreciation allowance = 12.5%; analysis period = 5 years; tractor costs are additional.

issues listed occur before the seed is sown, forward planning becomes one of the most important issues.

Cost Comparisons

No management analysis of a no-tillage system would be complete without an examination of the cost–benefits of choosing a drill or planter with different complexity, capability and cost. Economic studies (Baker, 1993a, b, c, 1994, 1995) show that, as the annual use of a seed drill increases, a point is reached where there is little difference in the ownership and operating costs between simple low-cost machines and large sophisticated (high technology) expensive machines. Table 15.2 shows a comparison of costs. While the absolute costs and taxation rates shown in Table 15.2 will not be generally applicable and will soon be out of date, the relative values between the various options are likely to be more nearly universal.

At annual use levels of 50–100 hectares, the large sophisticated drills are prohibitively expensive (US$95–182/ha) compared with simple low-cost machines (US$45–69/ha). However, from about 600 hectares per year upwards, the differences are negligible, at US$18–26/ha and may even favour the larger machines. The data in Table 15.2 can be considered conservative as they do not account for increased seedling establishment or yields likely to result from using the more sophisticated machines. The costs do, however, account for higher operating speeds and lower maintenance for the more advanced machines. Saxton and Baker (1990), for example, found an advanced no-tillage drill with winged openers increased wheat yields an average of 13%. Calculations using a higher marginal tax rate than 24% and/or lower interest rates than 11% will result in the larger machines becoming economic at a lower annual usage than 600 hectares per year.

Summary of Managing a No-tillage Seeding System

1. The failure risk of a no-tillage seeding system can be reduced by ensuring a high level of input for all factors, not just the seeding equipment.
2. Choose sites that will offer a high potential return from the no-tillage system.
3. Chemicals generally replace tillage as a means of weed control and must be selected and applied with care.
4. Crop rotation can be an effective management tool when used in conjunction with chemicals to control weeds, pests and diseases.
5. Some no-tillage seeding equipment permits a wide range of options for fertilizer application. Accurate analysis of soil fertility levels and crop requirements will make full use of this benefit.
6. Using excessive quantities of poor-quality seed to compensate for poor drill or planter design or technique can be costly and ineffective.
7. No-tillage requires that new operator skills be learned but also offers the opportunity for greater operator specialization.
8. An otherwise well managed and executed no-tillage seeding programme can fail from poor post-seeding crop observation and follow-up.
9. Good planning of all aspects of the no-tillage programme is a key part of risk management.
10. Advanced no-tillage drills become economic at about 600 hectares use per year.
11. No-tillage is a short cut compared with conventional tillage. Do not short cut the short cut!

16 Controlled-traffic Farming as a Complementary Practice to No-tillage

W.C. Tim Chamen

Removing vehicle-induced compaction from the cropped area liberates crops and soils from unnecessary stress, enhances their performance and sustains production with the minimum of inputs.

What is Controlled-traffic Farming?

Controlled-traffic farming (CTF) divides the crop area and traffic lanes into distinctly and permanently separated zones. All implements have a particular span (or multiple of it) and all wheel tracks are confined to specific traffic lanes. It should not be confused with tramline systems, which just provide guidance for chemical applicators but do not offer permanent separation of wheels and crops. Figure 16.1 shows the system based on existing technology. In the longer term, it is likely that more specialized equipment will be developed that will improve flexibility and further enhance efficiency of the system.

Why Adopt a CTF Regime within a No-tillage Farming System?

The benefits of CTF

Soils not only physically support crops; they are also the medium through which their roots grow and extract water, nutrients and air to sustain their development. Confinement or restriction of roots will almost invariably lead to a negative outcome. Removing vehicle-induced compaction improves and sustains the health of soils. More rainfall is absorbed and available to crop roots, which in turn are better able to explore and extract nutrients. Improved porosity also ensures effective gaseous exchange and drainage, both of which further improve the potential for optimum crop performance.

No-tillage improves many critical soil properties but some soils are still susceptible to wheel and hoof compaction, no matter how long they have been under no-tillage.

Machine performance is also improved by the avoidance of mechanically induced compaction. Variably compacted soils differ greatly in their strength and response to mechanical inputs. For example, this makes it difficult to achieve optimum performance of seed drill openers. Openers may work well in one condition or position on a drill and poorly or less well in others. A more homogeneous soil condition over the field provides greater machine precision. Soil responses are more predictable and vary less from point to point. Avoiding soil compaction diminishes the heterogeneity (variability) of soil properties both within and between soil types, making them easier to

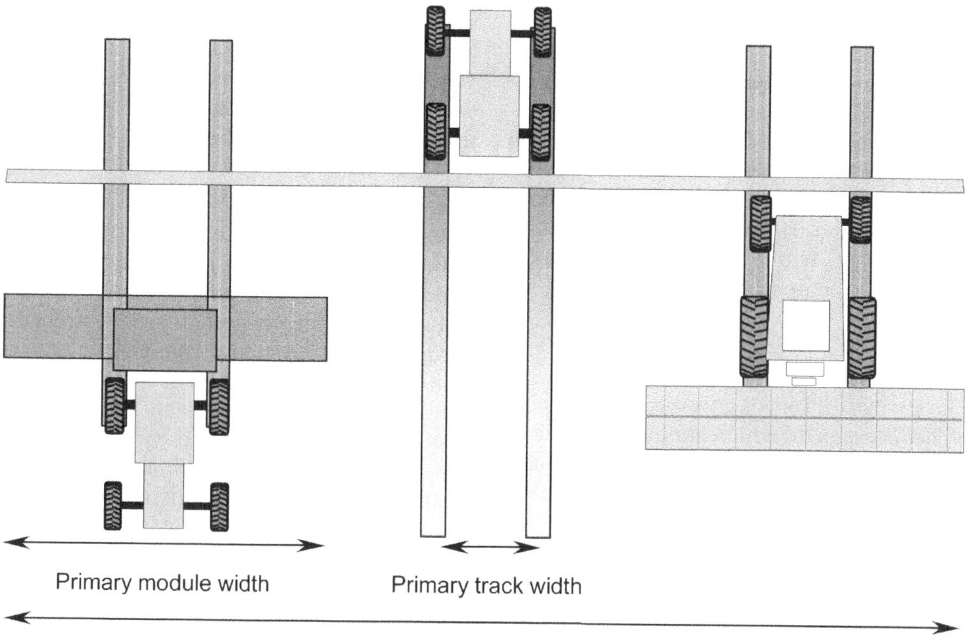

Fig. 16.1. A tractor-based controlled-traffic farming system showing drilling, spraying and harvesting. On this scale, all wheel tracks other than those used for chemical applications might be sown.

manage and more suitable for a wider range of crops under a no-tillage regime.

The effects of CTF on soil conditions

No-tillage farming systems may cause varying amounts of soil disturbance. Initially no-tillage concentrated on avoiding general tillage operations, but recent emphasis has added the importance of minimizing the disturbance created by the no-tillage tools (openers) themselves. Low-disturbance no-tillage is where drill and planter openers aim to disrupt the soil as little as possible – sufficient only to sow the seed and place the fertilizer, but otherwise leaving the soil almost as if it had not been drilled at all. Other forms of no-tillage involve aggressive shank, hoe or tined openers that leave the surface, and often deeper layers, in a disturbed state resembling the effects of minimum or reduced tillage.

Defining low-disturbance no-tillage is difficult. A general rule of thumb is that at least 70% of the original surface residues should remain undisturbed after passage of the drill. But, for openers operating at 750 mm row spacing, 30% disturbance allows 112 mm either side of each row to be disturbed, whereas, at 150 mm spacing, only 22.5 mm either side of the row is acceptable.

In general terms, the greater the compaction applied to the soil, the greater will be the need for repair. No-tillage provides a large measure of remedial action by reducing the traffic intensity, avoiding soil disturbance and allowing the soil to restructure. However, removing the traffic altogether will allow this to happen in greater measure and more quickly. Central to the creation and maintenance of an improved soil structure is the minimization of disturbance, and, as we have seen from the above, the more aggressive the opener, the more disturbance there will be.

Unlike randomly trafficked soils, where the openers may need to create a seedbed as well as sow the seed, non-trafficked soils tend to retain their seedbeds from one season to the next, so that only seed and fertilizer placement is required. From all points of view, the less the disturbance created during seeding within a no-tillage regime, the better, and CTF helps to make this possible. Where comparisons have been made between random trafficking and CTF, the research data often do not include details about the opener designs, and so the optimum no-tillage conditions for the trials may not always have been present, which may or may not have affected the comparisons.

Soil strength

The strength of soils is governed by a number of factors, some of which are interrelated and all of which have an impact on no-tillage. Compacted soils are stronger and have greater resistance to penetration than non-compacted soils, particularly when their water contents diminish (Blackwell *et al.*, 1985; Campbell *et al.*, 1986; Gerik *et al.*, 1987; Chamen *et al.*, 1990, 1992; Dickson and Campbell, 1990; Carter *et al.*, 1991; Unger, 1996; Radford *et al.*, 2000; Yavuzcan, 2000; Abu-Hamdeh, 2003; Radford and Yule, 2003).

In a 10-year experiment, one particular treatment subjected a moist (25–32% water content) Vertisol to a wheel load of 5 t in year 1 and 3 t annually thereafter for 5 years (Radford and Yule, 2003). Tillage to control weeds was used in the first 5 years of an arable rotation. At the end of the initial 5 years, no-tillage and controlled traffic were applied for a further 5 years to these same plots. The greater shear strength persisted in the 0–100 mm profile for over 3 years, while, in a treatment with repeated 5 t wheel loads in all of the first 5 years (compared with the 3 t after year 1), strength effects to 100 mm persisted for nearly 5 years after no-tillage was introduced.

These data suggest that randomly trafficked soils may exhibit high levels of variability in strength as a result of a history of indiscriminate wheeling. Although these differences may tend to diminish with time under a no-tillage regime, the natural amelioration in the top and most important few centimetres will tend to differ according to soil type, opener design and newly applied traffic. Added to this will be a general increase in soil strength arising from repeated wheel passes. On some soils this may not be completely counteracted by structural improvements resulting from lack of disturbance or by a greater concentration of organic matter in the surface layers.

EFFECTS OF SOIL STRENGTH ON NUTRIENTS AND SEEDLING GROWTH. Increased soil strength reduces a crop's ability to extract nutrients and as a result some will be lost from the soil system. With any particular soil, strength variation is dominated by changes in water content, but strength at a specific water content is determined by its state of compactness. Denitrification caused by compaction is a source of nitrogen loss, and restricted rooting may cause poor phosphorus uptake (Wolkowski, 1990, 1991). Potassium uptake is primarily affected by aeration. Below an oxygen concentration of about 10%, uptake is impaired.

Denitrification may lead to fertilizer loss with no-tillage in wet conditions (Torbert and Reeves, 1995). When the soil is dry, uptake of N can be compaction-impaired by limiting root growth. This effect has been the cause of N loss, particularly under no-tillage, following N fertilization and heavy rainfall (Ball *et al.*, 1999). Denitrification and methane production were identified as one of the main constraints to the improved environmental performance of no-tillage compared with reduced tillage (King *et al.*, 2004). King *et al.* attributed this to an increase in the bulk density of the topsoil and to poor aeration.

Soil strength directly above emerging seedlings may also be an issue. Addae *et al.* (1991) suggested the following relationship:

$$Y = 90.4 - 3.58X$$

where:

Y = seedling emergence, percentage
X = soil strength, kPa

The maximum force that a wheat seedling coleoptile can exert is around 30 g and only when resistance is less than 25 g can 100% emergence be expected (Bouaziz *et al.*, 1990).

Compaction of the soil above an emerging seedling therefore reduces emergence, particularly if the soil is wet. Variation in the time to emergence is also often associated with soil strength variations (Brown, 1997).

EFFECTS OF SOIL STRUCTURE ON SOIL STRENGTH. Increased soil strength can be attributed to changes in soil structure. It is a readily observable fact that compacted non-shrinking clay soils exhibit plasticity when moist and cloddiness when dry. They rarely display the friability and flow characteristics of non-compacted granular material. Consequently, randomly trafficked soils not only reveal large variations in penetration resistance, but they also react differently when disturbed. In some areas they will flow and in others they will smear or fracture into variably sized and often large aggregates. This is not easy to deal with when designing an opener to work consistently within a given soil type at a given moisture content. It is even more difficult when soil type changes across a given field. To overcome the problem of variable penetration depth, electromechanical control systems for no-tillage drills have recently been designed to cope with changes in soil strength and go a long way towards overcoming the problem (see Chapter 13).

One of the outcomes of tillage to remedy compaction, in an attempt to create a uniform but artificially structured seedbed, is interruption of natural soil structural-forming processes. This is despite the fact that the very mechanical processes being employed will themselves immediately render that soil more susceptible to the negative effects of random wheeling and other compacting influences. Therefore, although tillage temporarily makes the operation of seed drills relatively simple, it commits soil to a downward negative spiral of compaction and structural degradation and has never been a long-term answer.

Cockcroft and Olsson (2000) suggested that no-tillage and zero traffic could not avoid the problem of hard setting on some soils. Although biopores help the infiltration of water and more organic matter improves the situation, drainage and root growth can still be impaired. A sustainable solution has yet to be found for these soils.

EFFECTS OF SOIL STRENGTH ON DRAUGHT FORCES AND IMPLEMENT WEAR. Although no-tillage aims to minimize soil disturbance, the force required to displace soil during sowing is still directly proportional to its strength. Chamen *et al.* (1990) reported a 25% reduction in energy requirement for no-tillage in non-trafficked compared with trafficked soil, despite a slightly greater depth of operation (56 mm in the non-trafficked compared with 50 mm in the trafficked soil). This is similar to reported reductions in energy for tine tillage in trafficked and non-trafficked soil (Lamers *et al.*, 1986).

A further consequence of lower soil strength is proportional reductions in wear on soil-engaging components. Lower wear saves on replacements and also saves on labour and downtime to fit new components.

While in tilled soils and some untilled soils it is often found that openers working behind wheels require replacement more frequently than elsewhere, in other situations the reverse may be true. When operating no-tillage drills in long-term pasture with good load-bearing ability in New Zealand, often the surface disturbance resulting from wheel slip by the tractor tyres loosens rather than compacts the soil and wear of openers in those wheel marks is reduced [Eds].

Soil structure

Avoiding vehicle-induced soil compaction can have a major impact on the structure-related aspects of water and gas movement in and out of the soil. Much research has concentrated on these characteristics. McQueen and Shepherd (2002) concluded that some soils brought into cropping from permanent pasture could suffer from soil deformation caused by traffic. Compaction, even on no-tillage soils, reduced water infiltration (Ankeny *et al.*, 1990; Meek *et al.*, 1990; Li *et al.*, 2001), soil porosity, saturated hydraulic conductivity (Wagger and Denton, 1989), air-filled porosity and permeability (Blackwell *et al.*, 1985; Campbell *et al.*, 1986).

On the other hand, minimal-disturbance no-tillage openers operating in silty soils in New Zealand have been shown to leave most indices of soil health (including soil structure) in a similar state to the original permanent pasture. Even after 20 years of continuous double cropping with no-tillage and random and repeated trafficking, there was no obvious effect on such soils compared with their pasture equivalents (Anon., 2000; Ross *et al.*, 2000, 2002a, b; Ross, 2001, 2002) [Eds].

Both air capacity and available water are primarily affected by bulk density, organic carbon and clay content, the latter being relatively more important in subsoils. Variability in air capacity and available water is highly dependent on bulk density and soil texture. In a clay loam, available water has been halved with an increase in bulk density from 1.4 g/cm^3 to 1.75 g/cm^3 (Hall *et al.*, 1977).

Reduced infiltration due to traffic compaction can increase runoff and erosion. Wang *et al.* (2003) measured a twofold increase in runoff on trafficked compared with non-trafficked no-tillage plots and an approximate threefold increase in soil loss.

Environmental improvements associated with non-compacted soils also relate to gaseous losses to the atmosphere. Reduced air-filled porosity due to compaction leads to denitrification in clay soils. Similarly, no-tillage and controlled traffic appear to preserve CH_4 oxidation rates (Ball *et al.*, 1999).

There is also evidence of improved water availability to crops on some non-trafficked, albeit shallow-tilled (100 mm) clay soils. Changes in matric potential at 150 mm depth over a 48 h period showed large fluctuations on a trafficked soil compared with relatively small changes on non-trafficked soil. The latter reinforces the importance of promoting natural soil structure through both no-tillage and controlled traffic (Chamen and Longstaff, 1995).

Campbell *et al.* (1986) working on a sandy clay loam found that, in the absence of traffic, the soil could be reclassified from being unsuitable to being entirely suitable for no-tillage.

The implications of CTF for no-tillage operations

RESIDUES AND RESIDUE HANDLING. Residues are a critical issue in no-tillage systems because they are not incorporated into the soil before the next crop is drilled. Indeed, many of the benefits of no-tillage accrue from this fact. It is preferable to leave the residues *in situ* on the soil surface to decay slowly and for both the residues themselves and their decayed products to be gradually incorporated into the soil by fauna such as earthworms. This is also advantageous in terms of nitrogen, which is often temporarily locked up by rapid organic matter decomposition. Residue management prior to and during drilling is therefore particularly important if the crop is to be sown without interference or subsequent adverse effects on germination and seedling growth.

The additional precision afforded by controlled traffic (see next section) should allow crop residues to be manipulated and placed more precisely, if required. For example, the tendency to use wider equipment is already initiating the design of more accurate residue placement methods by harvesters. Working from permanent wheel ways created as part of carefully prescribed routes and where future sowing lines are predetermined, residues could specifically be placed to avoid the new crop row.

With random traffic systems, crop stubbles and residues are flattened in an arbitrary way, resulting in their variable orientation to drill openers. Some openers do not perform reliably in these conditions; while others not only perform reliably but they utilize random residues to control the seed micro-environment. Controlled traffic avoids random stubble trampling and its associated variability. It is possible, for example, to develop the system where small grains have been stripped from straw that remains standing following the harvester pass. Both manual and assisted-guidance methods could then allow sowing between the standing straw rows and into soil that may only have a covering of the chaff and light fraction (Fig. 16.2).

There will be additional effects on residues from increased earthworm activity in

Fig. 16.2. Soil and residue conditions following grain stripping. Global positioning systems (GPS) and other precision guidance methods allow sowing to take place between the rows in a CTF regime.

non-trafficked soils. Radford *et al.* (2001) recorded an increase in earthworm numbers from 2 to 41/m^2 when all compaction on a moist vertisol was avoided. Pangnakorn *et al.* (2003) found a favourable differential of 26% in numbers of earthworms in no-till compared with cultivated soils and an additional 14% increase when traffic was removed.

Compaction restricts oxygen supply, nutrient intake and physical movement. Although the effect of additional earthworm activity is unlikely to have a direct effect on the sowing operation in terms of residues, the reverse is often true. Residues encourage earthworms and they in turn may improve seedling emergence, particularly in wet soils, primarily as a result of improved porosity. (Chaudry and Baker, 1988; Giles 1994).

Considering that there are increasing levels of CO_2 in the atmosphere and this scenario is likely to continue, crop and weed residues and crop yields are likely to increase (Prior *et al.*, 2003). Improved management of residues will therefore be of increasing importance, not only to deal with the quantity, but also to avoid a temporary lock-up of nutrients and longer-term excessive acidity in the surface layers. This issue remains to be dealt with adequately.

WEED CONTROL. Traditional cultivation systems use a combination of cultural, chemical and tillage methods to achieve weed control. Weeds are always a threat to the sustainability of cropping, and they continuously evolve to overcome any particular means of control. The most recent example of this is the resistance of *Lolium rigidum* (annual ryegrass) to glyphosate (Wakelin *et al.*, 2004). Therefore, it can be argued that reducing the number of options for weed management is risky; but there are positive aspects too, some of which are aided by CTF, the most important of which is minimizing soil disturbance.

There are several approaches that improve weed control without tillage. One of the few defendable objectives of tillage is to stimulate weed seed germination so that the offending seedlings can be killed by a subsequent tillage operation. In the absence of such stimulation, the most widely practised weed control measure is to blanket spray with either selective or non-selective herbicides. CTF will make this more efficient because a greater proportion of weed seeds are likely to germinate during the inter-crop period. Seeds lying on a friable soil surface are more likely to germinate through intimate soil contact or by burial, either through their own activities (e.g. wild oats, *Avena fatua*) or external forces such as rainfall, frost, wind or the activities of soil fauna. After spraying, the aim is to avoid further weed seed germination, and crucial to the success of this is the minimization of soil disturbance by the no-tillage openers.

This approach has been effective in New Zealand. Troublesome weeds such as wild turnip had forced many farmers to stop growing forage brassicas by conventional tillage because of the difficulty in controlling volunteer wild turnip plants, the seeds of which may remain dormant in undisturbed soil for up to 40 years. Even vigorous no-tillage openers often disturbed sufficient soil within the rows to create rows of the weeds where none had existed before drilling. But use of the disc version of winged no-tillage openers or double disc openers, either of which minimizes surface disturbance, avoids the problem [Eds].

After drilling, it may be possible to utilize the close precision of CTF to target inter-row weeds that will either germinate as a function of their own activity (as described above) or be prompted to do so by shallow inter-row tillage with a light implement. Inter-row flaming, steaming, mowing and non-selective herbicides can then be applied where there is sufficient room between the rows. Vision guidance methods for doing this are now fast and reliable.

The efficiency of spray booms is likely to be improved by CTF systems. Most CTF systems use extended track widths and it is anticipated that future developments will provide additional boom support even further from the boom centre. Improved stability reduces roll and allows booms to be positioned closer to the crop or ground without fear of contact. The auto-guidance systems generally associated with CTF also reduce boom yaw, a feature associated with manual overcorrection of steering. Reduction of roll and yaw improve the application accuracy while diminishing the risk of drift.

OPENER DESIGN AND PERFORMANCE. The main implication of CTF for no-tillage opener designs involves the general reduction in soil strength in the absence of vehicle-induced compaction. This reduces the penetration and draught forces required between wheeled and non-wheeled areas. Chamen *et al.* (1990) found that a triple disc opener pressed into non-trafficked no-tillage soil by rubber buffers penetrated too deeply. A solution was to use a traditional single disc opener designed for cultivated soils. Thus it may be seen that no-tillage seeding on non-trafficked soils can be carried out with significantly lighter and less robust machines.

Non-trafficked soils tend to present a more friable seedbed regardless of the soil moisture regime. This can have negative as well as positive effects. The positive effects are obvious and important, but hairpinning with discs may be a greater problem with CTF because there is less soil resistance to the vertical cutting of residues. Setting the discs deeper is unattractive because draught forces and soil disturbance are greatly increased. Other options include managing the residues to avoid their presence in the sowing line (Fig. 16.2) and using openers that do not create hairpinning or deliberately separate the seed from contact with hairpinned residues. The disc version of a winged opener places seeds to one side of any hairpins that the central disc may create, and eliminates this problem. The more friable nature of the soil under CTF will have largely a neutral effect on hairpinning with this opener [Eds].

Wide spacing of narrow tines works well in dry conditions but becomes unacceptable in moist soils because of the large wedges of residue left as the tines eventually clear themselves (see Chapter 10). Punch planters show promise if hairpinning can be avoided, but their potential has been limited by the high strength of trafficked soils. The greatest problems will be with moist clays, when fine soil and residues cling to every part of the opener. Experience of these conditions within a CTF regime is still limited and further use and customized opener development are needed.

On a more general note, the more friable seedbed structure associated with CTF should ensure that the firming devices of seed openers operate more effectively. As suggested by Baker and Mai (1982b) and Addae *et al.* (1991), firming should be around or under the seed, not above it. With CTF, a more homogeneous soil condition is likely to be presented to the opener and there will therefore be less need for compromises in depth settings between individual openers and less variation in seed covering. There will also be less wear, lower overall draught and reduced power and traction demands.

Figure 16.3 shows how two disc-type openers on the same machine can perform very differently, depending on whether

Fig. 16.3. Performance of two closely adjacent openers on the same drill working in a moist clay soil. The opener was behind the tractor wheels in the left photo and the seed is clearly visible on the surface, while on the right it was operating correctly in less compacted soil.

they are behind wheels or in between. In the absence of differential rutting from wheels, the soil surface will also be smoother. This reduces the potential for differences in opener performance, particularly where they are mounted in gangs. Openers mounted individually on parallel linkages will be less prone to depth variation where ruts are present, but a more level surface will still have a positive influence on their performance.

Consistent sowing depth is vital to avoid too shallow planting in dry conditions or too deep in others, and Kirby (1993) noted that the time to emergence was extended as sowing depth increased. Heege (1993) found that, in the range of cereal seeding depths from 25 to 45 mm, field emergence dropped from 82% when the depth varied by around 6 mm, to 50% when the variation increased to 20 mm. Heege and Kirby both found that rate of emergence affected subsequent growth, as did Benjamin (1990). They all suggested that differences in date of emergence were perpetuated and even exacerbated in subsequent growth. Although these differences may not be large enough to create differences in yield, they do make it more difficult to estimate crop growth stage for chemical applications. Additionally, this means that a larger proportion of the crop will be treated at the wrong growth stage and, as a result, suffer a greater setback.

In summary, fewer differences in soil strength and a more level surface will both help to make sowing depth more consistent. This minimizes crop emergence time and makes subsequent management easier and more effective.

The implications of CTF for soils and crops

AGRONOMY. Provided that severely compacted soils are loosened before introducing CTF, it seems certain that the problem of poor initial crop growth and loss of nitrogen through denitrification will be reduced, particularly in the early years of no-tillage. Improved initial growth will be promoted by the lack of a compacted surface layer and encourages crop root growth, which explores and extracts nutrients from a greater proportion of the profile.

Australian farmers have found that row cropping is a natural extension to controlled traffic. This is possible because the position of each crop row can be planned in advance and achieved in practice with precision guidance techniques.

Seed rates have often been increased slightly with no-tillage, although rates of several crops have been actually reduced with advanced no-tillage openers (Baker *et al.*, 2001). Regardless, controlled traffic makes seeding more reliable and works in favour of lowering seed rates because the surface is more level and there is less compaction variation across the drill width. Without compaction, many soils form a stable fine crumb at the surface, which readily accepts seeds with minimal disturbance. This makes drill setting easier, reduces irregularities in performance and avoids the need for increased 'insurance' seed rates.

A no-tillage farmer in the UK (Hollbrook, 1995) found that spring barley sown 3–4 mm deep was noticeably healthier than the crop sown at 40–50 mm. Shallow sowing resulted in the first node emerging from the coleoptile when it was 20–30 mm above ground rather than at the surface. This precluded the incidence of eyespot (*Cercosporella*) and subsequent weakness of straws, which often resulted in crop lodging.

Slugs (*Deroceras reticulatum*) have frequently been a problem with cropping systems that retain surface residues, and particularly those with cloddy seedbeds and smeared and open sowing slots (Moens, 1989). Slugs attack crops in two ways – below ground, where they eat the seeds, and above ground, where they eat the young leaves. Openers that produce small clods mean that slugs can access seeds more readily, while open or smeared sowing lines allow them to move unhindered from one seed to the next. CTF has the potential to address these problems through the avoidance of 'cloddiness' and smeared sowing lines.

CROP YIELDS. Most research comparing trafficked and non-trafficked soils has been with systems using cultivation, but work on no-till in Scotland found that, even with fairly modest wheel loads, no-tillage yields were reduced. This occurred in the early years of no-tillage, but differences were absent by the fourth season, despite no actual reduction in bulk density on the trafficked soil (Campbell *et al.*, 1986). In the USA and in Argentina, soybean yields in no-tillage systems were reduced by between 10% and 39% with repeated but often quite modest wheel loads. Even where no-tillage had been practised for 7 years it was still possible to reduce yields as a result of newly imposed wheel loads (Flowers and Lal, 1998; Botta *et al.*, 2004).

RANGE OF CROPS. Although we have concentrated primarily on small-grain cropping, the introduction of CTF should make it possible to grow a wider range of crops with no-tillage. No-tillage establishment of cotton, for example, was successful even in the presence of wheel compaction. Lint yields for no-tillage were only reduced in one year out of three, while those for transplanted tomatoes, albeit with strip tillage, were comparable at two sites in 2002. Strip tillage for melons resulted in marginally lower yields than the traditional method, but, with both tomatoes and melons, 'cloddy' soil conditions at planting/sowing were partly responsible for the poorer crop performance. A vegetable producer in Australia growing tomatoes, zucchini, melons, onions and broccoli predicted that CTF would allow him to establish these crops with no-tillage. Potatoes have also been grown successfully with deep mulches and no-tillage (Lamarca, 1998; Mitchell *et al.*, 2004a, b; Ziebarth, 2003, personal communication).

The possible constraints on cropping within a no-tillage CTF regime arise from a number of sources:

- Soil structure/crop interactions.
- Inexperience and perception.
- Machinery.

Because completely non-trafficked soil has until recently been largely unknown within farming systems, it is difficult to predict how some crops will react to these no-tillage conditions. Equally, there are very few data that might be used to determine whether crops such as carrots, sugar beet and potatoes will perform adequately in non-trafficked, non-tilled soil.

The only way that this might be determined is through the comparison of a number of soil parameters, such as bulk density, penetration resistance and porosity. For example, does the bulk density of a given non-trafficked non-tilled soil exceed that of its cultivated counterpart for a particular crop? In addition, within what soil environment will a root crop perform equally to that of the cultivated norm? Many of these questions remain unanswered. We shall also have to be aware that considerable soil disturbance is often experienced during the harvest of root crops. Although this would at least partly interrupt the no-tillage cycle, it would still be advantageous for the remainder of the rotation and for establishment of the root crop. Controlled traffic would also minimize the repair needed after harvesting and ensure a quick and effective return to no-tillage.

The crops that we can probably grow now under a CTF regime with a proven agronomy based on no-tillage include:

• Wheat	• Barley
• Oats	• Rye
• Sorghum	• Millet
• Oilseed rape	• Maize
• Soybean	• Dry peas
• Field beans	• Linseed
• Dryland rice	• Cotton

This range is necessarily more limited than mentioned previously, and further technological developments and in-field experience are needed before more crops can be considered. However, given the characteristics of these crops and the typical climatic conditions under which they have been successful, it would be quite rational to extrapolate to other crops and climates in locations where CTF no-tillage farming has not been extensively attempted.

Making CTF Happen

Basic principles

There are several principles involved in CTF:

1. Forward planning.
2. Matching of vehicle track widths.
3. Matching of single (primary) or multiples of implement widths.
4. Discipline.

These principles will be outlined in the following sections, but far greater detail can be found in *Tramline Farming Systems*, published by the Department of Agriculture, Western Australia (Webb *et al.*, 2000), in conjunction with the Grains Research and Development Corporation.

Forward planning and machinery matching

Planning is probably the most important aspect of conversion to CTF, because it ensures, amongst other things, that the cost is kept to a minimum. Some farms may be able to convert within 12 months; others may require planning and change extending over several years. In the context of this book, it is assumed that the end point of transition is a no-tillage crop establishment system, but the starting point could be mouldboard ploughing, secondary cultivation and drilling. There must therefore be an initial commitment to a significantly lower input system. In some ways, changing from an extensive machinery system makes the economics easier because the excess machinery can be sold and appropriately sized new or second-hand equipment purchased, probably at little additional cost. It will also entail a reduction in labour. The economics, however, will be dominated by the change to no-tillage rather than to CTF. If a minimum- or no-tillage system is already being used, the transition may have to be planned more carefully and over a longer timescale because fewer costs will be lost from the system, but returns will still be improved.

The width-matching process

The objective is to match all implement working widths, on the one hand, and machine track widths, on the other. The purpose is to minimize costs and number of wheel tracks per unit area. The cost factor means that most transitions will start with examination of existing equipment to consider its adaptability. As an example, take a small farm growing grain crops with a minimum-tillage system that has a 3.5 m wide cultivator, a 5 m wide roll and a 3 m drill; the cereal harvester is 6.1 m wide and chemicals are applied with 12 m booms. Tractors are on track widths varying from 1.5 to 1.8 m and trailers have a track of around 1.8 m; the harvester is on 2.8 m. In effect, nothing matches up with anything in terms of controlled traffic (Fig. 16.4, left). The tractor wheel track settings can, however, be changed to 1.8 m (to match the trailers) relatively easily.

Two challenges remain – the track width of the harvester and the choice and width of no-tillage drill. If the 6.1 m harvester is to be retained, the drill should be 6 m wide (to ensure that the harvester gathers the entire crop on most occasions) and the cost of this will need to be budgeted, with allowance made for the second-hand value of the existing drill, cultivator and the rolls (the economics of CTF will be studied more closely in a later section). It may also be possible to sell one tractor, but one of the remaining tractors must be capable of pulling the proposed replacement drill or a new (larger) tractor will have to be purchased.

The harvester track width cannot easily be changed and these wheels will be the one set that extend outside the primary track width. Their position, however, is known and they will not necessarily cause damage every season because soils are often drier at this time (and therefore able to withstand more weight) than at sowing. If compaction and surface rutting occur, they can be repaired with a subsoiler having tines positioned so that they loosen just the additional width imposed by the harvester. A 6 m system as described will create wheel ways that cover around 16% of the area, depending on the width of the tyres used. Providing the wheel ways are well maintained, it may be possible in the longer term to fit narrower tyres.

On a larger farm, an alternative might be to use a 'twin-track' CTF system. This largely eliminates the harvester problem, while maintaining wheel track settings more or less standard. Figure 16.5 shows that the system works by straddling the harvester across adjacent passes of the primary

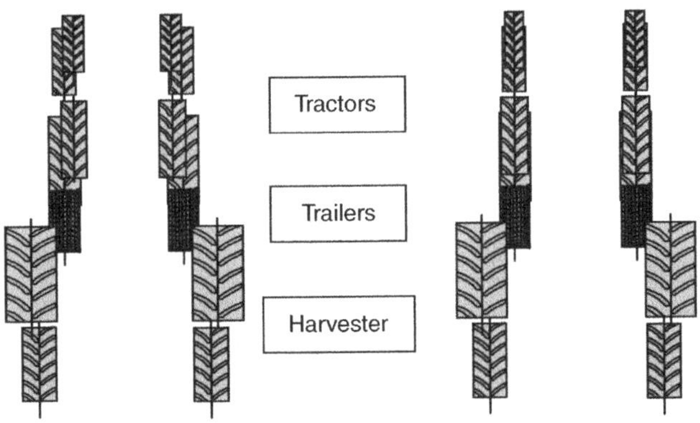

Fig. 16.4. Placing all the equipment in the example around a common centre line (left) shows that it is only the harvester that has a significantly different track width. Available settings on the tractors will allow them to be aligned with the trailers, as indicated on the right, with only the cost of time.

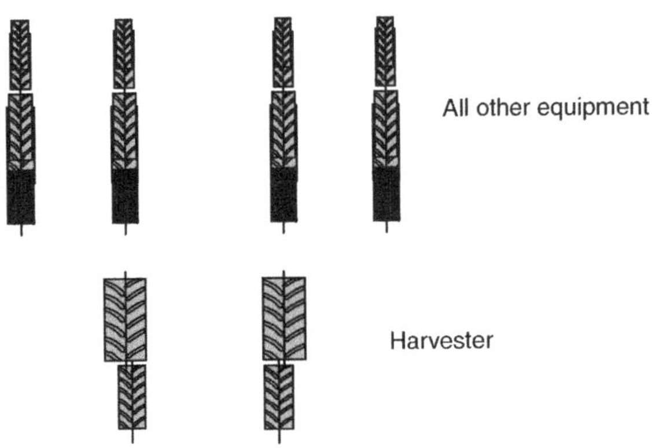

Fig. 16.5. Twin-track CTF system, where the harvester straddles single tracks of adjacent pairs of tractor tracks. Primary implement width is determined by the addition of the tractor and harvester track widths.

tracks. The primary implement width is determined from the simple addition of the common track width of the tractors, trailers and chemical application equipment, plus the harvester track width. In the example above, primary implement width would be: 1.8 + 2.8 = 4.6 m. The harvester cutting width can be any multiple of this; in this instance, the most practical would be 4.6 or 9.2 m. The drilling width, however, can only be odd multiples of the primary implement width and this probably limits it to a single multiple. Chemical applications can be any multiple of the primary implement width if the primary tracks are used, e.g. 4.6 m, 9.2 m, etc. If the chemical application equipment is on a wide axle and runs on the harvester tracks (to improve the stability of the applicator), the width of the chemical application equipment can only be even multiples of the primary implement width.

Presently none of the implement widths quoted above is standard, so some adjustment to the primary track width might be needed even in a twin-track system. For example, if the primary track were narrowed to 1.7 m, this would correspond with available harvester widths (9 m) and chemical application equipment (18 m, 27 m, 36 m). Alternatively, the track settings could be 2 m and 3 m, giving a primary implement width of 5 m. The harvester cutting width should be slightly wider than the calculated width to ensure capture of the entire crop in all circumstances.

A further method of matching is to align all field machinery on the same track width as the harvester because, as previously mentioned, this machine is difficult to alter. Unfortunately, the harvester is probably the machine with the widest track, and with current designs this will mean a primary track width of around 3 m for all vehicles and implements. This is common practice in Australia (Fig. 16.6), where there may be less need to drive on highways and where rural areas have relatively low population densities. In Europe and other parts of the world with high population densities and often-narrow roads, much greater difficulties are likely. However, because no-tillage reduces the number of field operations and future spray vehicles may have ‘on the move’ variable track widths, the extent of the problem should diminish considerably. It may only be the harvester and sowing machine and associated tractor that have the 3 m track setting on the road. The advantage of this system is that there are few constraints in terms of primary implement widths. With very wide machines, some means of extending the harvester’s unloading auger may be needed to ensure that the transport unit can be reached in the adjacent traffic lane.

A further alternative similar to ‘twin-track’ for smaller farms is for the harvester

Fig. 16.6. Example of an Australian system with a 9 m primary module and a 3 m primary track width (Webb *et al.*, 2000).

to span between the same wheels of adjacent tractor passes, as shown in Fig. 16.7. The basis for this is:

Primary implement width	=	harvester track width
Primary track width	=	harvester track width/2
Harvester cutting width	=	harvester track width × 1.5
Chemical application width	=	any multiple of primary implement width

This system potentially introduces a large number of wheel tracks, but some of these may only be used once a year for crop sowing and most could be sown, as described later.

So far, we have dealt primarily with the systems used for small-grain crops, but the principles of no-tillage can be applied equally to most other crops. Although little research has been done on no-tillage for vegetable crops, improved potential may exist within CTF systems, as discussed later.

Field layout and system management

Orientation and layout of controlled traffic wheel ways are all part of the planning process, and each individual area or block of land needs to be considered independently. Detailed field maps are an essential part of this planning, by measurements, historical records or aerial photographs. Topographic data are also valuable, particularly on farms with significant slopes. Changes in soil type across a property are likely to be of lesser significance than with random traffic systems, but it will still be useful to know these boundaries, particularly with respect to drainage. With regard to drainage, it is essential that any installed drainage systems are operating properly or problems corrected before installing a CTF system. This is also true for soil structural remediation. If inspection reveals a pan layer, fissuring of the profile should be attempted according to the guidelines suggested by Spoor *et al.* (2003).

The principal aspects to consider in any CTF layout are:

- Orientation of permanent wheel ways in relation to:
  - length of run;
  - slope and water movement;

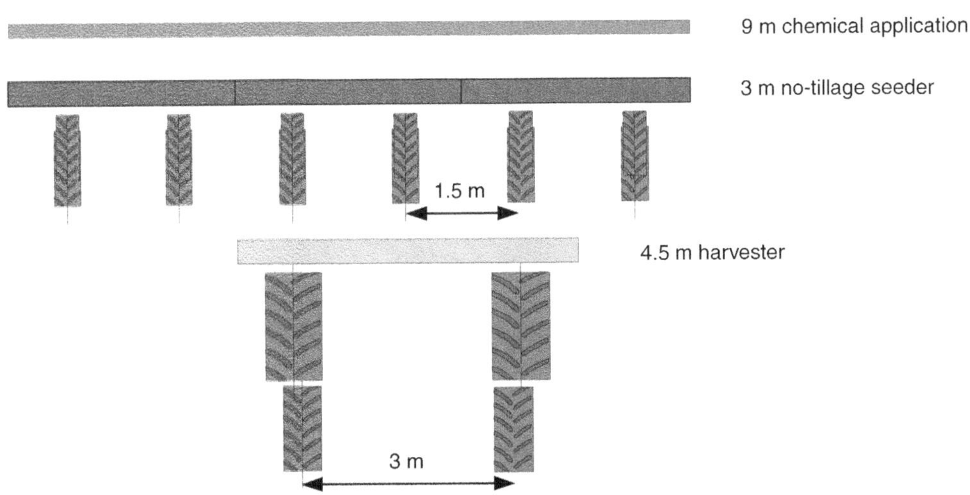

Fig. 16.7. A controlled traffic machinery system for small farms. The 1.5 m primary track width is spaced at 1.5 m intervals and thus any pair of wheel tracks can be used by all equipment other than the harvester.

  - field shape and short rows;
  - extraneous objects (trees, pylons, ponds, etc.);
  - field drainage system.
- Wheel-way management and field access.

Orientation of permanent wheel ways

In most situations the longest length of the area being considered is chosen for the orientation because this improves field efficiency by reducing the number of end turns. The length of run that this creates must also be considered in respect of any significant field slope. Although water infiltration on the soil 'beds' is likely to be improved significantly compared with traditionally managed fields, water will still tend to run along and erode the wheel ways, particularly if they run uninterrupted over long distances and are orientated up and down slopes. In Australia, where CTF is widely practised and where rainfall events can be very heavy, orientation of operations has become more flexible with CTF. Both up and down and across the slope can work, whereas with random traffic across-slope or contour layouts predominate.

CTF orientation must also consider the presence of any drainage system, and particularly one that involves mole channels. The latter will run predominantly up and down slopes and the aim with a controlled traffic system is to run parallel to them. The danger with repeated wheeling across the mole channels is that they may collapse prematurely. Running parallel to the moles will mean crossing the drains themselves, but it is unlikely that these will be damaged, partly due to their depth but also because they are often backfilled with gravel. If the wheel ways run parallel to the mole channels, although there is a danger that some will be coincident with and may damage them, the overall effect on the drainage system within a field is likely to be minimal. Running parallel will also ensure that the mole channels can be redrawn without complete disruption of the wheel ways. For more information on drainage systems, see Spoor (1994).

A similar approach is adopted with a row of pylons going across a field; in this case, they may be used for the orientation and as a line to set up the first wheel tracks. Unlucky indeed would be the farmer who has both a drainage system and pylons with completely different orientations! The compromise would have to be with the pylons. Experience with either drainage systems or field 'infrastructure' is limited, because CTF has yet to be adopted in areas where these situations occur extensively.

Wheel-way management

The potential for wheel-way erosion can be countered in a number of ways. As a first principle, the wheel ways need active management from the outset; they cannot be allowed to sink or rut differentially. They should be filled as required by drawing in soil from the surrounding area, particularly in the early days of establishment, and particularly if the soil has been deep-loosened recently. Within a tillage regime, these recommendations could be met coincidentally during the creation of a false seedbed for weeds. However, in the context of no-tillage, a customized narrow unit (Fig. 16.8) might be used if rutting or plastic flow of soil out of the wheel ways has occurred. This implement should not be used too frequently, however, as the edges of the beds may become rounded and cause uneven sowing depth.

If weed or erosion pressures on bare soil become unacceptable or, due to machinery constraints, the wheel tracks take up a large proportion of the area, crop may be established within them (in general, this applies only to those tracks that will not be used after crop sowing). The roots of plants established in these tracks will often explore laterally and find their way into the main crop bed. As a result and although they perform less well, they do mature in unison with and add significantly to the main crop yield. This is not the case for sown wheel ways that are used subsequently for crop management. In these the plants are often

Fig. 16.8. Rolling maintenance tool used to deal with plastic flow of soil out of the wheel ways. This would not normally be used on more than an annual basis (J. Grant, 2001, personal communication).

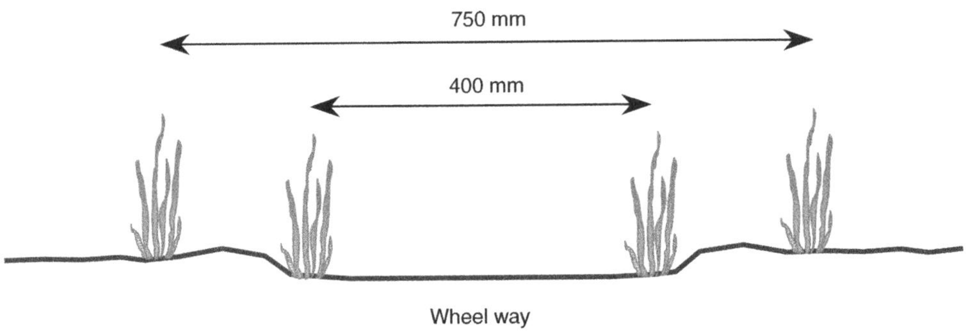

Fig. 16.9. Example of how a cereal crop might be sown on a wheel way. The nominal 250 mm spacing is modified to 400/175 mm to encourage roots of the plants in the wheel way to access the adjacent bed.

dwarfed by repeated wheeling and are late to mature. Where wheel ways are sown within a narrowly spaced crop (300 mm or less), the row spacing may be altered slightly, as illustrated in Fig. 16.9. The openers will need to be set very specifically to deal with this situation and the wear rate on them is likely to be higher. To date there is limited experience with this technique and growers will need to use some field experimentation initially, but this technique has the added advantage of temporarily marking the wheel ways.

In some instances, further active management of the wheel ways might be needed on slopes to ensure that water gathered within them does not reach erosive potential. This could be achieved by introducing diagonal channels at regular intervals, which divert water into the beds alongside.

The second principle of wheel-way management is to avoid water standing in or flowing along them. To a large extent, the first of these problems can be avoided by attending to active management, but low spots in the field or areas of poor natural drainage can also create this situation. Orientation should aim to avoid low spots, but this will not always be possible and an alternative in the form of modifying the wheel-way edges, as described above, may need to be introduced.

Wheel-way erosion may also be reduced by a buffer strip part-way downslope.

This might also provide an area for beneficial insects and, if sited correctly, address 'short row' issues.

Guidance systems

Fundamental to any controlled-traffic farming system is a means of ensuring that the wheel ways are not only orientated but also positioned correctly at the outset. Traditionally, positioning has been achieved with machine-mounted hardware that provides an adjacent parallel marking line offset by the required distance. The driver then uses this line on the next pass to position the machinery correctly. This works well with modest machinery widths but, when these approach 10 m or more, the size, strength and durability of the equipment become a significant factor. Offset loads can also be a problem if the marker engages with the soil, and maintenance costs can also be high. It is an even greater problem under no-tillage because the marker has to make a visible line in untilled and often residue-covered soil, and this is difficult. Markers have relatively low precision and introduce errors that are cumulative pass to pass. An alternative, but still with cumulative errors and poor precision, is to place a closed-circuit video camera on the extremity of the implement, with pictures relayed to a screen in the driver's cabin. This requires that the driver continuously monitor the screen to keep the machinery on course, as he or she would with a marked line.

An increasingly available and attractive alternative is electronic systems based on a differential global positioning system (DGPS) using satellite signals. There is a wide range of available costs, depending upon the degree of accuracy delivered. With CTF systems, an accuracy of ± 3 cm is desirable, with a peak error of ± 5 cm if wide-row crop operations are planned. Such systems can also be coupled directly to the vehicle's steering to provide auto-steer capability for both straight-line and curved parallel tracking. Automatic steering allows drivers to concentrate on the implement operation, relieves them of the constant stress of driving to a mark and also avoids excessive steering corrections, which can adversely affect machinery operation. A further advantage, particularly with wide equipment, is that any pass can be driven in any sequence, because the positioning is absolute, since it does not rely on a mark from the previous pass. Drivers can skip every other pass, for example. This makes turning at the end easier and has the added advantage that field completion can be at the start point, which is normally the point of field access.

It is also important to note that the implement lateral offset feature found with a number of satellite guidance systems cannot be used with CTF. This feature compensates for an implement that does not trail centrally behind the tractor by shifting the tractor appropriately on adjacent passes. If this were used with CTF, it would move the vehicle off the permanent wheel ways. Any misalignment in a CTF system must therefore be dealt with physically on the tractor or implement and this can create a significant challenge on side slopes. Trailed equipment may need some wheel steer to overcome this problem.

Economics

There are a number of ways in which the economics of CTF systems can be assessed and all will give different answers. Every property, circumstance and range of machinery will be unique and the economics of change will be very specific. The aim in this chapter, therefore, will be to establish the principles and the cost/revenue centres rather than entering into detailed cost analyses that provide only a single hypothetical solution. This approach also concentrates on the transition from no-tillage seeding in the presence of random traffic to a similar but controlled traffic system.

Economics centre on:

- Planning and transition costs and their timescale.
- Fixed and variable costs of the CTF system employed.

- Change in output.
- Management costs.

Transition costs and timescale for change to CTF

Planning is the key to minimizing costs. And yet the cost of planning itself is difficult to quantify. A typical consultancy fee for CTF conversion in Australia is around US$75 per hour. There will, however, be many growers who will study the subject carefully and put a plan together themselves, the costs of which will be absorbed within normal overheads. But serious consideration should be given to employing the services of experts to determine the most efficient field layouts. Changing a layout after installation is not an attractive proposition and is very wasteful of time and resources, as well as resulting in a loss in productivity.

The planning process will involve taking stock of existing farm equipment and how much it can be applied within the new regime. A clear picture of the new CTF cropping and machinery regime needs to be clearly identified at this stage before transition costs can be estimated. The transition costs fall into three main categories: (i) those associated with changing the implements or machines; (ii) those associated with changing wheel-track settings; and (iii) those associated with guidance:

1. Changing machinery might include buying new, as well as discarding old equipment. If a change to no-tillage is being made at the same time as adoption of CTF, the equipment requiring attention will be greater, but an opportunity exists to integrate the full range rather than just parts. With CTF, the no-tillage drill will experience lower penetration and draught forces and as a result there will be lower power demands on the tractor, so some longer-term savings may be possible. Centralization of the harvester cutting platform may also be necessary because many are offset to assist unloading. The other main aspect to consider is the matching of implement widths, on the one hand, and wheel-track settings, on the other.

2. The cost of changing wheel tracks may be in the range US$750 to US$4000 (Webb *et al.*, 2000) and reflects considerable diversity in machine designs, axle configurations and wheel equipment. This cost will also vary considerably depending upon the type of system adopted, for example single- or twin-track, as described earlier. For single-track systems, the cost is likely to be greater because all equipment will probably have to be matched to the wider track setting of the harvester. Such conversions are now available for some tractors, with total costs for front and rear axles being in the region of US$10,000. Most other equipment can be modified locally or in the farm workshop. For twin-track systems, the costs may be confined to the labour required to alter the position of rims on centres or swapping wheels from side to side, for example.

3. Costs for guidance systems can be as little as the time required to make up marker arms from existing farm equipment to around US$50,000 for a satellite system delivering auto-steer with a mean offset error of around ± 3 cm. The market and therefore the cost structure for these satellite-based systems is changing rapidly and to such an extent that the full cost of the system may not necessarily be attributable only to conversion to CTF. Many farmers are now purchasing these systems within conventional practice as a means of improving the accuracy of their operations, as well as establishing tramlines for chemical applications.

Not only does the latter give greater flexibility, but it also precludes the need to establish marks within the crop. Traditionally these have been installed by special equipment on the drill that leaves lines unsown at the required intervals.

To introduce CTF, an existing system might be upgraded from perhaps ± 25 cm manual to ± 3 cm auto-steer. The additional cost of this would be in the region of US$17,000.

The timescale for change will depend on the investment that has been calculated;

the greater the investment, the shorter should be the timescale. This is because the greatest benefits will only be realized when a complete CTF system is in place. These benefits are dealt with in the section on outputs.

Fixed and variable costs

Fixed costs are generally considered to be regular labour, machinery, rent and general overheads, while chemicals, seeds, fuel, wearing parts, contractors and casual labour are considered to be variable (Nix, 2001). With CTF, we would expect the main impact to be a lowering of fixed costs, and particularly those related to labour and machinery. The marginal labour benefit from CTF will be less than but additional to the marginal labour benefit from changing to no-tillage in the first place.

Although it would be easy to attribute CTF with improvements in field efficiency due to better guidance, this can now be achieved equally within conventional practice using 'tramline' systems on drills or through satellite guidance, and is therefore not considered as a CTF benefit. The main impact of CTF on labour in a no-tillage system will be a reduced demand during drilling, which could be slightly faster (conditions permitting) as a result of lower draught forces on the drill. Unless a contractor is employed for this task, the farmer is only likely to experience a timeliness benefit in the short term. In the longer term it may be possible to increase the land farmed with a given labour force or lose some labour costs if several are employed during drilling.

Changes in variable costs centre on seeds, fuel, wearing parts and chemicals, all of which should be reduced. Typically, power demands for drilling at any particular speed are reduced by up to 25%, including the lower rolling resistance that can be attributed to working on the permanent wheel ways rather than on the crop bed. Due to the improved soil conditions, lower seed rates may be possible with less risk, although this issue should also be handled by improving the drilling methodology rather than relying on CTF alone to make up for deficiencies in drilling equipment or technique. Savings on wearing parts are more difficult to predict but will increase the longer the soil is under no-tillage.

Chemical costs are likely to be reduced principally through greater precision and the ability to inter-row band-spray with non-selective chemicals while simultaneously applying selectively to the crop row. Although such a system is not exclusive to CTF, the well-maintained wheel ways offer greater potential. If one considers that the cost of protection chemicals for wheat grown in a temperate region approaches 50% of the total cost of seed, fertilizer and spray (Nix, 2001), then any saving on these chemicals is likely to have significant cost implications. Equally, a reduction in chemical inputs, or at least input of less environmentally damaging chemicals, is an added benefit. We may also presume that fertilizers applied in a CTF regime will be more efficiently utilized and, although this may not be a cost saving to the farm, it will result in an improved yield (discussed below) and a lower risk of off-farm pollution of watercourses.

Change in output

Reviewing research undertaken over the past 30–40 years on soil compaction with 17 different crops showed that yields under CTF in both tilled and non-tilled conditions had increased in the range 9–16% compared with random traffic. The less extensive data quoted for no-tillage systems suggest a more modest level of improvement; a safe figure may be around 10%. Soil type, cultural practices, crop rotations and the percentage area of land taken up by permanent wheel tracks will obviously moderate these percentages, and the crop row spacing will further influence the effect. To determine what happens in practice, each individual case needs to be considered and the following suggests an approach that might be taken.

Taking the 8 m system considered earlier and a close-spaced row crop such as wheat (250 mm in this case), the following is assumed to apply:

Primary implement width = 8 m
Primary track width = 3 m
Chemicals applied at 24 m width
Two out of every three primary tracks are sown

Assuming that crop yield is improved by 10% only on the non-trafficked area and that the harvester will have wheels around 750 mm wide, the number of rows affected by wheels will be 3 × 2 × 4 = 24 rows out of a total of 96. There will be no improvement in yield on these rows and therefore the net improvement will be 7.5%. This is actually a conservative estimate because conventional systems usually have tramlines where at least two rows will be missing within a 24 m width.

In-field management costs

The main ongoing management cost to sustain field operations is likely to be that associated with the permanent wheel ways. As indicated earlier, a customized small implement (Fig. 16.8) may suffice for this task, but within a no-tillage regime this represents an additional pass, usually carried out after harvest. Experience suggests that this may be needed in the early years of conversion and when any operation has to be carried out in wet conditions. In some instances, this may only be needed on the chemical application wheel ways.

Summary of costs and returns

Table 16.1 provides an overview of the aspects considered in the foregoing text and attempts to quantify a number of the variables. As stressed earlier and confirmed by Uri (2000), the variables are so numerous that any fully calculated example involving conservation or no-tillage systems will only provide a specific solution unique to a particular situation. It is better, therefore, to have the tools and a procedure to calculate rather than to give a single answer.

The magnitude of these costs can be put into context by examining some of the benefits. A world price for wheat of US$100/t and an average yield of 4 t/ha increased by 7.5% on 500 ha, equates to an additional income of US$15,000 per annum. At 2001 prices, a 20% reduction in tractor size from 134 kW would give a saving of around US$17,000. The net benefit from these two items on 500 ha is US$32,000 at the end of the first year.

Detailed analyses on a regional basis are offered by a number of authors and the reader is referred to these specific studies for further information. Gaffney and Wilson (2003), for example, suggest a net benefit of US$15–25/ha for a change to CTF within a no-tillage regime on a vertisol in Queensland, while Mason *et al.* (1995) for the same scenario in the South Burnett of Australia suggest a net improvement of US$75/ha.

Summary of Controlled-traffic Farming as a Complementary Practice to No-tillage

1. Controlled-traffic farming (CTF) is a crop production system in which the crop zone and traffic lanes are distinctly and permanently separated. In practice, it requires:

a. use of the same wheel tracks for all field operations;
b. all machines to have the same wheel-track setting;
c. all implements to have a particular span or multiple of it.

2. CTF relies on good guidance systems to install and keep the permanent wheel ways in the same place from year to year. The main systems used to do this are:

a. physical markers, which provide a means of positioning the next pass, which, if integrated with seeding, may be used to introduce guide rows for later use;
b. closed-circuit television (CCTV) video cameras with an associated display in the driver's cabin;

Table 16.1. Factors and variables that impact on the economics of changing from a random traffic to a controlled traffic no-tillage seeding system, their likely magnitude and level following transition.

Factor/variable	Costs, US$	Savings/benefits, %
Consultancy for CTF field layout	75/h	
Drill price (from Uri, 2000)	6,400	11
DGPS guidance with ± 25 cm pass-to-pass accuracy	2,400[a]	
DGPS guidance upgrade from ± 25 cm to ± 3 cm accuracy[c]	15,400[b]	
DGPS guidance to ± 3 cm with automatic steering[c]	5,400–10,200	
Axle conversions to 3 m:		
Tractors – per tractor with full warranty	750–4,000	
Drill, chasers or trailers, per item	5,000–7,000	
Self-propelled chemical applicators with full warranty		17–25[d]
(Not needed if tractor mounted. Also, many North American special-purpose vehicles are now available with 3 m axles)		5[d]
Lower-power tractor for hauling drill		20
Labour		15
		20
Variable costs:		10
Seed		
Fuel	3/ha	7.5
Wearing parts – soil-engaging elements		
Chemicals		
Wheel-way maintenance		
Crop yield		

[a]Additional cost to the ± 25 cm system, i.e. total cost would be 6400 + 2400 = US$8800.
[b]Additional cost to the ± 3 cm system, i.e. total cost would be 8800 + 15,400 = US$24,200.
[c]This option has an annual US$1330 correction signal fee.
[d]Tractor power or labour reduction, not both – see 'Fixed and variable costs' in main text.

c. differential global positioning systems using satellites;

d. automatic steering controlled by the guidance system.

3. CTF should liberate the full potential of no-tillage seeding by avoiding soil compaction damage in the cropping zone. This is likely to result in:

a. improved crop yields from the outset;

b. better nutrient use efficiency achieved through greater root proliferation;

c. improved soil porosity, which provides better water infiltration, drainage and gaseous exchange;

d. reduced threat of denitrification, particularly in the presence of organic residues;

e. lower draught forces and wear on seed openers;

f. reduced labour and fuel inputs, particularly during seeding operations;

g. lower power demand for drilling, allowing a smaller tractor to be used for a given output;

h. more reliable and consistent operation of seed openers in a wider range of conditions and soils;

i. the potential for a wider range of crops to be grown with no-tillage.

4. In other situations, many of these advantages will come from the change to no-tillage, which reduces, but seldom eliminates, the additional gains to be had from CTF. In most cases, combining CTF and no-tillage achieves a greater potential from no-tillage [Eds].

5. CTF allows farmers to anticipate greater levels of precision in all operations so that they may:

a. increase the flexibility and effectiveness of weed control;

b. spray the crop row and inter-row independently;

c. use non-selective chemicals in the inter-row;

d. perhaps position and manage residues to allow their manipulation to greater benefit.

6. The cost of converting to CTF need not be great, providing it is carefully designed and part of the forward-planning process. If properly planned, the benefits are likely to far outweigh the costs.

7. There are a number of ways that CTF can be achieved and all will vary in terms of cost. Field layout is a particularly important aspect because it needs to account for field drainage, slope, operating efficiency and permanent obstacles.

8. Permanent wheel tracks within a CTF regime need to be managed to ensure optimum performance. Management is likely to include:

a. regular infilling, preferably as an integral part of normal field operations;

b. engineering their drainage down slopes and in low areas;

c. sowing with crop in particular circumstances and in a particular way.

9. Additional environmental benefits can be achieved by no-tillage in combination with CTF.

17 Reduced Environmental Emissions and Carbon Sequestration

Don C. Reicosky and Keith E. Saxton

While tillage agriculture contributes significant greenhouse gases detrimental to the atmosphere, no-tillage agriculture will reduce them by both storing new SOM and reducing the oxidation of existing SOM.

Introduction

Agriculture affects the condition of the environment in many ways, including impacts on global warming through the production of 'greenhouse gases', such as CO_2 (Robertson *et al.*, 2000). In 2004, the US Environmental Protection Agency (EPA) estimated that agriculture contributed approximately 7% of the US greenhouse gas emissions (in carbon equivalents, CE), primarily as methane (CH_4) and nitrous oxide (N_2O). While agriculture represents a small but relevant source of greenhouse gas emissions, it has the potential, with new practices, to also act as a sink by storing and sequestering CO_2 from the atmosphere in the form of soil carbon (Lal, 1999). Estimates of the potential for agricultural conservation practices to enhance soil carbon storage range from 154 to 368 million metric tons (MMTCE), which compare to the 345 MMTCE of reduction proposed for the USA under the Kyoto Protocol (Lal *et al.*, 1998). Thus, agricultural systems can be manipulated for the dual benefits of reducing greenhouse gas emissions and enhancing carbon sequestration. The influence of agricultural production systems on greenhouse gas generation and emission is of interest as it may affect potential global climate change. Agricultural ecosystems can play a significant role in production and consumption of greenhouse gases, specifically, CO_2.

Conservation tillage reduces the extent, frequency and magnitude of mechanical disturbance caused by the mouldboard plough, reduces the air-filled macropores and slows the rate of carbon oxidation. Any effort to decrease tillage intensity and maximize residue return should result in carbon sequestration for enhanced environmental quality.

Tillage-induced Carbon Dioxide Emissions

Tillage or soil preparation has been an integral part of traditional agricultural production. Tillage is also a principal agent resulting in soil perturbation and subsequent modification of the soil structure with soil degradation. Intensive tillage can adversely affect soil structure and cause excessive breakdown of aggregates, leading to potential soil movement via erosion. Intensive tillage causes soil degradation through carbon loss and tillage-induced greenhouse gas emissions, mainly CO_2, which have an impact on productive capacity and environmental quality.

Intensive tillage decreases soil carbon. The large gaseous losses of soil carbon following mouldboard ploughing compared with relatively small losses with no-tillage have shown why crop production systems using mouldboard ploughing have resulted in decreased SOM and why no-tillage or direct-seeding crop production systems are stopping or reversing that trend (Reicosky and Lindstrom, 1993). Reversing the trend of decreased soil carbon with less tillage intensity will be beneficial to agriculture as well as the global population through better control of the global carbon balance (Reicosky, 1998).

Emission measurements

The tillage studies reported in this chapter were conducted in west central Minnesota, USA, on rich soils high in soil organic carbon (Reicosky and Lindstrom, 1993, 1995; Reicosky, 1997, 1998). The CO_2 flux from the tilled surfaces in these studies was measured using a large, portable chamber, described by Reicosky (1990) and Reicosky *et al.* (1990), in the same manner as described by Reicosky and Lindstrom (1993) and Reicosky (1997, 1998). Measurements of CO_2 flux were generally initiated within 1 minute after the tillage pass and continued for various times. The CO_2 flux from the soil surface was measured using the large, portable chamber described by Reicosky and Lindstrom (1993).

Briefly, the chamber, with mixing fans running, was placed over the tilled surface or the no-tilled surface, the chamber lowered and data collected for 1 s intervals for a total of 60 s to determine the rate of CO_2 and water vapour increases inside the chamber. The chamber was then raised, calculations completed and the results stored on computer floppy disk.

The data included the time, plot identification, solar radiation, photosynthetically active radiation, air temperature, wet bulb temperature, output of the infrared gas analyser measuring CO_2 and water vapour concentrations in the same airstream. After the appropriate lag and mixing times, data for a 30 s calculation window were selected to convert the volume concentrations of water vapour and CO_2 to a mass basis and then regressed as a function of time using linear and quadratic equations to estimate the gas fluxes. These fluxes represent the rate of CO_2 and water vapour increase within the chamber from a unit horizontal land area as differentiated from a soil surface basis caused by differences in soil roughness. Only treatment differences in respect of tillage methods, tillage type or experimental objectives are described, with the results.

Tillage and residue effects

Recent studies, involving the dynamic chamber described above, various tillage methods and associated incorporation of residues in the field, indicated major carbon losses immediately following intensive tillage (Reicosky and Lindstrom, 1993, 1995). The mouldboard plough had the roughest soil surface, the highest initial CO_2 flux and maintained the highest flux throughout the 19-day study. High initial CO_2 fluxes were more closely related to the depth of soil disturbance that resulted in a rougher surface and larger voids than to residue incorporation. Lower CO_2 fluxes were caused by tillage associated with low soil disturbance and small voids, with no-tillage having the least amount of CO_2 loss during 19 days.

The large gaseous losses of soil carbon following mouldboard ploughing (MP) compared with relatively small losses with no-tillage (NT) or direct seeding have shown why crop production systems using mouldboard ploughing have decreased SOM and why no-tillage or direct-seeding crop production systems are stopping or reversing that trend. The short-term cumulative CO_2 loss was related to the soil volume disturbed by the tillage tools. Lower CO_2 fluxes were caused by tillage associated with low soil disturbance and small voids, with no-tillage having the least amount of CO_2 loss during 19 days. Similarly, Ellert and Janzen (1999) used a single pass with a heavy-duty cultivator that was relatively shallow and a small dynamic chamber to show that fluxes from 0.6 hours after tillage were two- to fourfold above the pre-tillage values and rapidly declined within 24 hours of cultivation. They concluded that short-term influences on tillage and soil carbon loss were small under semi-arid conditions, in agreement with Franzluebbers *et al.* (1995a, b).

On the other hand, Reicosky and Lindstrom (1993) concluded that intensive tillage methods, especially mouldboard ploughing to 0.25 m deep, affected this initial soil flux differently and suggested that improved soil management techniques can minimize the agricultural impact on global CO_2 increase. Reicosky (2001b) further demonstrated the effects of secondary tillage methods and post-tillage compaction in decreasing the tillage-induced flux. Apparently, severe soil compaction decreased porosity and limited the CO_2 flux after plough tillage to that of the no-tillage treatment.

This concept was further explored when Reicosky (1998) determined the impact of strip tillage methods on CO_2 loss after five different strip tillage tools were used in row-crop production and no-tillage. The highest CO_2 fluxes were from mouldboard plough and subsoil shank tillage. Fluxes from both slowly declined as the soil dried. The least CO_2 flux was measured from the no-tillage treatment. The other forms of strip tillage were intermediate, with only a small amount of CO_2 detected immediately after the tillage operation. These results suggested that the CO_2 fluxes appeared to be directly and linearly related to the volume of soil disturbed. Intensive tillage fractured a larger depth and volume of soil and increased aggregate surface area available for gas exchange, which contributed to the vertical gas flux. Narrower and shallower soil disturbance caused less CO_2 loss, suggesting that the volume of soil disturbed must be minimized to reduce carbon loss and the impact on soil and air quality. The results also suggest that the environmental benefits and carbon storage of strip tillage compared with broad-area tillage need to be considered in soil management decisions.

Reicosky (1997) reported that average short-term CO_2 losses 5 hours after the use of four conservation tillage tools were only 31% of that of the mouldboard plough. The mouldboard plough lost 13.8 times as much CO_2 as the soil area not tilled, while different conservation tillage tools lost an average of only 4.3 times. The benefits of residues on the soil surface to minimize erosion and smaller CO_2 loss following conservation tillage tools are significant and suggest progress in developing conservation tillage tools that can enhance soil carbon management. Conservation tillage reduces the extent, frequency and magnitude of mechanical disturbance caused by the mouldboard plough and reduces the large air-filled soil pores to slow the rate of gas exchange and carbon oxidation.

Reicosky *et al.* (2002) have shown that removal of maize stover as silage for 30 years of continuous maize, compared with returning the residue and removing only the grain, resulted in no difference in the soil carbon content after 30 years of mouldboard ploughing. Fertility level had no observable effect on CO_2 losses. The tillage-induced CO_2 flux data represented the cumulative gas exchange for 24 h for all treatments.

The pre-tillage CO_2 flux from the same area not tilled averaged 0.29 g $CO_2/m^2/h$ for the high-fertility plots at the start of measurements. This contrasts with the largest cumulative flux after tillage of 45 g $CO_2/m^2/h$ on a low-fertility grain plot. The CO_2 flux showed a relatively large initial flux immediately after tillage and then rapidly decreased 4 to 5 hours after tillage. The CO_2 flux decrease continued as the soil lost CO_2 and dried out to 24 hours, when values were lower but still substantially higher than those from the no-tillage treatment. The flux 24 h after tillage on the same plots above was approximately 3 g $CO_2/m^2/h$, considerably higher than the pre-tillage value.

The temporal trend was similar for all treatments, suggesting that the physical release controlled the flux rather than the imposed experimental treatments. The consistency of the C : N ratio across all four treatments suggests little effect of residue removal or addition and that mouldboard ploughing masked the effects of residue removal as silage or grain removal and above-ground stover returned. Intensive tillage with the mouldboard plough overshadowed any residue management aspects and resulted in essentially the same lower carbon content at the end of 30 years. The results suggest that intensive tillage with a mouldboard plough may overshadow any beneficial effect of residue management (return or removal) that might be considered in a cropping system.

Strip tillage and no-tillage effects on CO_2 loss

The impact of broad-area tillage on soil carbon and CO_2 loss suggests possible improvements with mulch between the rows and less intensive strip tillage to prepare a narrow seedbed, as well as no-tillage. Reicosky (1998) quantified short-term tillage-induced CO_2 loss after the use of strip tillage tools and no-tillage. Various strip tillage tools, spaced at 76 cm, were used and gas exchange measured with a large portable chamber. Gas exchange was measured regularly for 6 hours and then at 24 and 48 hours. No-tillage had the lowest CO_2 flux during the study and mouldboard ploughing had the highest immediately after tillage, which declined as the soil dried. Other forms of strip tillage had an initial flush related to tillage intensity, which was intermediate between these extremes, with both the 5 and 24 hour cumulative losses related to the soil volume disturbed by the tillage tool.

Reducing the volume of soil disturbed by tillage should enhance soil and air quality by increasing soil carbon content. These results suggest that soil and environmental benefits of strip tillage should be considered in soil management decisions. Limited tillage can be beneficial and do much to improve soil and air quality, minimize runoff to enhance water quality and minimize the greenhouse effect. The energy savings represent an additional economic benefit associated with less disturbance of the soil. The results suggest environmental benefits of strip tillage over broad-area tillage, which need to be considered when making soil management decisions.

The CO_2 flux as a function of time for each tillage method for the first 5 hours showed that mouldboard ploughing had the highest flux, which was as large as 35 g $CO_2/m^2/h$ and then rapidly declined to 6 g $CO_2/m^2/h$ 5 hours after tillage. The second largest CO_2 flux was 16 g $CO_2/m^2/h$ following subsoil shanks, which also slowly declined. The least CO_2 flux was measured from the no-tillage treatment, with an average flux of 0.2 g $CO_2/m^2/h$ for the 5 hour period. Other forms of strip tillage were intermediate and only a small amount of CO_2 was detected immediately after some tillage operations, which ranged from 3 to 8 g $CO_2/m^2/h$ and gradually declined to approach no-tillage values within 5 hours. These results suggest a direct relationship between the magnitude of the CO_2 flux that appears to be related to the volume of soil disturbed.

The cumulative CO_2 losses calculated by integrating the flux as a function of time for both 5 and 24 h periods showed similar trends. The values for 24 hours may be subject to error due to the long time between the last two measurements and tillage-induced drying, which may have caused the tilled treatments to dry out faster than the no-tillage treatments. The cumulative flux for the first 5 hours after tillage for mouldboard ploughing was 59.8 g CO_2/m^2, decreasing to 31.7 g CO_2/m^2 for the subsoil shank to a low of 1.4 g CO_2/m^2 for the no-tillage treatment. The strip tillage methods had slightly more CO_2 loss than no-tillage. Similarly, the cumulative data for the 24 h period reflect the same trend, the maximum release by mouldboard ploughing,

159.7 g CO_2/m^2, decreasing to 7.2 g CO_2/m^2 for no-tillage. The other forms of strip tillage were intermediate between these, which paralleled the 5 hour data. The results suggest that cumulative CO_2 loss was directly related to the soil volume disturbed by the tillage tool. The narrower and shallower soil disturbance caused less CO_2 loss.

The cross-sectional areas of the soil disturbed by the tillage were estimated from field measurements drawn to scale, using graphical techniques. The drawings were then cut out and run through an area meter. The cumulative CO_2 fluxes for 24 hours were then plotted as a function of these soil areas disturbed and showed a nearly linear relationship between the 24 hour cumulative CO_2 flux and the soil volume disturbed by tillage. These results suggest that intensive tillage fractured a larger depth and volume of soil and increased aggregate surface area available for gas exchange. This increased soil porosity and area for gas exchange contributed to the vertical flux, which was largest following mouldboard ploughing.

The results of short-term CO_2 loss from the strip tillage study for row crops suggest that, to minimize the impact of tillage on soil and air quality, the volume of soil disturbed must be minimized. Tilling the soil volume necessary to get an effective seedbed and leaving the remainder of the soil protected and undisturbed to conserve water and carbon to minimize soil erosion and CO_2 loss should be the preferred strategy. Limited tillage can be beneficial and do much to improve soil and air quality, minimize runoff to enhance water quality and minimize the greenhouse effect. The energy savings represent an additional economic benefit associated with less disturbance of the soil (West and Marland, 2002; Lal, 2004). The results suggest that the environmental benefits of strip tillage over broad-area tillage need to be considered when making soil and residue management decisions.

The concept that each soil has a finite carbon storage capacity is being revisited. This has important implications for soil productivity and the potential of using soil to enhance soil carbon storage and reduce greenhouse gases in the atmosphere. Most agricultural and degraded soils can provide significant potential sinks for atmospheric CO_2. However, soil carbon accumulation does not continue to increase with time with increasing carbon inputs but reaches an upper limit or carbon saturation level, which governs the ultimate limit of the soil carbon sink (Goh, 2004). The relation between no-tillage and conservation tillage in the way they affect soil carbon stocks is open to further debate and definition of carbon pools.

The relationship between tillage-induced changes in soil structure and subsequent effect on carbon loss was reviewed by Six *et al.* (2002) within the framework of a newly proposed soil C-saturation concept. They differentiated SOM that is protected against decomposition by various mechanisms from that which is not protected and discussed implications of changes in land management for processes that affected carbon release. This new model defined a soil C-saturation capacity, or a maximum soil carbon storage potential, determined by the physicochemical properties of the soil, and was differentiated from models that suggested soil carbon stocks increased linearly with carbon inputs. Presumably, this carbon saturation capacity will be soil-, climate- and management-specific. This causes a change in the thinking about carbon sequestration and that a soil-dependent natural limit may exist in both natural and managed systems.

Superimposed on this analysis is the role of glomalin, a sticky substance produced by fungal hyphae that helps glue soil aggregates together (Nichols and Wright, 2004). No-tillage is one management practice that has been successful in increasing the hyphal fungi that produce glomalin. The next researchable challenge will be to determine if the carbon saturation and glomalin over the entire profile in no-tillage and conservation tillage systems are substantially different. Presumably with less tillage-induced breakdown of soil aggregates, no-tillage may have an advantage over other forms of conservation tillage. The final answer awaits further research.

Carbon Sequestration Using No-tillage

Conservation agriculture is receiving much global focus as an alternative to the use of conventional tillage systems and as a means to sequester soil organic carbon (SOC) (Follett, 2001; Garcia-Torres *et al.*, 2001). Conservation agriculture can work under many situations and is cost-effective from a labour standpoint. More importantly, the practices that sequester soil organic carbon contribute to environmental quality and the development of a sustainable agricultural system. Tillage or other practices that destroy SOM or cause loss and result in a net decrease in soil organic carbon do not result in a sustainable agriculture. Sustainable agricultural systems involve those cultural practices that increase productivity while enhancing carbon sequestration. Crop residue management, conservation tillage (especially no-tillage), efficient management of nutrients, precision farming, efficient management of water and restoration of degraded soils all contribute to a sustainable agriculture.

Kern and Johnson (1993) calculated that conversion of 76% of the cropland planted in the USA to conservation tillage could sequester as much as 286 to 468 MMTCE over 30 years and concluded that US agriculture could become a net sink for carbon. Lal (1997) provided a global estimate for carbon sequestration from conversion of conventional to conservation tillage that was as high as 4900 MMTCE by 2020. Combining economics of fuel cost reductions and environmental benefits derived by converting to conservation tillage are positive first steps for agriculture towards decreasing carbon emissions into the atmosphere.

Soil tillage practices are of particular significance for the carbon status of soils because they affect carbon dynamics directly and indirectly. Tillage practices that invert or considerably disturb the surface soil reduce soil organic carbon by increasing decomposition and mineralization of biomass due to increased aeration and mixing plant residues into the soil, exposing previously protected soil organic carbon in soil aggregates to soil fauna, and by increasing losses due to soil erosion (Lal, 1984, 1989; Dick *et al.*, 1986a, b; Blevens and Frye, 1993; Tisdall, 1996). Conversely, long-term no-tillage or reduced tillage systems increase soil organic carbon content of the soil surface layer as a result of various interacting factors, such as increased residue return, less mixing and soil disturbance, higher soil moisture content, reduced surface soil temperature, proliferation of root growth and biological activity and decreased risks of soil erosion (Lal, 1989; Havlin *et al.*, 1990; Logan *et al.*, 1991; Blevens and Frye, 1993; Lal *et al.*, 1994a, b).

Cambardella and Elliott (1992) observed for a loam soil that the soil organic carbon content in the 0 to 20 cm depth was 3.1, 3.5, 3.7 and 4.2 kg/m^2 for bare fallow, stubble mulch, no-tillage and native sod, respectively. They observed that tillage practices can lead to losses of 40% or more of the total soil organic carbon during a period of 60 years. Edwards *et al.* (1992) observed that conversion from mouldboard plough tillage to no-tillage increased soil organic carbon content in the 0 to 10 cm layer from 10 g/kg to 15.5 g/kg in 10 years, an increase of 56%. Lal *et al.* (1998) stated:

> A summary of the available literature indicates that the soil organic carbon sequestration potential of conversion to conservation tillage ranges from 0.1 to 0.5 metric tons ha^{-1} yr^{-1} for humid temperate regions and from 0.05 to 0.2 metric tons ha^{-1} yr^{-1} for semi arid and tropical regions.

They further estimated that the soil organic carbon increase may continue over a period of 25 to 50 years, depending on soil properties, climate conditions and management.

Carbon sequestration in the soil has benefits beyond removal of CO_2 from the atmosphere. No-tillage cropping reduces fossil fuel use, reduces soil erosion and enhances soil fertility and water-holding capacity. Beneficial effects of conservation tillage on soil organic carbon content, however, may be short-lived if the soil is ploughed, even after a long time under conservation tillage (Gilley and Doran, 1997;

Stockfisch *et al.*, 1999). Stockfisch *et al.* (1999) concluded that organic matter stratification and accumulation as a result of long-term minimum tillage were completely lost by a single application of inversion tillage in the course of a relatively mild winter. Tillage accentuates carbon oxidation by increasing soil aeration and soil residue contact, and accelerates soil erosion by increasing exposure to wind and rain (Grant, 1997). Several experiments in North America have shown more soil organic carbon content in soils under conservation tillage compared with plough-tillage seed beds (Doran, 1980; Doran *et al.*, 1987; Rasmussen and Rohde, 1988; Havlin *et al.*, 1990; Tracy *et al.*, 1990; Kern and Johnson, 1993; Lafond *et al.*, 1994; Reicosky *et al.*, 1995).

Similar to the merits of no-tillage reported in North America, Brazil and Argentina (Lal, 2000; Sa *et al.*, 2001), several studies have reported a high potential for soil organic carbon sequestration in European soils. In an analysis of 17 European tillage experiments, Smith *et al.* (1998) found that the average increase of soil organic carbon, with a change from conventional tillage to no-tillage, was 0.73 ± 0.39% per year and that soil organic carbon may reach a new equilibrium in approximately 50 to 100 years. Analysis of some long-term experiments in Canada (Dumanski *et al.*, 1998) indicated that soil organic carbon can be sequestered for 25 to 30 years at a rate of 50 to 75 g carbon/m^2/year, depending on the soil type in well-fertilized Cherozem and Luvisol soils cropped continuously to cereals and hay. Analysis of these Canadian experiments focused on crop rotations, as opposed to tillage, and is unique in that it considered rates of carbon sequestration with regard to soil type.

On a global basis, West and Post (2002) suggested that soil carbon sequestration rates with a change to no-tillage practices can be expected to have a delayed response, reach a peak sequestration rate in 5 to 10 years, and then decline to nearly 0 in 15 to 20 years, based on regression analysis. This agrees with a review by Lal *et al.* (1998), based on results from Franzluebbers and Arshad (1996) showing that there may be little or no increase in soil organic carbon in the first 2 to 5 years after a change in management practice, followed by a large increase in the next 5 to 10 years. Campbell *et al.* (2001) concluded that wheat rotation systems in Canada will reach an equilibrium, following a change to no-tillage, after 15 to 20 years, provided average weather conditions remained constant. Lal *et al.* (1998) estimated that rates of carbon sequestration may continue over a period of 25 to 50 years. The different estimates of carbon sequestration may be expected partly based on different rotations and rotation diversity.

Nitrogen Emissions

Cropping systems and nitrogen fertilization affect plant biomass production, partially controlling input of organic carbon to the SOM stocks. Agriculture alters the terrestrial nitrogen cycle as well. Through nitrogen fertilization, annual cropping, monocropping and improper water management, nitrogen is more prone to being lost to both ground- or surface water and the atmosphere. N_2O, a common emission from agricultural soils, is a potent greenhouse gas (310 times more potent than CO_2), which has increased its atmospheric concentration by 15% during the past two centuries (Mosier *et al.*, 1998). Reductions can be achieved through improved nitrogen management, as well as with irrigation water management, because N_2O is generated under both aerobic conditions (where nitrification occurs) and anaerobic conditions (where denitrification occurs) in the soil.

Due to the tightly coupled cycles of carbon and nitrogen, changes in rates of carbon sequestration and terrestrial ecosystems will directly affect nitrogen turnover processes in the soils and biosphere–atmosphere exchange of gaseous nitrogenous compounds. Some data suggest that increasing N_2O emissions may be closely linked to increasing soil carbon sequestration (Mosier *et al.*, 1991; Vinther, 1992; McKenzie *et al.*, 1998;

Robertson *et al.*, 2000). If no-tillage is a truly viable management practice, it must mitigate the overall impact of no-tillage adoption by reducing the net global warming potential determined by the fluxes of all the greenhouse gases, including N_2O and CH_4.

Six *et al.* (2004) assessed potential global warming mitigation with the adoption of no-tillage in temperate regions, by compiling all available data reporting differences in fluxes of soil-derived C, N_2O and CH_4 between conventional tillage and no-tillage systems. Their analysis indicated that, at least for the first decade, switching from conventional tillage to no-tillage would generate enhanced N_2O emissions for humid environments and somewhat lower emissions for dry environments, which would offset some of the potential carbon sequestration gains; and that, after 20 years, N_2O emissions would return to or drop below conventional tillage fluxes. They found that N_2O emissions, with a high global warming potential, drive much of the trend in net global warming potential, suggesting that improved nitrogen management is essential to realize the full benefits from carbon storage in the soil for the purposes of global warming mitigation. They suggested caution in the promotion of no-tillage agriculture to reduce greenhouse gas emissions and that the total radiative forcing needs additional consideration beyond just the benefit of carbon sequestration. They suggested that it is critical to investigate the long-term as well as short-term effects of various nitrogen management strategies for long-term reduction of N_2O fluxes under no-tillage conditions. These results suggest the need for more basic research on N_2O emissions during the transition from conventional tillage to no-tillage and after equilibrium conditions have been achieved to adequately quantify the carbon-offsetting effects in global warming potential.

In Brazil, most, but not all, studies indicate that the introduction of zone tillage increases SOM (Bayer *et al.*, 2000a, b; Sa *et al.*, 2001). Sisti *et al.* (2004) evaluated changes in soil carbon in a 13-year study comparing three different cropping rotations under zone tillage and conservation tillage in a clayey Oxisol soil sampled to 100 cm. They found that, under a continuous sequence of winter wheat and summer soybean, the stock of soil carbon to 100 cm under zone tillage was not significantly different from that under conservation tillage. However, in rotations with a vetch crop, soil carbon stocks were significantly higher under zone tillage than under conservation tillage. They concluded that the contribution of nitrogen fixation by the legume crop was the principal factor responsible for the observed carbon accumulation in the soil under zone tillage. The results demonstrate the role of diverse crop rotations, especially including legumes supplying organic nitrogen under zone tillage, in the accumulation of soil carbon. The dynamic nature of the carbon : nitrogen ratio may require additional organic nitrogen to increase carbon sequestration at depth. Sisti *et al.* (2004) found that much of the nitrogen gain was at depths below the plough layer, suggesting that most of the accumulated soil carbon was derived from crop root residues.

Further work in Brazil reflects the importance of soil and plant management effects on soil carbon and nitrogen losses to 1 m depth (Diekow *et al.*, 2004). They evaluated carbon and nitrogen losses during a period of conventional cultivation that followed on native grassland and 17-year no-tillage cereal- and legume-based cropping systems with different nitrogen fertilization levels to increase carbon and nitrogen stocks. With nitrogen fertilization, the carbon and nitrogen stocks of the oat/maize rotation were steady with time. However, they found increased carbon and nitrogen stocks due to higher residue input in the legume-based cropping systems. The long-term no-tillage legume-based cropping systems and nitrogen fertilization improved soil carbon and nitrogen stocks of the previously cultivated land to the original values of the native grassland. Nitrogen and legume residues in a rotation were more effective for building soil carbon stocks than inorganic nitrogen from fertilizer applied to the grass crop in the rotation. In addition, legume nitrogen does not require the cost of using fossil fuel to manufacture nitrogen fertilizer. The dominant soil change took place in the surface layer; however, deeper layers were important for carbon and nitrogen storage, which leads to improved soil and environmental quality.

The literature holds considerable evidence that intensive tillage decreases soil carbon and supports the increased adoption of new and improved forms of conservation tillage or direct seeding to preserve or increase SOM (Reicosky *et al.*, 1995; Paul *et al.*, 1997; Lal *et al.*, 1998). Based on the soil carbon losses with intensive agriculture, reversing the decreasing soil carbon trend with less tillage intensity should be beneficial to agriculture and the global population by gaining better control of the global carbon balance (Houghton *et al.*, 1983; Schlesinger, 1985). The environmental and economic benefits of conservation tillage and direct seeding demand their consideration in the development of improved management practices for sustainable production. However, the benefits of no-tillage for soil organic carbon sequestration may be soil- or site-specific, and the improvement of soil organic carbon may be inconsistent on fine-textured and poorly drained soils (Wander *et al.*, 1998). Six *et al.* (2004) indicated a strong time dependency in the greenhouse gas (GHG) mitigation potential of no-tillage agriculture, demonstrating that greenhouse gas mitigation by adoption of no-tillage is much more variable and complex than previously considered.

Policy of Carbon Credits

The increase in greenhouse gas concentrations in the atmosphere is a global problem that requires a global solution (Kimble *et al.*, 2002; Lal, 2002). Concern about negative effects of climate warming resulting from increased levels of greenhouse gases in the atmosphere has led nations to establish international goals and policies for reductions of these emissions. Initial targets for reductions are stated in the Kyoto Protocol of the United Nations Framework Convention on Climate Change, which allows trading credits that represent verified emission reductions and removal of greenhouse gases from the atmospheres (United Nations Framework Convention on Climate Change Secretariat, 1997).

Emissions trading may make it possible to achieve reductions in net greenhouse gas emissions for far less cost than without trading (Dudek *et al.*, 1997). Storing carbon in soils using conservation agriculture techniques can help offset greenhouse gas emissions while providing numerous environmental benefits, such as increasing site productivity, increasing water infiltration and maintaining soil flora and fauna diversity (Lal *et al.*, 1998; Lal, 2002). Storing carbon in forests may also provide environmental benefits resulting from increased numbers of mature trees contributing to carbon sequestration (Row *et al.*, 1996). While carbon is a key player for agriculture in solving the problem of global warming, a critical caveat is that other greenhouse gases change with changes in land use, including CH_4 and N_2O. We must look at the net global warming potential, not only for carbon in future trades but global warming potential credits, rather than carbon credits alone.

As interest in soil carbon sequestration grows and international carbon trading markets are developed, it is important that appropriate policies be developed that will prevent the exploitation of soil organic carbon and at the same time replace the lost carbon and establish its value (Walsh, 2002). Policies are needed that will encourage the sequestration of carbon for all environmental benefits that will evolve (Kimble *et al.*, 2002). Making carbon a commodity necessitates determining its market value and doing so with rational criteria.

Both farmers and society will benefit from sequestering carbon. Enhanced soil quality benefits farmers, but farmers and society in general benefit from erosion control, reduced siltation of reservoirs and waterways, improved air and water quality and biodegradation of pollutants and chemicals. Farmers need to be compensated for the societal benefits of carbon sequestration and the mechanisms that develop will allow for carbon trading and maintaining property rights. One important criterion in developing the system is the measurement and verification of the carbon options for sequestration that must be developed and the importance of making policymakers aware of these procedures and the technical difficulties. The use of international carbon credit market mechanisms is intended to help meet the challenge of climate change and future carbon constraints, which enable sustainable development and at the lowest social cost.

Carbon credit accounting systems must be transparent, consistent, comparable, complete, accurate and verifiable (IPCC, 2000). Other attributes for a successful system include global participation and market liquidity, linking of different trading schemes, low transaction costs and rewards for early actions to voluntarily reduce emissions before regulatory mandates are put in place. Characterizing the relationships between soil carbon and water quality, air quality and all the other environmental benefits should be an easy sell to get social acceptance of this type of agriculture. The largest impediment is the educational processes directed at the policymakers and food-consuming public, which require further enhancement.

A growing number of organizations around the world are implementing voluntary projects that are climate-beneficial as a means to improve efficiency and reduce operating costs and risk. Businesses and institutions throughout the world are realizing that the benefits of good environmental management far outweigh the cost, both now and in the future, of good corporate management, which includes strategies to reduce greenhouse gas emissions, risk exposure and costs and to enhance overall competitive operations. Multinational organizations are participating in carbon energy credit trading markets in order to avoid future compliance costs and to protect their global franchise in the face of increasing concern over global warming (Walsh, 2002). In the evolution towards a global economy and as concerns over global environmental impacts increase, CO_2 emission management will become a factor in the planning and operations of industrial and government entities all over the world, creating challenges and opportunities for those who are able to recognize and capitalize on them.

The global ecosystem services provided by farmers and other landowners could provide a source of carbon-emission credits to be sold to carbon emitters and hence provide an additional source of income for farmers, particularly no-tillage farmers. Trade in carbon credits has the potential to make conservation agriculture more profitable and enhance the environment at the same time. The potential for carbon credits has attracted considerable attention of farmers and likely buyers of the carbon credits. However, it is difficult to stay fully informed about developing carbon credits because of their technical complexity and the pace of development on this subject. Rules for trading in carbon credits are not yet agreed upon, but international dialogue is under way to develop a workable system and rules for trading. The number of organizations working on developing a carbon trading system suggests that some type of international mechanism will evolve and that carbon credit trading will become a reality.

Information is rapidly becoming available on publicly traded carbon credits; however, little information is available on privately traded contracts. A great deal of uncertainty exists at this time as to which companies will emerge as reliable sources of high-quality information and entities that can handle trading in a fair and reliable manner. Potential suppliers and buyers of carbon credits are urged to proceed with caution because many of the issues central to carbon credit markets and trade are yet to be clarified. We must convince policymakers, environmentalists and industrialists that soil carbon sequestration is an additional important benefit of adopting improved and recommended conservation agricultural production systems. This option stands on its own, regardless of the threat of global climate change from fossil fuels.

Conservation agricultural practices (especially no-tillage) can help to mitigate global warming by reducing carbon emissions from agricultural land and by sequestering carbon in the soil through regulatory, market incentive and voluntary or educational means (Lal, 2002). Public policy can encourage adoption of these practices. For the present, there is a degree of uncertainty for investors and potential investors in forest-related carbon sinks over the specific rules that will apply to implementation of the sinks provisions of the Kyoto Protocol. Investors and potential investors in carbon sinks need to be aware that there is uncertainty at the international level. Administration and transaction costs could play a key role in determining the success of any carbon credit trading system. Costs in these areas are expected to be minimized through improved techniques and services for measuring and reporting sequestered carbon, private-sector consultants, economies of scale and the emergence of market mechanisms and strategies such as carbon pooling or aggregating. There are risks involved in selling carbon credits in advance of any formalized international trading system and those participating in early trading need to clarify responsibilities and obligations. However, care should be taken in the design of these policies to ensure their success, to avoid unintended adverse economic and environmental consequences and to provide maximum social benefit.

Summary of Reduced Environmental Emissions and Carbon Sequestration

While we learn more about soil carbon emissions, soil carbon storage and their central roles in environmental benefits, we must understand the secondary environmental benefits of no-tillage and what they mean to sustainable production agriculture. Understanding these environmental benefits directly related to soil carbon and getting the conservation practices implemented on the land will hasten the development of harmony between humans and nature while increasing production of food, fibre and biofuels.

Reducing soil carbon emissions and increasing soil carbon storage can increase infiltration, increase fertility, decrease wind and water erosion, minimize compaction, enhance water quality, impede pesticide movement and enhance environmental quality. Increased levels of greenhouse gases in the atmosphere require all nations to establish international and national goals and policies for reductions. Accepting the challenges of maintaining food security by incorporating carbon storage in conservation planning demonstrates concern for our global resources and our willingness to work in harmony with nature. This concern presents a positive role for no-tillage, which will have a major impact on global sustainability and our future quality of life.

18 Some Economic Comparisons

C. John Baker

The long-term economics of no-tillage will be determined more by maximizing crop performance and net cash returns than by minimizing the inputs costs.

In this chapter we look at some economic comparisons of tillage versus no-tillage. But, no matter how the comparisons are analysed, in the end, crop yield will affect the results at least as much as input costs.

Comparisons between different levels of no-tillage are also important. For example, a relatively inexpensive no-tillage drill costing half as much as a more advanced alternative will only need to cause a 4–5% reduction in crop yield to become a bad investment.

But the most common comparison is between no-tillage and tillage. Opinions abound about whether it is cheaper to use no-tillage or tillage. Comparisons are often misleading for the following reasons:

1. Farmers who consider changing from tillage to no-tillage often compare the cost of engaging a no-tillage contractor (custom driller) with the cost of undertaking their own tillage. Many only include direct costs (such as fuel) as the cost of undertaking tillage since they already own the equipment, which they consider has already been paid for. The real issue is not apparent until these farmers have to replace their worn-out tillage equipment. None the less, we attempt to analyse this situation by comparing the cost of used tillage equipment with used no-tillage equipment.

2. Understandably, even if farmers are determined to make a switch to no-tillage, they will often keep their tillage equipment for a few years as a form of insurance – 'in case no-tillage does not work out' – while also paying for a no-tillage contractor. Thus, for a period, they are paying twice, but not by as much as they might imagine, as shown later by the analyses.

3. Many comparisons penalize no-tillage by imposing expected reductions in crop yields and/or increases in seeding and/or fertilizer rates for the first few years. This no longer applies when using modern no-tillage equipment and methodologies. Recent experience has repeatedly shown that using advanced no-tillage machinery and systems will produce crop yields at least comparable to tillage in year 1, and probably significantly better with time. Seeding rates of some crops and pastures have actually been reduced, not increased – some by up to 50%. On the other hand, if lower technology no-tillage systems and equipment are used, temporary yield reductions may well be applicable.

4. Economic comparisons should, but seldom do, factor in no-tillage reductions in labour, tractor numbers, tractor hours, fuel use and depreciation. One US farmer, for example, using modern no-tillage methods, recently reported that he now uses more fuel to harvest his crops than to grow them – an unheard-of scenario using conventional tillage (D. Wolf, 2005, personal communication).

5. Tractors often clock only one-quarter of the annual hours using no-tillage compared with tillage and thus last considerably longer. Therefore, the annual depreciation, interest and insurance costs can be reduced and machinery replacement intervals lengthened.

6. Some farmers already have a permanent labour force and no alternative function for that labour when the demand at seeding is reduced; thus there is seemingly little to be gained by adopting no-tillage. On the other hand, enterprising farmers have used the freed-up time to increase the area cropped each year. The economics of this are hard to factor into any analysis.

7. The amount of capital recovered from the sale of second-hand tillage equipment will diminish as no-tillage increases in popularity. The market for second-hand tillage equipment will shrink and this has certainly been a factor for some farmers when making the change.

So how do the figures stack up on both sides of tillage versus no-tillage? We provide answers to this question from two perspectives. The first was to examine four possible scenarios of ownership (C.J. Baker, 2000, unpublished data). We use the costs of equipment in New Zealand because that country has some of the more expensive and capable no-tillage options available, as well as cheaper alternatives. The second analysis was to review the results of charges made by a contractor in England to a client over two seasons. The first season (2002/03) was for tillage and minimum tillage. The second season (2003/04) was for no-tillage (J. Alexander, 2004, personal communication).

In both analyses we assume that crop yields are the same for both tillage and no-tillage. Such an assumption is only realistic if advanced (and usually more expensive) no-tillage equipment is used. If less advanced (cheaper) no-tillage equipment is used, it is likely that crop yields will be depressed below tillage, which will add an effective additional cost to the no-tillage. The comparisons quoted below may therefore require adjustment for less advanced equipment.

Obviously the actual figures will require adjustment for other countries and years. But readers are encouraged to change the input data to those applying locally and recalculate the figures. In most cases the relative values will remain approximately the same, regardless of how the actual figures change over time and location.

New Zealand Comparisons

- Scenario A: Economics of using a tillage contractor or a no-tillage contractor.
- Scenario B: Economics of purchasing new tillage or new no-tillage equipment.
- Scenario C: Economics of retaining used tillage equipment or purchasing either new or used no-tillage equipment.
- Scenario D: Economics of retaining used tillage equipment or engaging a no-tillage contractor.

Assumptions

1. Farmed area 300 hectares – 150 hectares cropped twice annually. (The cropped area could increase substantially with no-tillage but this is not included.)

2. With no-tillage, glyphosate, slug bait and chlorpyrifos are used in spring for weed and pest control.

3. For tillage, glyphosate is applied prior to spring ploughing (at a lighter rate than for no-tillage) but is omitted for autumn sowing.

4. All values are shown in 2004 New Zealand dollars.

Scenario A: Economics of using a tillage contractor or a no-tillage contractor

Establishing 150 hectares of spring wheat (Table 18.1), followed by 150 hectares of autumn forage crop (Table 18.2). Table 18.3 summarizes the pre-tax costs.

Table 18.1. Spring cropping using contractors.

Item	Tillage	No-tillage
Glyphosate (including application)	$55/ha[a]	$65/ha
Chlorpyrifos (applied with glyphosate)		$40/ha[b]
Slug bait (applied with drill)		$40/ha
Contractor	$250/ha	$100/ha
Seed and fertilizer	Same	Same
Total	$305/ha	$245/ha
Crop yield	Same	Same
× 150 hectares	$45,750	$36,750

[a]Glyphosate is applied at a lower rate for tillage.
[b]The chlorpyrifos cost would reduce to $8/ha when there was lighter pest pressure.

Table 18.2. Autumn cropping using contractors.

Item	Tillage	No-tillage
Glyphosate		
Chlorpyrifos		
Slug bait		
Contractor	$150/ha	$100/ha
Seed and fertilizer	Same	Same
Total	$150/ha	$100/ha
Crop yield	Same	Same
× 150 hectares	$22,500	$15,000

Table 18.3. Summary of total annual pre-tax costs.

	Tillage	No-tillage
Costs	$68,250	$51,750
Costs/ha	$227/ha	$172/ha
Difference (in favour of no-tillage)		$16,500 ($55/ha)

CONCLUSIONS

1. On a contractor basis, costs (and therefore gross margins) for the year favour no-tillage by $16,500 or $55/ha.

2. Even if glyphosate is omitted from tillage in the spring (at $55/ha), the comparison still favours no-tillage by $8250 per year or $27.50/ha for the whole year.

3. No allowance has been made in this analysis for the benefits of establishing autumn crops or pasture using advanced no-tillage methods immediately after harvest, nor for the additional spring utilization of land that comes from no-tillage. These factors alone can be valued at an additional $440/ha in favour of no-tillage (W.R. Ritchie, 2003, unpublished data).

NOTES

1. When sowing brassicas, peas or other broadleaved crops in spring, the chlorpyrifos cost for no-tillage can be reduced to $8/ha, which reduces the per-hectare cost of no-tillage in spring to $213/ha (overall cost $140/ha), increasing the overall difference between the two to $87/ha in favour of advanced no-tillage.

2. Contract tillage varies by district from $250/ha to $500/ha. The conservative lower figure was used.

3. Contract no-tillage with advanced equipment varies from $100/ha to $150/ha, depending on contour, size of field, etc. The conservative lower figure was used.

4. If using cheaper no-tillage equipment, drilling costs will be reduced, but crop yields are likely to be reduced by more than the saving in costs.

5. Herbicides and pesticides are often unnecessary in autumn with no-tillage. Some or all may be necessary in other situations, in which case their cost at reduced application rates should be added to no-tillage.

6. Autumn tillage in New Zealand (NZ) usually involves minimum tillage.

Scenario B: Economics of purchasing new tillage or new no-tillage equipment

Establishing 150 hectares of spring wheat followed by 150 hectares of autumn forage crop. The capital costs associated with purchasing all new equipment are shown in Table 18.4. The annual pre-tax operating costs of the two systems are shown in Table 18.5.

Table 18.4. Pre-tax capital costs of purchased new equipment.

Item	Tillage	No-tillage
1 × 170 hp tractor		$170,000
1 × 120 hp tractor	$120,000	
1 × 80 hp tractor	$80,000	
Sprayer	$6,000	$6,000
Plough (5 furrow)	$28,000	
Power harrow (3 m)	$23,000	
Roller	$6,000	
Leveller	$3,000	
Drill	$34,000	$120,000
Total capital cost	$300,000	$296,000
Difference		Negligible

Table 18.5. Annual pre-tax operating costs of new equipment.

Item	Tillage	No-tillage
Depreciation[1]		
(tractors)	$10,000	$4,250
(other equipment)	$2,500	$3,150
Interest[2] (9%) on average investment	$20,250	$19,980
Maintenance[3] (tractors @ 5%/year)	$10,000	$8,500
Maintenance[3] (soil-engaging equipment @ 7%/year)	$6,580	$8,400
Maintenance[3] (non-soil-engaging equipment @ 3%/year)	$180	$180
Fuel		
(50 l/ha spring tillage) @ 65c/l	$4,875	
(25 l/ha autumn tillage) @ 65c/l	$2,438	
(15 l/ha spring and autumn no-tillage) @ 65c/l		$2,925
Labour		
(4 h/ha spring tillage) @ $15/h	$9,000	
(2 h/ha autumn tillage) @ $15/h	$4,500	
(1 h/ha spring and autumn no-tillage) @ $15/h		$4,500
Total annual operating cost	$70,323	$51,885
Cost per hectare	$234	$172
Difference (in favour of no-tillage)		$18,438 (or $61/ha)

[1,2,3] See 'Notes' on pp. 271–272.

CONCLUSIONS

1. The capital cost of advanced no-tillage equipment was very similar to new tillage equipment.

2. With new equipment, annual savings in operating costs of approximately $18,000 per year ($61/ha) will be achieved by purchasing advanced no-tillage equipment rather than tillage equipment.

NOTES

1. Depreciation was calculated on a straight-line basis as:

Tillage tractors: Annual depreciation = new price minus trade-in price (50% of new price) divided by service life (10 years).

No-tillage tractor: Annual depreciation = new price minus trade-in price (50%

of new price) divided by service life (20 years).

All other equipment: Annual depreciation = new price minus trade-in price (50% of new price) divided by service life (20 years).

2. Interest was calculated on the average investment (new price plus trade-in price divided by 2) × 0.09.

3. Maintenance was from published data (Bainer *et al.*, 1955).

4. Actual total cost of labour will probably be closer to $20/hour if allowance is made for downtime, travel, maintenance, etc.

Scenario C: Economics of retaining used tillage equipment or purchasing either new or used no-tillage equipment

Establishing 150 hectares of spring wheat followed by 150 hectares of autumn forage crop. The capital costs associated with purchasing new or used no-tillage equipment, compared with retaining ownership of used tillage equipment, are shown in Table 18.6. The annual pre-tax operating costs of new or used no-tillage equipment versus used tillage equipment are shown in Table 18.7.

CONCLUSION. Capital costs are virtually halved by owning second-hand equipment (tillage or no-tillage) compared with new equipment. Some $95,000–$97,500 in capital cost is saved by purchasing second-hand tillage or no-tillage equipment.

NOTE

1. The value of used equipment was assumed to be two-thirds of its new value and the equipment is halfway through its service life. The trade in value remains at 50% of the new value at the end of its service life.

Table 18.6. Pre-tax capital costs of new no-tillage and used tillage and no-tillage equipment.

Item	Tillage (used)[1]	No-tillage (new)	No-tillage (used)[1]
1 × 170 hp tractor		$170,000	$114,000
1 × 120 hp tractor (3300 h)	$80,000		
1 × 80 hp tractor (3300 h)	$54,000		
Sprayer	$4,500	$6,000	$4,500
Plough (5 furrow, used)	$19,000		
Power harrow (3 m, used)	$15,500		
Roller (used)	$4,500		
Leveller (used)	$4,500		
Conventional drill (used)	$23,000		
No-tillage drill		$120,000	$80,000
Total capital cost	$205,000	$296,000	$198,500
Difference (in favour of used equipment – see Scenario B above)	$95,000		$97,500

Table 18.7. Annual pre-tax operating costs of new and used no-tillage and used tillage equipment.

Item	Tillage (used)	No-tillage (new)	No-tillage (used)
Depreciation[1] (tractors)	$6,800	$4,250	$2,900
Depreciation[1] (other equipment)	$2,100	$3,150	$2,150
Interest[2] @ 9% (tractors and equipment)	$15,975	$19,980	$15,592
Maintenance[3] (tractors @ 5% new price/year)	$10,000	$8,500	$8,500
Maintenance[3] (soil-engaging equipment @ 7% new price/year)	$3,360	$8,400	$8,400
Maintenance[3] (non-soil-engaging equipment @ 3% new price/year)	$180	$180	$180
Fuel			
(50 l/ha spring tillage) @ 65c/l	$4,875		
(25 l/ha autumn tillage) @ 65c/l	$2,438		
(15 l/ha spring and autumn no-tillage) @ 65c/l		$2,925	$2,925
Labour			
(4 h/ha spring tillage) @ $15/h	$9,000		
(2 h/ha autumn tillage) @ $15/h	$4,500		
(1 h/ha spring and autumn no-tillage) @ $15/h		$4,500	$4,500
Total annual operating cost	$59,228	$51,885	$45,147
Cost per hectare	$197	$173	$150
Difference (in favour of no-tillage)		$7,343 (or $24/ha)	$14,081 (or $46/ha)

[1,2,3] See 'Notes' on p. 273.

CONCLUSIONS

1. Annual costs of owning and operating used tillage equipment ($59,228/year) were approximately $11,000 lower than for new tillage equipment ($70,323/year – Scenario B).

2. The annual costs of owning and operating used tillage equipment ($59,228/year) were approximately $7000 (or $24/ha) greater than owning and operating new advanced no-tillage equipment ($51,885/year) and approximately $14,000 (or $46/ha) greater than used advanced no-tillage equipment.

NOTES

1. Depreciation was calculated on a straight-line basis as follows:

Tillage tractors: Annual depreciation = used price minus trade-in price (50% of new price) divided by remaining service life (5 years).

No-tillage tractor: Annual depreciation = new or used price minus trade-in price (50% of new price) divided by remaining service life (20 years for new or 10 years for used).

All other equipment: Annual depreciation = new or used price minus trade-in price (50% of new price) divided by remaining service life (20 years for new or 10 years for used).

2. Interest was calculated on the average investment (used or new price plus trade-in price divided by 2) × 0.09.

3. Maintenance was from published data (Bainer *et al.*, 1955).

4. The maintenance costs shown for used equipment are conservative because maintenance could be expected to increase with age of machines.

Scenario D: Economics of retaining used tillage equipment or engaging a no-tillage contractor

Establishing 150 hectares of spring wheat followed by 150 hectares of autumn forage crop. The annual pre-tax costs of operating used tillage equipment versus hiring a no-tillage contractor are shown in Table 18.8.

Table 18.8. Costs of used tillage equipment versus hiring a no-tillage contractor.

Item	Tillage	No-tillage
Annual operating costs of used tillage equipment (from Scenario C)	$59,228	
Glyphosate in spring (from Scenario A)	$8,250	
Annual cost of contractor including glyphosate and pesticides (from Scenario A)		$51,750
Totals	$67,478	$51,750
Cost per hectare	$225	$172
Difference (in favour of no-tillage)		$15,728 ($52/ha)

CONCLUSION

1. Ownership of used tillage equipment was more expensive (by approximately $15,000 per year or $52/ha) than engaging a contractor with advanced no-tillage equipment.

Summary and conclusions

The A–D scenarios outlined above are summarized in Table 18.9.

General conclusions

1. It made little difference whether such comparisons were made between new or used equipment, hiring contractors, or combinations of these options. No-tillage was less expensive than tillage for all situations.

2. For 150 hectares cropped twice per year, it was cheaper to use advanced no-tillage equipment in any form than to use any form of tillage ($7000–$18,000/year, or $24–$61/hectare).

3. The smallest difference was ownership of used tillage versus ownership of new no-tillage equipment ($24/ha).

4. The largest difference was ownership of new tillage versus ownership of new no-tillage equipment ($61/ha).

5. All other comparisons result in an approximate $50/ha saving using no-tillage.

6. Hiring a no-tillage contractor with advanced equipment is most often accompanied by a high level of specialist expertise.

7. The only valid economic argument for not adopting advanced no-tillage is if a farmer does not have access to an advanced no-tillage drill. Substandard crop yields will be likely, if not a regular occurrence, with less advanced no-tillage equipment. Tillage is more forgiving of substandard equipment.

8. If a farmer chooses to continue ownership of the used tillage equipment while hiring a no-tillage contractor with advanced equipment on a trial basis (a sensible practice), the costs of depreciation and interest on the tillage equipment will remain although it is not being used ($80/hectare, Scenario C). Since the use of

Table 18.9. Summary of Scenarios A–D.

Scenario	Tillage ($/year)	Tillage ($/ha)	No-tillage ($/year)	No-tillage ($/ha)	Differences $/year	Differences $/ha
Scenario A (contractors)	68,250	227	51,750	172	16,500	55
Scenario B (own new equipment)	70,323	234	51,885	173	18,438	61
Scenario C (own used equipment)	59,228	197	45,145–51,885	150–173	7,343–14,081	24–47
Scenario D (own used equipment versus contractor)	67,478	225	51,750	172	15,728	53

a no-tillage contractor is less than a tillage option ($53/ha, Scenario D), the net cost of trying out advanced no-tillage for a year will be about $27/ha ($80–$53), which is a modest price to pay with the prospect of saving $24–61/ha/year for every year thereafter with the adoption of no-tillage.

European Comparisons

In these comparisons, an English tillage contractor provided the following figures for a client who cropped 404 hectares (1000 acres) per year. The tillage and minimum-tillage figures were actual charges made to the farmer in previous years. The advanced no-tillage figures were quotations for 2004.

Two scenarios are compared: plough-based tillage versus no-tillage, and minimum tillage versus no-tillage. The tillage and minimum-tillage programmes are outlined in Tables 18.10 and 18.11 and are considered typical for many English properties.

The no-tillage quote was for an advanced and more expensive no-tillage drill (which would assure crop production with at least equal yield to the tillage systems), as reflected in the higher per-hectare charge rate. As with the New Zealand comparison, substituting a less advanced no-tillage drill for the advanced no-tillage drill might have had the potential to reduce the costs of no-tillage but it also had the potential to reduce the no-tillage crop yield.

Scenario (A) Comparison of no-tillage with full plough-based tillage

Establishing cereal grain on a 404 hectare (1000 acre) farm using a plough-based tillage system, compared with advanced no-tillage (contractor charges). Comparative costs are shown in Table 18.10.

Table 18.10. Comparison of tillage and no-tillage costs in England.

	Cost/ha	Area	Total
Tillage machines			
Subsoiler, with packer roller	£31.75	404	£12,827.00
Ploughing	£36.00	404	£14,544.00
'Cultipress'	£14.20	404	£5,736.80
Rolling	£10.75	404	£4,343.00
Power harrow	£25.60	200	£5,120.00
Fertilizing	£7.50	404	£3,030.00
Combination conventional drill	£29.75	304	£9,044.00
Cultivator-drill	£30.00	100	£3,000.00
Spraying	£7.00	404	£2,828.00
Total			£60,472.80
No-tillage machines			
Advanced no-tillage drill	£55.00	404	£22,220.00
Spraying	£7.00	404	£2,828.00
Total			£25,048.00
Difference			£35,424.80
Difference per hectare			£87.68/ha

Scenario (B) Comparison of no-tillage with minimum tillage

Establishing cereal grain on a 404 hectare (1000 acre) farm using a minimum-tillage system, compared with advanced no-tillage (contractor charges). Comparative costs are shown in Table 18.11.

Table 18.11. Comparison of minimum tillage and no-tillage costs in England.

	Cost/ha	Area	Total
Minimum-till machines			
Subsoiler, with packer roller	£31.75	202	£6,413.50
Tillage train	£35.00	404	£14,140.00
'Cultipress'	£14.20	404	£5,736.80
Rolling	£10.75	404	£4,343.00
Fertilizing	£7.50	404	£3,030.00
Cultivator-drill	£30.00	404	£12,120.00
Spraying	£7.00	404	£2,828.00
Total			£48,611.30
No-tillage machines			
Advanced no-tillage drill	£55.00	404	£22,220.00
Spraying	£7.00	404	£2,828.00
Total			£25,048.00
Difference			£23,563.30
Difference per hectare			£58.32/ha

Conclusions

1. On a contractor basis, minimum tillage was cheaper than tillage by £29/ha.
2. On a contractor basis, advanced no-tillage was cheaper than plough-based tillage by £87/ha.
3. On a contractor basis, advanced no-tillage was cheaper than minimum tillage by £58/ha.
4. These comparisons may not have been valid if less advanced no-tillage machines had been used.
5. Comparisons between tillage, minimum tillage and no-tillage are machine-dependent, since no-tillage drill designs have the potential to influence crop yields markedly.

Summary of Some Economic Comparisons

1. The most common economic comparison is between no-tillage and tillage but such comparisons are often misleading for any one of a number of reasons and assumptions.
2. Several possible scenarios provide economic examples of tillage versus no-tillage, but the items and figures will require changing for other countries and years.
3. Machine costs involved with changing from a tillage to a no-tillage system are a major consideration.
4. Maintaining ownership of tillage machines for a period after beginning no-tillage adds some costs to the transition but may be a comforting and affordable choice for many farmers.
5. Economics of using a tillage contractor or a no-tillage contractor favours using a no-tillage contractor.
6. Economics of purchasing new tillage or new advanced no-tillage equipment showed similar capital costs in either case but significantly lower operating costs for no-tillage.
7. Economics of retaining used tillage equipment or purchasing either new or used no-tillage equipment showed that capital costs are virtually halved by owning second-hand equipment (tillage or no-tillage), compared with new equipment, but again operating costs are in favour of no-tillage.
8. Economics of retaining used tillage equipment or engaging a no-tillage contractor showed that ownership of used tillage equipment was more expensive than hiring a contractor with advanced no-tillage equipment.
9. It made little difference whether comparisons were made between new or used equipment, hiring contractors, or combinations of these options. No-tillage was less expensive than tillage for all situations.
10. Hiring a no-tillage contractor with advanced equipment is most often accompanied with a high level of specialist expertise.
11. A US farmer who recently converted from tillage to no-tillage reports a 'win–win' situation with advanced no-tillage equipment. He has not only recorded his best crop yields ever with no-tillage, but he now also uses less fuel to grow his crops than to harvest them.

19 Procedures for Development and Technology Transfer

C. John Baker

Measuring the mechanical performance of no-tillage machines is far less important than measuring their biological performance.

One of the distinguishing aspects of experiments conducted with agricultural tillage machines is that there are very few common experimental techniques and standardized instruments that can be universally applied. The designs and functions of most agricultural machines are quite diverse; thus the techniques used to evaluate them are tailor-made for specific purposes and to answer specific questions.

This situation contrasts with experiments with plants, for example, in which the most common procedure is to grow plants in pots or plots of soil, each with a designated treatment. Since all plants perform essentially the same functions of utilizing the sun's energy to convert nutrients from the soil, atmosphere and water into biomass, there is a high degree of commonality of plant experiments.

In the study of no-tillage drills, planters and openers, design scientists have sought knowledge not only about resulting plant growth, using well-established experimental procedures, but also about their mechanical performance and, perhaps most importantly, about the interactions between infinite design variations of the machine components, the soil, surface residues, pests and the plants.

Described here are some of the experimental procedures and techniques used by the authors and their colleagues to gain knowledge about the functions and performance of no-tillage components and subsequently to develop new no-tillage technologies, designs and practices. Many of the techniques developed are specific to no-tillage but should be useful to others pursuing similar investigations. Some were unique experiments, while others followed well-established common procedures.

This is not an attempt to provide a comprehensive review of all techniques used by scientists in this field, although the results of much relevant work by a wide range of scientists are reported elsewhere in this book. The technique descriptions and instrumentation given here are restricted to those used or devised by the authors. We explain how many of the experiments were conducted in some detail because they were designed to address a variety of questions about how plants and soil interact with no-tillage machines, and because there were no known methodologies for those purposes available at the time.

The techniques and procedures described examined the following subjects:

1. Plant responses to no-tillage openers in controlled conditions.
2. The micro-environment within and surrounding no-tillage seed slots.
3. Soil compaction and disturbance by no-tillage openers.
4. Locating seeds in the soil.
5. Seed travel within no-tillage openers.
6. Drag on a disc opener.
7. Accelerated wear tests of no-tillage openers.
8. The effects of fertilizer banding.
9. Prototype drills and management strategies.

Plant Responses to No-tillage Openers in Controlled Conditions

It is often assumed that most seeds will germinate and grow satisfactorily if sown into moist soil followed by favourable climatic conditions. Unfortunately, under no-tillage this assumption is not always correct. Early experience with no-tillage had suggested that, as the soil and climatic conditions became less favourable, seed, seedling and plant performance often suffered more than where seeds were sown into tilled seedbeds.

Thus, it became important to develop a fundamental procedure to evaluate the biological performance of different no-tillage openers under controlled conditions. The aim was to create a facility where scientists could put stress on the no-tillage system by superimposing unfavourable soil moisture conditions followed by unfavourable climatic conditions without the risk of intervention by unpredictable weather.

Sowing seeds in the field was considered too impractical and imprecise to control the soil moisture and climate. Conventional 'rainout' shelters, which involve large movable transparent canopies covering several plots of soil, were expensive and would have limited the experiments to one site. This contrasted with tillage experiments, where the soil beneath a 'rainout' shelter can be re-tilled several times to repeat several experiments on the same site.

The scientists also did not have the convenience of being able to place seeds in disturbed soils that had been prepared in pots or trays so that they could later be transported into glasshouses or other artificially controlled climate laboratories. For no-tillage experiments, the soils had to have been truly undisturbed for at least 12 months, and preferably longer, and to remain this way throughout the experiments.

A new technique was developed to transport untilled soil in bins to an indoor climatically controlled facility. This involved removing large 2.0 m × 0.7 m × 0.2 m blocks of soil weighing approximately 0.5 t from the field in an undisturbed state, controlling pre-drilling soil moisture content, drilling with openers arranged to duplicate their performance on a field drill or planter and then controlling the post-drilling climate and soil moisture content for the duration of the experiment (Baker, 1969a, 1976a, b).

Rectangular steel bins were constructed with both ends open. The front end of each bin was able to be attached to the rear of a stirrup-shaped soil cutter, which was itself attached to and pulled through the soil by a tractor (Fig. 19.1). The horizontal blade of the cutter was hollow, with exit ports drilled along its rearmost edge. Water was pumped into the hollow blade during extraction of the 0.5 t soil blocks to create a thin slurry on the underside of each soil block and thus temporarily lubricate it as it slid along each of the 2 metre bins. The base of each bin was lined with a veneer of stainless steel to assist this process.

In practice, it was found that 2 m was about the maximum slice length that a 200 mm deep undisturbed soil slab could be expected to slide without becoming compressed and perhaps buckled. Increasing the depth beyond 200 mm may have permitted longer blocks to be extracted, but such bins would have been difficult to handle because of their added weight and length.

Although a 200 mm soil depth could not be expected to sustain plant growth for long periods before roots reached the stainless steel bases, all of the studies that utilized

Fig. 19.1. A stirrup-shaped soil cutter with bin attached for extracting undisturbed soil blocks (from Baker, 1969a).

these bins concentrated on the germination and seedling emergence phases of crop production, since these were considered to be the most critical phases obstructing reliable no-tillage. It was also considered that machine influences on plant growth were likely to be greatest at the germination and emergence phases and thereafter would be of less influence than other factors, such as weather, soil and management effects.

The soil remained in its bin throughout each experiment. Bins were transported from the field to the laboratory using heavy lifting equipment on a tractor (Fig. 19.2). The moisture content of the soil in each bin was manipulated either by covering each bin with clear plastic and leaving it to air-dry or by irrigating it from above by sprinkler or from below by placing the perforated bins in shallow troughs containing a predetermined quantity of water.

Two processes were used to drill these undisturbed blocks of soil with a variety of no-tillage openers. Where measurements of the drilling process itself were to be made or multiple openers were to be tested in each bin, five bins were placed end to end on the raised bed of a 'tillage bin' arrangement, which also had a tool carrier on a moving gantry that straddled the line of bins and could be moved forwards or backwards at infinitely variable speeds from 0 to 8 km/h (0 to 5 mph) (Fig. 19.3).

Where drilling took place indoors, the openers on test were usually arranged at 150 mm row spacing with three rows to a bin. This resulted in 200 mm of clearance between the outside rows and the edges of the bins. The slightly larger distance in this zone was to avoid soil disturbance at the bin edges. All openers were mounted on parallel drag arms attached to a subframe. The vertical angle was variable to alter the opener pitch for any geometrical arrangement. Downforce was applied by adding weights to individual openers and draught forces were measured by a load cell mounted within the drag arm attachment subframe.

Mounting openers on parallel arms and applying downforce by application of weights were not a true duplication of common field practice. Weights ensured that the downforce applied to any one opener remained constant regardless of its position in the vertical plane. This seldom happens in practice. But the objective was to remove most ancillary functional differences between openers and their modes of operation to

Fig. 19.2. A filled soil bin being transported.

Fig. 19.3. The 'tillage bin' with soil bins arranged end to end ready for drilling (from Baker, 1969a).

evaluate differences associated with their actions in the soil and the shape of the slots they created.

Individual seeds were metered by a modified vacuum seeder designed by Copp (1961). As drilling was usually conducted at slow speeds, a visual count was made of the seeds entering the soil by observing them as they passed down a clear plastic delivery tube at bench height. In this manner, the exact number of seeds sown was known to make accurate counts of germination percentages. With the 'tillage bin' elevated to bench height, this allowed instrumentation to be inserted from beneath or beside the soil to monitor variables such as vertical and/or lateral soil forces resulting from the passage of individual openers.

It was occasionally necessary to test openers operating on actual field drills. In this case, the open-ended steel bins were left embedded in the soil after pulling them in with a tractor and the stirrup-shaped cutter. A field drill was then operated over them while they were *in situ*, taking care to avoid contact with the steel side walls of the bins. The soil bins could then be removed to controlled climate facilities.

The 'tillage bin' facility successfully allowed an accurate measure of how different shapes of no-tillage openers and slots respond to different soil conditions in terms of their abilities to promote satisfactory seed germination and seedling emergence. Almost all previous no-tillage experiments had used field conditions reporting successful establishment, but the results may have been as much a function of favourable conditions as of mechanical performance. While field experiments served to demonstrate that no-tillage seeding could work, there was a need to identify and eliminate the causes of failures. This required precise control to be exercised over the seeding conditions.

The tillage bin facility, because of its moving gantry, was also used for a variety of other related experiments. Among these were a study of spray droplet dissipation in pasture (Collins, 1970; see Chapter 12), monitoring of seed spacing from precision spacing planters (Ritchie and Cox, 1981; Ritchie, 1982; Carter, 1986; see Chapter 8) and the transplanting of cabbage seedlings into untilled soil (Pellow, 1992).

The micro-environment within and surrounding no-tillage seed slots

To learn the environmental requirements of seeds and seedlings within the seed slot, the following variables were tested to define the effects of opener designs: (i) soil moisture regime within the slot; (ii) soil-air humidity within the slot; (iii) soil oxygen within and around the slot; and (iv) soil temperature within the slot.

No attempt was made in these experiments to monitor the presence of allelopathic substances from decaying residue or other root material in the slot, since this was being well researched by Lynch and others at the time (Lynch, 1977, 1978; Lynch *et al.*, 1980). However, later experiments on wet soils by the authors and their colleagues added knowledge about these effects and how they might be avoided through opener design (see Chapter 7).

Soil moisture regime within the slot

Most non-destructive devices for measuring the liquid water content of soil sample a reasonably large soil volume. This is necessary to average the variations inherent in small soil volumes. The slot zone left by a no-tillage opener represents a relatively small volume of soil, which has made monitoring of liquid-phase moisture particularly difficult.

Gypsum blocks and most other physical absorption-based devices work best at the wet, low-tension, end of the moisture range, which made them unsuitable for experiments with dry soils. Early designs of dew-point psychrometers were tried, but the steep temperature gradients at or near the soil surface made them unreliable. Eventually, recourse was had to destructive gravimetric sampling, in which miniature cores of soil (20 mm diameter × 10 mm deep) were removed from the slot zone and oven-dried to provide a measurement of the liquid-phase soil moisture content on a differential weight basis. More sophisticated instruments have become available since these experiments were conducted.

The research showed that the liquid-phase water content of the soil in and around contrasting slot shapes did not greatly differ, at least in the short term, even when there were marked differences in seedling emergence between openers in otherwise relatively dry soils. While this at first seemed anomalous, it was decided that exhaustive testing of further alternative devices for measuring liquid-phase soil water was not justified. Rather, attention shifted to the measurement of slot humidity, or vapour-phase soil water.

Soil-air humidity within the slot

Soil physics shows that the atmosphere (air) in soil macropores and voids forms an equilibrium water vapour pressure with the liquid water contained in the surrounding soil pores. At a given temperature, the vapour-phase water in these soil spaces represents soil-air humidity. Since soil temperature at seeding depth does not change rapidly and is easily measured, soil humidity became a reasonably reliable way to measure the water-vapour pressure of the soil atmosphere.

Choudhary (1979) first monitored soil-air humidity within no-tillage slots using an aspirator to slowly draw quantities of air from the slot and pass these through a dew-point hygrometer for a direct reading of the relative humidity of the air sample. While this method produced interesting figures, the scientists were conscious that the removal of air from the slot inevitably resulted in its being replaced with air drawn predominantly from the atmosphere above the soil surface. Thus, the slot air samples only partly reflected the humidity within the slot.

The accuracy of the method relied on the removal rate of the slot air and the diffusion resistance of the slot cover, which controlled the rate that atmospheric air replaced that being removed. A high diffusion resistance of the slot cover, for example, might result in the removed slot air sample being replaced by additional slot air from further down the slot, while a low diffusion resistance might contain a larger proportion of atmospheric air. As it turned out, this diffusion resistance was later identified as an important variable in seed/seedling survival, but in the meantime a method was found that sampled the relative humidity *in situ* without removing air from the slot.

A modified direct-reading humidity probe was inserted into the slot and allowed to equilibrate with the undisturbed slot atmosphere for at least 2 minutes. The probe selected was originally designed to monitor relative humidity between sheets of newsprint. As such it was flat and thin in shape. The point was removed and a small piece of fibreglass filter material was wrapped over the end to prevent soil from falling into the sensitive probe. The filter was left behind in the soil when the probe was withdrawn and was not reused. Figure 19.4 shows a humidity probe being inserted into a dry no-tilled soil that is contained within a climate-controlled room.

This method yielded a direct reading of relative humidity, approximating what the seeds experienced in the slot. The information gathered with this technique had far-reaching consequences. The experiments showed that no-tilled seeds could germinate in a high-humidity slot atmosphere, i.e. without access to substantial amounts of liquid-phase water, a fact that was later confirmed by Martin and Thrailkill (1993) and Wuest (2002).

More importantly, subsurface seedlings could survive beneath the soil for several weeks if the slot atmosphere was maintained at or near 100% relative humidity. The latter observation was shown to be a function of the diffusion resistance of the slot cover and the humidity gradient between the slot air and the ambient air outside the slot. Slot cover was itself a function of slot shape, the presence of surface residues over the slot and the design of the opener.

Being able to monitor slot atmosphere humidity was one thing, but being able to control and vary that humidity for the purposes of experimentation was quite another matter. Even rain-protection covers were not satisfactory since they were unable to alter the ambient humidity of the day. Utilizing a multi-room controlled-climate facility, the 0.5 t blocks of soil in their steel bins were moved after drilling into climate-control rooms in groups of three. Each room had an artificial climate in which the temperature, humidity, light intensity, light spectrum, day length, nutrients and, if necessary, wind speed and direction could be controlled. In this way, the effects of high and low ambient humidity levels and/or

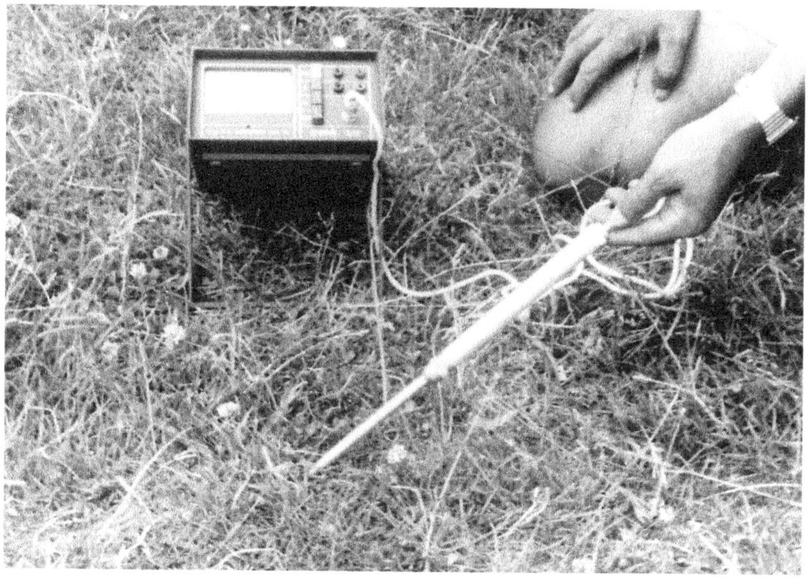

Fig. 19.4. Sampling soil humidity in the field.

temperatures were varied and the effects on the establishing seedlings measured (see Chapter 6).

Soil oxygen within and around the slot

The main consequence of a no-tilled soil becoming very wet after drilling is restriction of oxygen supply to the germinating seeds and embryonic roots. In a tilled soil, there is much artificial loosening, which exaggerates the oxygen regime around the seeds for a time. In an untilled soil, seeds rely almost entirely on the ability of the soil to remain adequately oxygenated in its natural state. To test a range of opener designs to provide varying oxygen conditions with wet soil conditions, variables of oxygen diffusion rates, earthworms, infiltration and soil temperatures were considered.

Several scientists have described an oxygen-diffusion measurement technique involving pushing a small platinum electrode into the soil and measuring the current passing between this electrode and a reference electrode. The current has the effect of reducing electro-reducible material, in this case oxygen, at the platinum surface. The size of the current is governed by the rate of oxygen diffusion from within the soil to the surface of the electrode and thus gives an indication of the oxygen diffusion rate (ODR) within the soil.

Most scientists agree that the ODR values obtained with platinum electrodes are only an approximation of what a root might experience, but the technique provides a relative measure of the difference between a range of soil conditions. The advantages are that it is cheap, non-destructive, quick, easy and capable of sampling very small zones of soil in the vicinity of the slot.

Chaudhry (1985) sampled ODR in a grid pattern around the basal area of a range of slots in a wet soil and used a computer program to draw iso-ODR lines reflecting the contrasting oxygen regimes generated by the passage of no-tillage openers and the presence or absence of surface residues and earthworms (see Chapter 7).

Earthworm activity was a likely contributor to the soil slot oxygen status. Mai (1978), Chaudhry (1985) and Giles (1994) monitored the numbers of earthworms present in the general plot soil and those around a seed slot. Cylindrical cores of soil centred on the slot were extracted and earthworms counted and weighed. Chaudhry also monitored earthworm activity on the soil surface by estimating the percentage of a given area of soil that was covered with earthworm casts. He termed this the 'casting index'.

Water infiltration into the slot zone was another potential factor in providing oxygen exchange. Relative infiltration rates were monitored by rectangular metal boxes (infiltrometers) inserted into the soil surface centred on the slot (Chaudhry, 1985; Baker *et al.*, 1987).

Exhaustive temperature comparisons were made by Baker (1976a) within a range of slot configurations. Temperature is relatively easy to measure in small discrete zones using miniature thermometers or electronic thermocouples. Short-term readings were by simple mercury thermometers, while thermocouples were used for continuous readings, such as diurnal ambient fluctuations.

Soil Compaction and Disturbance by No-tillage Openers

It had long been thought that a logical result of no-tillage openers operating in untilled soils would be progressive compaction and restricted root growth in the slot zone. Therefore, several studies centred on monitoring these aspects. The parameters measured were: (i) soil strength; (ii) instantaneous soil pressure (stress); (iii) instantaneous and permanent soil displacement; (iv) soil bulk density; and (v) smearing.

Soil strength

Soil strength is traditionally assessed by measuring the force required to push a probe (penetrometer) into the soil. To more closely resemble the actions of a root, the probe ends are usually conical in shape so that the force dissipation is radial as well as longitudinal. Such probes, however, are usually designed to sample reasonably large volumes of soil and, because of the natural heterogeneity of soil, repetitive sampling with a single probe is common.

To get the benefits of multiple soil probing within the confines of the slot zone, a miniature multi-point penetrometer was designed (Dixon, 1972; Baker, 1976a; Baker and Mai, 1982b). This device consisted of 20 1 mm diameter stainless steel probes mounted in a common horizontal press bar in such a way that the vertical position of each probe with respect to the bar could be adjusted and clamped individually. The press bar could be angled at any desired position from horizontal to vertical and was attached to a threaded shaft that acted as the thrust mechanism, together with a sensitive ring-shaped force-measuring device (known as a 'proving ring'). Two different displacement-measuring devices have been used to monitor the changes in diameter of the ring. Initially, a micrometer sufficed, but in later tests a displacement transducer was substituted to facilitate recorded results. The multi-point penetrometer is shown in Fig. 19.5.

Because soil tends to flow as a plastic body to a limited extent for several seconds after a rigid probe is inserted, it was necessary to insert the probes at a predetermined and constant speed and to read the force applied at a standard time interval after the probe penetration had been stopped at the desired depth (when plastic flow had ceased). The probes were inserted at a constant speed of penetration by rotating the threaded shaft at a constant speed, using a slow-speed electric motor drive, which was immediately disconnected upon reaching the desired depth, and then waiting 10 seconds before reading the gauge.

To accommodate the irregularities of the soil surfaces, the press bar was positioned parallel to the chosen surface and each probe was slipped through the bar until it lightly contacted the soil surface, then clamped in that position. Care was taken to ensure that an equal number of probes on each side of the central threaded shaft contacted the soil to ensure, as nearly as possible, symmetry of forces about the central point when all of the probes were pushed into the soil. Even then, a single probe would occasionally contact a stone, greatly distorting the symmetry, and the readings were discarded.

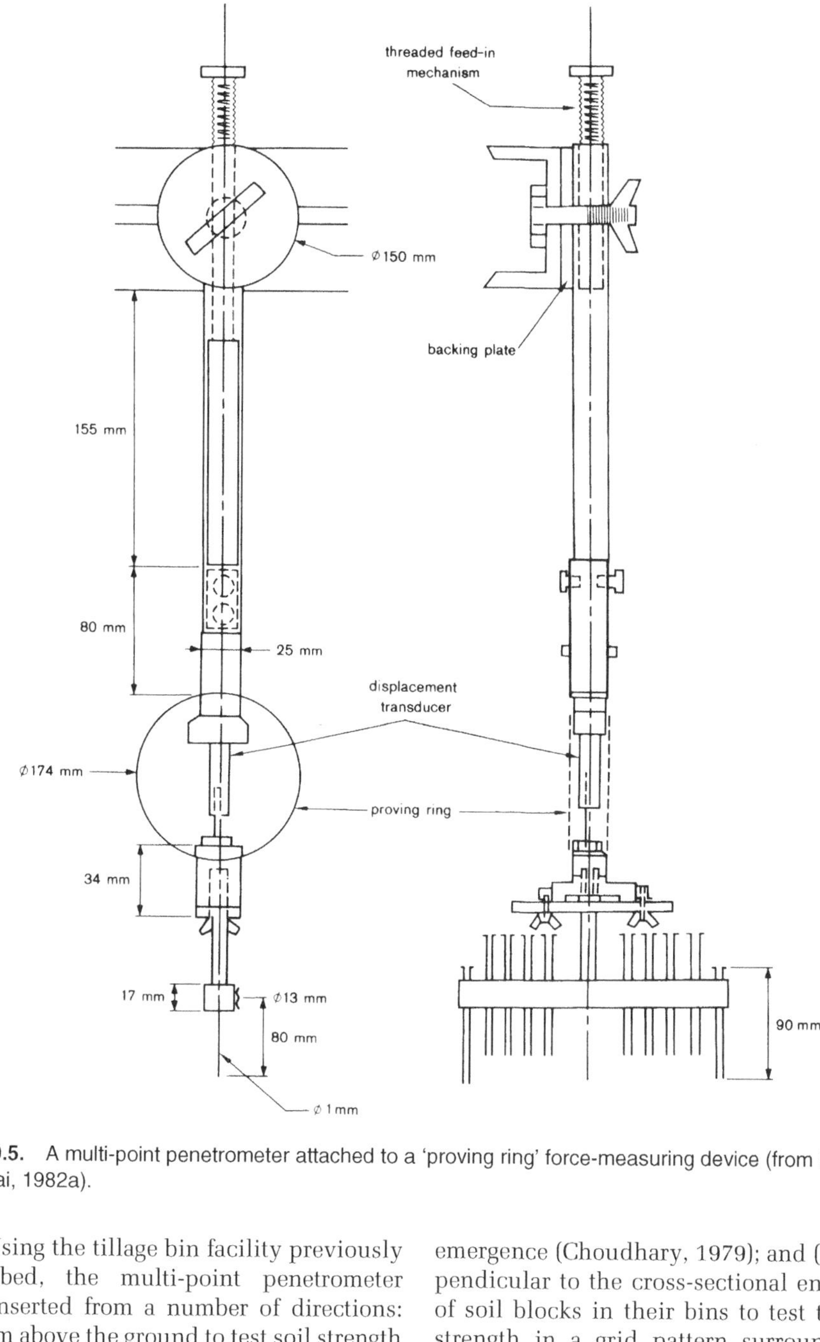

Fig. 19.5. A multi-point penetrometer attached to a 'proving ring' force-measuring device (from Baker and Mai, 1982a).

Using the tillage bin facility previously described, the multi-point penetrometer was inserted from a number of directions: (i) from above the ground to test soil strength vertically downwards at the base of slots (Baker and Mai, 1982b); (ii) from the side perpendicular to the side walls of slots (Mai, 1978; Baker and Mai, 1982b); (iii) from beneath the bins pushing upwards to measure the resistance of slot cover to shoot emergence (Choudhary, 1979); and (iv) perpendicular to the cross-sectional end faces of soil blocks in their bins to test the soil strength in a grid pattern surrounding a cross-section of the slots (Mitchell, 1983).

The penetrometer was not usable in the field as its high sensitivity required a very stable base from which to derive the penetration force. This could only realistically be provided by the tillage bin supported on a concrete floor. Even then, a person pressing on one of the bins could cause the penetrometer reading to deflect.

Instantaneous soil pressure (stress)

As the opener passes through the soil, pressures are created to move the soil aside, with multiple potential consequences from compaction to smearing. These pressures were measured using a specially designed diaphragm pressure pad (Mai, 1978). A small length of 9.5 mm diameter brass tube had a rubber diaphragm attached to one end. The other end had a sensitive electronic miniature pressure transducer attached. The tube was filled with water to act as a non-compressible liquid and a small bleed screw was used to expel all air. These tubes were inserted through holes in the side walls and base of the steel bins into close-fitting pre-bored holes in the soil so as to position the rubber diaphragm in intimate contact with soil a set distance (as close as 10 mm) from the expected pathway of a no-tillage opener to be tested.

Since each opener travelled a well-controlled pathway on the tillage bin tramway, it was possible to very accurately predetermine the side position of the soil-stress devices. The depth of penetration of each opener was somewhat less predictable, despite common ground-gauging wheels being used with each opener, because the ground surface of each bin did not finish exactly the same distance from the base of its steel bin during the field extraction process. Thus, somewhat more latitude was allowed for vertical positioning.

Even so, the water-filled tubes were used to protect the expensive miniature pressure transducers in the event of mechanical contact with a passing opener. The brass tubes and their rubber diaphragms were considered expendable in the event of an accident. The expensive pressure transducers were not. Figure 19.6 shows one such tube. In this manner, the contrasting instantaneous soil stresses created by a range of passing openers in an untilled soil were monitored and reported (Baker and Mai, 1982a).

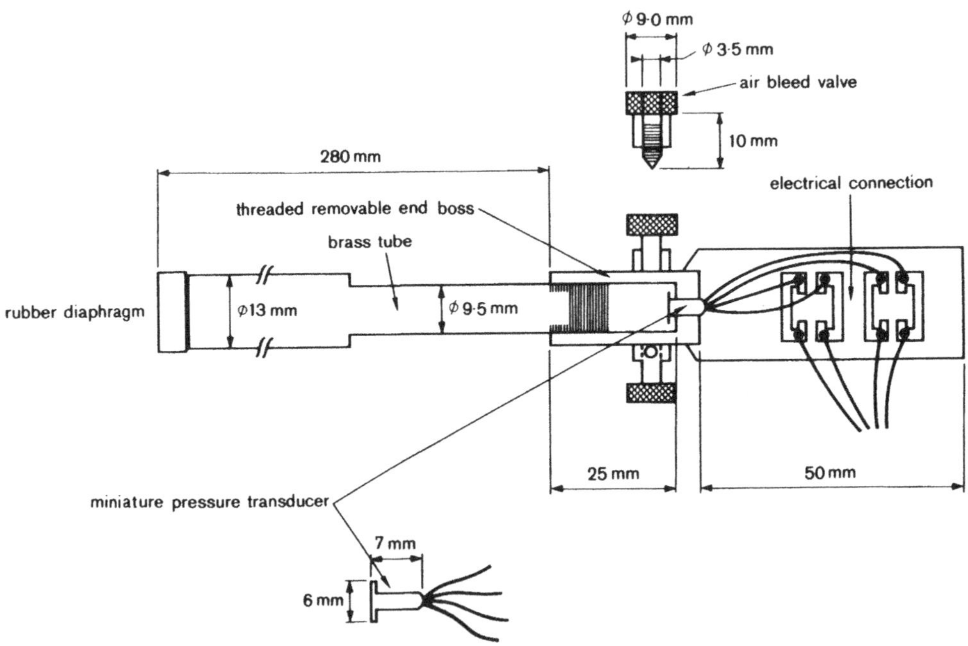

Fig. 19.6. A soil pressure measuring tube (from Baker, 1969a).

Instantaneous and permanent soil displacement

This was measured by placing small vertical probes in the soil at predetermined distances from the anticipated pathway of an opener to be tested in the soil bins on the tillage bin (Mai, 1978). A light non-stretchable thread was attached at one end to each probe and at the other end to a small electronic displacement transducer, which recorded both the instantaneous horizontal displacement of the soil as the opener passed and the permanent displacement after it had passed. The displacement data gave a measure of the direction in which an opener displaced the soil, as well as the plasticity of the soil and how it had responded to the mechanical action of that particular opener.

Soil bulk density

This was measured by extracting small soil cores (10 mm × 10 mm) from the slot zones in a location and pattern required by the specific experiment (Mai, 1978; Chaudhry, 1985). The cores were weighed and a standard procedure was used to calculate soil bulk density as the weight per unit volume of soil.

Smearing and compaction

This was a difficult parameter to accurately quantify, since smearing, in particular, was often confined to a layer less than 1 mm thick. It was determined that smearing in any case only affected root growth when it was allowed to dry and become a crust. Other environmental parameters determine slot drying, as previously described. Thus, no effort was made to develop a direct method to accurately quantify smears. It appeared that the difference between a smear and a compacted layer was only a matter of thickness.

Locating Seeds in the Soil

Three aspects of seed position within the soil were considered important to the design of no-tillage seed drills and planters (Ritchie, 1982): (i) seed spacing along the row; (ii) seed depth; and (iii) lateral position of the seed relative to the centre line of the slot.

Seed spacing

Measuring seed spacing is relatively simple. At least, it is if no account is taken of seed bounce in the slot and other soil factors, such as cloddiness. Accurate measurement can be achieved by simulated drilling, which involves moving a seeder over a sticky plate or paper so that the seeds dropped from the seeder are immediately fixed on the paper as the machine moves forward. The tillage bin and moving gantry described earlier were ideal for this function (Ritchie, 1982; Carter, 1986). Seed spacing can also be determined directly by measuring the distance along the surface of the soil between emerged seedlings. The latter method takes no account of displacement of shoots from the original positions of the seeds (by, for example, weaving around soil clods or stones) or of failure of seeds to germinate or of seedlings to emerge.

Seed depth

Measuring seeding depth is a deceptively difficult problem. For obvious reasons, the position of seeds in the vertical plane in the soil can only be determined after they have been sown, unlike horizontal seed spacing, which can be simulated on sticky paper without the opener having to penetrate the soil.

The problem is that when scientists excavate the soil to find individual seeds, it is almost inevitable that other seeds in the vicinity will be disturbed. In recent years, scientists have used one of four approaches:

Manual excavation (Hadfield, 1993; Thompson, 1993)

Despite the disadvantages, careful excavation of the soil in the field to expose individual seeds is still the most common method. This method has the problem that inherent errors are difficult to quantify and correct. With tilled soils, the seeds are approached from above, but, because of the lack of disturbance and the relative stability of some untilled soils and slots, it is sometimes possible to cut a trench alongside and approach the seeds from the side, which reduces the risk of disturbing other seeds.

Scoop sampling

A semi-cylindrical horizontal core of undisturbed soil, which centres on a drilled row, is removed with a specially shaped scoop, and then carefully split open on a bench in a laboratory to expose the seeds (Baker, 1976a). This technique can only be used with untilled soils because tilled soils are too friable and the cores collapse. It is somewhat more accurate than manual excavation from above because the seeds are approached from the side. It is also more convenient than field sampling from the side because the operator works mostly at bench height and the soil samples can be laid on their sides on the bench. The technique removes relatively short lengths of row at a time, and transports these to a laboratory. It is more time-consuming than other methods. It is more useful for locating and counting seeds and seedlings in a given length of row than for accurately recording their positions relative to the soil surface.

Tracing down seedlings

After emergence of seedlings, careful tracing down from the emerged shoots to the seed position will establish the original position of sown seeds within the soil (Stibbe *et al.*, 1980; Pidgeon, 1981; Allam and Weins, 1982; Choudhary *et al.*, 1985). This procedure has been mechanized for automatic recording to provide measurements for relatively large numbers of seedlings. But, because it only measures the emerged seedlings, it fails to record any position for non-emerged seeds. Since identifying disadvantaged seeds was one of the more obvious aims of locating them in the soil for no-tillage studies, the technique has had limited application.

X-ray imagery of seeds

By coating seeds with red lead oxide (a common bird repellent) prior to sowing, images of the seeds can be recorded by X-raying samples of soil removed from the field in metal boxes using a veterinary X-ray facility (Campbell, 1985; Choudhary *et al.*, 1985; Praat, 1988; Campbell and Baker, 1989; D. de Kantzow, 1985, 1993, personal communication). Both aluminium and steel are suitable for the boxes, as X-rays readily pass through these metals without an image. The technique is non-injurious to the seeds (they will germinate after X-raying) and it positively identifies seeds beneath the soil without disturbing them. It is also largely unaffected by soil type, moisture content or organic matter levels, but it is best suited to large seeds and relatively small numbers of samples because it is time-consuming and relatively expensive.

X-rays are derived from a point source on the X-ray machine; thus, as the X-rays scan a sample, a parallax error is created at all positions except those directly beneath the point source. This parallax error increases towards the extremities of the sample and affects the accuracy of quantifying the distances between individual seeds or between seeds and the surface of the soil. Campbell (1985) derived a mathematical correction for this error. He also used a strip of lead soldering wire to indicate the position of the soil surface in the X-rays. Figure 19.7 shows pea seeds coated with lead oxide X-rayed beneath the soil after seeding.

Lateral position of seeds relative to the centre line of the slot

As with seed depth, manually locating the lateral position of seeds after they have been drilled presents problems arising from the possibility of inadvertently displacing them before their positions can be recorded. Both scoop sampling and X-ray imagery were used on the few occasions this parameter was studied.

To date, no totally satisfactory method has been devised to positively, cheaply

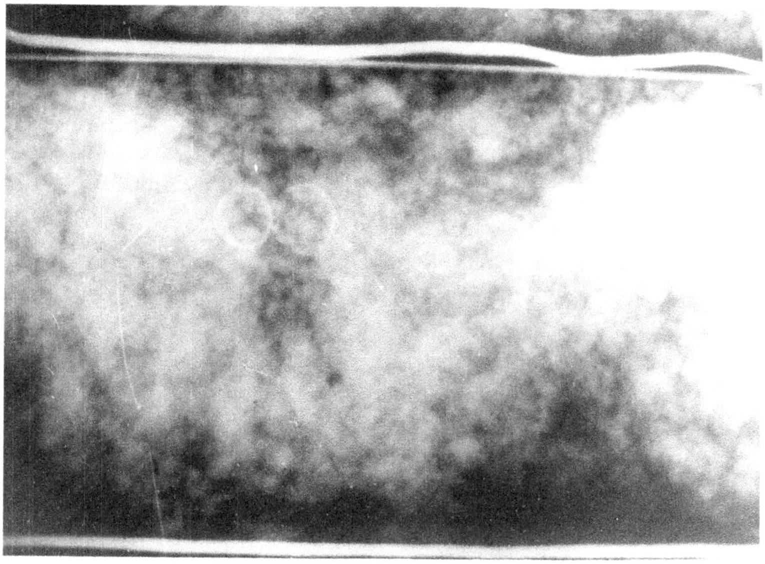

Fig. 19.7. Pea seeds coated with lead oxide X-rayed beneath the soil after seeding (from Campbell and Baker, 1989).

and repeatably identify the final three-dimensional position of seeds in the soil. Perhaps this accounts for why most designers of furrow openers and seed drills seem to satisfy themselves with defining how well their openers follow the ground surface, with the implied assumption that final seed placement is solely related to this capability.

Seed Travel within No-tillage Openers

The pathway seeds are required to travel through and from no-tillage openers is often more tortuous and less predictable than with simpler openers for tilled soils. Thus, it has been important to monitor seed travel and to analyse the causes of blockage or disruption to the flow.

All of the techniques adopted by the authors have involved use of video camera and slow replay facilities. Ritchie (1982) studied discharge of seeds from precision singulation seeders, together with a range of delivery tubes, by videotaping the seeds as they fell. He calculated the delay times between passage of successive seeds past a grid and the resulting potential variations in horizontal spacing along the row. The video was then replayed on a frame-by-frame basis against a background grid calibrated on both a time and distance basis. Figure 19.8 shows seed ejection being monitored in this manner using the tillage bin moving gantry as the source of seeder movement.

One study of seeds within the disc version of a winged opener involved substituting a clear Plexiglas disc for the normal steel disc on the opener and videotaping the seed pathway through the transparent disc. This opener is somewhat unique in that much of the internal pathway for the seeds involves a three-sided tube in close proximity to a revolving disc. The rotation of the disc forms one wall of this delivery tube and moves continuously. Scientists wanted to study the influence of this moving wall and the geometric shape of the stationary walls on seed drop and ejection from the opener. Figure 19.9 shows the seed flowing through such an opener.

To date, no satisfactory technique has been found for viewing seeds as they emerge from an opener beneath the soil, although knowledge of such action would assist greatly in designing openers with improved

Fig. 19.8. The ejection of seeds from a no-tillage opener being filmed on video. Four individual maize seeds can be seen dropping from the precision seeder at the centre right of the photograph.

Fig. 19.9. Seed flow being monitored through a clear Plexiglas disc.

seed ejection and depth control qualities. The advent of endoscopes and laparoscopes appeals as a possibility, but dust collection on the lens while operating beneath the soil would seem to be inevitable, and continuous dust removal, by, for example, a small jet of air, might interfere with the seed ejection process itself. None the less, there is potential for innovative design in the pursuit of this objective.

Drag on a Disc Opener

The disc version of winged openers, in particular, operates on the principle of a central vertical disc with a number of other components rubbing on it, creating a drag on the disc, resisting turning. Contact between the disc and some of these components, e.g. the left- and right-hand side blades and scrapers, is essential to the residue-handling and seed-placement functions of the opener. So, too, is continued and uninterrupted rotation of the disc. Thus, it became important to be able to quantify the magnitude of the various torsional drag forces opposing continuous rotation of the disc so that those that are unnecessary might be eliminated and those that are useful could be minimized.

The method adopted consisted of designing a special test stand in which a single opener was mounted in such a way as to allow each of the components contributing to torsional drag to be individually attached and removed without otherwise affecting the function of the opener (Javed, 1992). The test stand with opener attached was pulled through a range of test soils at a constant and known ground speed. The disc had a modified motorcycle disc brake assembly attached to it, which was capable of stopping the disc, resulting in 100% disc slip in the soil. The force required to achieve any intermediate and predetermined degree of braking of the disc was recorded by an electronic force transducer mounted between the disc brake assembly and the frame of the test stand. The speed of the disc, in revolutions per minute, was indicated by a tachometer and was directly proportional to disc slip in the soil at any given forward speed. Figure 19.10 shows the disc drag test stand and opener.

The free disc, i.e. without any torsionally dragging components attached, was first braked down to a predetermined speed, representing a set amount of disc slip in the soil. Then each of the components thought to cause torsional drag was added to the opener individually and measurements were taken of the residual braking found necessary to achieve the same set amount of disc slip. The difference between this and the original reading represented the torsional drag on the disc attributable to the added component. Variability of the soil that provided the tractive forces driving the disc required that a large number of recordings

Fig. 19.10. A test stand for monitoring disc drag of a no-tillage opener.

be made to develop accuracy. These were made using a high-speed electronic data logger, which recorded some 10,000 individual readings per test.

Accelerated Wear Tests of No-tillage Openers

The disc version of the winged opener was quite different from other seed drill openers for either tilled or untilled seedbeds. Thus, little was known about the relative wear rates of its essential components, although Baker and Badger (1979) had studied aspects of wear on earlier simple winged openers. The two most important areas of wear on this opener were considered to be the soil-to-metal wear on the outside of the side blades and their wings and the metal-to-metal wear between these side blades and the rotating disc.

Indeed, it had not yet been determined whether the side blades actually rubbed on the disc (metal-to-metal contact) or were held fractionally clear of the disc by a fine film of soil passing between the two components, in which case the contact would result in metal-to-soil-to-metal wear. The question of possible contact between the side blades and the disc was important because, if there was no direct contact, it would allow the side blades to be manufactured from material of considerably greater wear resistance. If there was direct contact, hard side blades might have eroded the discs themselves, which would have been unacceptable.

A technique was developed to examine both questions (Brown, 1982; Brown and Baker, 1985). A single opener was assembled in such a manner as to electrically isolate the side blades from the disc. It was then operated in the soil with leads connected to both the disc and side blades through a 12-volt battery to complete a circuit if the two made electrical contact and monitored by a meter or resistance light bulb. In the soils tested, a thin film of soil continually isolated the blades from the disc. Subsequent field experience confirmed that the hardness of blades had no effect on the life and integrity of the face of the disc, and that the abrasion patterns on both the disc and insides of the blades are consistent with metal-to-soil-to-metal wear.

None the less, the thin film of soil wears both components at this interface. A further technique was developed to accelerate wear testing of alternative strategies for prolonging the life of the side blades. The opener was modified so that the axle of the disc could be powered, causing it to rotate when the opener was stationary. The modified opener was arranged so that the base of the disc and blades were immersed in an open box of crushed (and, in one case, slurried) soil at normal sowing depth. The side blades were held against the disc with springs to simulate the forces experienced in the field if the opener was proceeding forwards. The test stand was left to run continuously in this manner for extended periods so as to monitor the pattern of wear at the interface between the blades and the disc. Figure 19.11 shows the accelerated wear box and test opener.

Where normal field wear patterns on the outside of the blades and wings were being studied (soil-to-metal wear), there was no substitute for continuous field drilling. By definition, the openers were required to experience continuously undisturbed soil; thus, re-drilling the same area repeatedly was not an option. In one test, a single-row drill was constructed and 16 hectares of undisturbed land were drilled in single rows. The opener covered some 500 km, which was equivalent to 225 hectares of continuous drilling with a 4.5 metre (15 foot) wide drill. Wear of the various blade treatments was measured both dimensionally and as weight loss (Brown, 1982; Brown and Baker, 1985).

Effects of Fertilizer Banding in the Slot

A number of experiments were conducted to determine the most appropriate position to place fertilizer separately from seed. Apart from the more common field experimentation techniques (which are not described in detail here), a number of

Fig. 19.11. An accelerated wear box for testing a no-tillage opener.

specialized experimental facilities were developed.

Horizontal, vertical or diagonal separation directions were compared using modified disc-version winged openers with side-blade combinations as follows:

1. The side blades were on opposite sides of the disc and of equal length (horizontal separation).
2. The side blades were on opposite sides of the disc but the fertilizer blade was 20 mm longer (diagonal separation).
3. One side blade was extended below the disc to create a deep band beneath and to one side of the seed (deep banding).
4. A short and a long side blade were both positioned on the same side of the disc (vertical separation).

Crop performance and seed damage were compared with field trials of these combinations. The horizontal option performed better than the diagonal or vertical options in all respects (see Chapter 9). This was fortunate, because the vertical option would have been difficult to implement on a field scale because the placement of two blades on one side of the disc would have been a difficult engineering task for other than experimental purposes. Figure 9.4 (Chapter 9) shows the experimental vertical placement opener.

Surprisingly, the extended diagonal option did not seem to interfere with the ability of the opener to handle surface residues, but it did cause undesirable wear patterns on the inside edges of the blades because each blade contacted the disc in the gullet zone for approximately half of the time, whereas contact was continuous if above the gullets. Longer blades also resulted in an increase in torsional drag on the disc because of the extended contact zone between the two. Since there was no benefit for the longer, more complicated, fertilizer blades, the option was not pursued.

Afzal (1981) studied vertical versus horizontal placement of fertilizer relative to seed without using an opener by extracting small blocks of undisturbed soil from the field and placing these in pots and boxes. For vertical placement, he bored small holes vertically into the soil, placed a preweighed amount of fertilizer in the base of the hole and replaced a known quantity of loose, tamped soil on top.

For horizontal separation he repeated the process described above but bored the vertical hole only to the seeding depth and covered the seeds with the plug of undisturbed soil. He then bored a horizontal hole from the side of the pot or box to position the fertilizer a predetermined distance from but at the same height as the seed. This hole was also closed using a plug of undisturbed soil, but in this case without surface residues.

Prototype Drills and Management Strategies

As part of the logical development of a new field technology, laboratory developments eventually need to be tested on a field scale. With seed drills and planters, this can only be partially achieved using small experimental machines. For example, one of the most important functions of no-tillage drills is the ability to handle surface residues. A single-row experimental machine might suggest how well an opener would perform this task, but only a machine with multiple openers would experience interactions of adjacent openers over a field with variable residue amounts and configurations. Thus, it is important to observe opener and drill performance on a field scale along with monitoring component wear and durability.

It is also necessary to compare different opener design performances on a field basis, but only after testing their biological performance in controlled laboratory conditions. When laboratory details are complete, appropriate field comparisons are possible using a test machine with several openers.

Operation in the field offers opportunities to monitor farmer reaction to the new technologies and to learn from farmers the constraints imposed by their management systems. It also allows the scientists, working with innovative farmers, to evolve new management strategies based on the increased capabilities of no-tillage and related emerging new technologies.

The development sequence involves testing: (i) single-row test drills; (ii) universal toolbars for field-testing several different designs of openers at the same time; (iii) plot-sized field drills and planters; and (iv) field-scale prototype drills and a drilling service for farmers.

Single-row test drills

A range of single-row drill designs were constructed for three objectives. First, they were a facility to test the mechanical performance of prototype openers in a field soil. Usually, the scope of such tests was focused on quantifying the mechanical functioning in different soil or residue conditions. Occasionally, as previously described, they may be used to drill an extended area for accelerated wear tests.

Generally, these single-row test drills consist of an opener rigidly mounted in a subframe attached to a tractor three-point linkage, with the downforce provided by removable ballast. In this manner, the tractor three-point linkage acted as the articulating drag arms for the opener, although the geometries of such linkages were seldom adjustable to form a perfect parallelogram. Within the limited range of vertical movement required of the test machines when the opener was in the ground, the tractor linkages were considered acceptable.

Secondly, single-row units were used for seeding purposes, at which time simple seed and fertilizer distribution systems were added to the basic machines. These simple drilling units offered field experience for verifying the laboratory biological performance of seed and fertilizer placement.

Thirdly, they became a convenient, although limited, machine to demonstrate the new opener capabilities to farmer groups without the need to transport heavy multirow machines to the field. But developers learned that, even with the aid of being able to see how each opener operated on the single-row demonstration drills, few observers were able to visualize the capabilities of a full-sized multi-row drill operating in the same circumstances. Consequently, the single-row demonstration concept played only a minor role in the wider technology transfer process, but was important in the engineering development process.

The single-row no-tillage drill concept was extended to become a commercially available machine as a plot drill for experimental stations; as a commercial drill for establishing edible shrubs by no-tillage on steep and erodible land; and as a commercial drill for small farmers in developing nations. The adaptability was further enhanced with the provision of a wheeled front steering frame to ensure that the wing angle on the opener remained correct and to facilitate turning corners when draught animals were used. A platform was added to the rear to allow an operator to step on or off to act as the downforce ballast. Figures 19.12, 19.13 and 19.14 illustrate several single-row test machines used to test and/or demonstrate the disc version of winger openers.

Simultaneous field testing of several opener designs

It is difficult to conduct a valid test of contrasting openers on a field scale without the ability to control the soil and climatic conditions. Almost invariably, such tests reveal the dominance of one opener over others being compared in that particular set of conditions, only to have the order altered in different conditions. The field conditions must be carefully identified under which any one opener is dominant, to learn the strengths and weaknesses of contrasting designs.

Often several parameters may vary, making it very difficult to isolate the reasons for one or more openers being superior for that particular set of conditions, without results from laboratory experiments that provide the biological capabilities of various no-tillage openers. And, unless the openers require very similar toolbar controls or are self-controlled, a single setting of height, down-pressure or speed may not be appropriate to all openers, biasing the results towards those openers that benefit most from the test settings.

It is interesting that, when people are asked to comment on the pros and cons of various no-tillage machines, many believe that such judgements cannot be made until several machines are lined up beside each other and tested in the same field. This seemingly obvious answer, however, is flawed because such field tests do not usually identify, let alone isolate, the individual causal processes of any differences that do arise. It

Fig. 19.12. A commercially available single-row no-tillage drill.

Fig. 19.13. An early single-row demonstration unit.

Fig. 19.14. A single-row machine for testing the residue-handling capability of a no-tillage opener.

is doubtful if any scientifically useful purpose has ever been served by field comparisons of multiple no-tillage machines.

Field toolbars are useful as an intermediate stage in the engineering field testing and development of prototype openers before any are considered sufficiently promising to incorporate into either a multi-row drill or planter, or even a self-contained single-row drill.

Figure 19.15 shows a universal field toolbar for evaluating a variety of openers,

Fig. 19.15. An example of a universal plot seed drill.

as designed by the University of New England, NSW, Australia (J. Scott, 1992, personal communication).

Plot-sized field drills and planters

Once the capabilities of an opener, e.g. the disc version of winged openers, are published or made public, it is common that other research organizations will design and construct plot-sized drills and planters equipped solely with these openers to sow test plots and fields for evaluation. In general, most designs of the plot machines have been an attempt to duplicate the mechanical arrangements of commercial field machines as faithfully as possible while at the same time incorporating facilities to more accurately monitor seed and fertilizer application rates, clean the product boxes between plots and adjust various mechanical options. These machines are made convenient to be easily transported to remote plots or farm field demonstrations. Such plot-sized drills have been an important intermediate stage of development before full-sized field prototype machines are contemplated. Figure 19.16 shows a selection of typical plot drills based on the disc version of winged openers.

Several designs of plot drills were used for plant-breeding purposes where plot sizes were small and the quantity of seed available was limited. Innovative mechanisms were introduced to delay release of the seed from the front gang of openers so that both the front and rear gangs began and ended seeding on the plot edges.

Field-scale prototype drills and a drilling service for farmers

The ultimate objective of any seed drill development programme is to produce a field-capable machine that can prove itself in normal commercial operation. One of the problems in developing effective no-tillage drills was that the drilling requirements were largely unknown and highly variable in this new style of farming, and few users could identify the causes of success or failure. Thus, field demonstration and proving took on a new dimension.

At first, a prototype drill was transported to a series of farmers' properties who were willing to try it on their farms, but this

RESEARCH–SCALE CROSS SLOT DRILLS

USDA-1, Pullman, WA

USDA-2, Pullman, WA

WSU Variety Testing – Pullman, WA

WSU Extension – Dayton, WA

WSU Research – Lind, WA

NDSU Research – Williston, ND

Fig. 19.16. Several plot drills based on the disc version of the winged opener.

often required modifying the hitches and hydraulic fittings each time a new farmer and tractor was involved. The problem of the incompatibility of hydraulic couplings was at first solved by equipping the test drill with a self-contained hydraulic system operated by a stationary petrol engine mounted on the drill itself, but this did not solve the other problems outlined above. It was also difficult to find a serious commitment from farmers to manage the no-tilled crops in a manner to provide reliable data on production and economics useful for field analyses.

A successful example of prototype testing and evaluation was a fully self-contained tractor, drill and truck developed and transported around New Zealand (Ritchie and Baker, 1987). That country offered a wide variety of agricultural enterprises, microclimates, farming systems and soil types representative of many of the agricultures of the world within a convenient travelling distance.

A charge was made to the farmers to both fund the operation and involve the participating farmers in a more committed and meaningful way. Thus, what was still primarily a field testing operation for the scientists also became a contract drilling service for the farmers ('custom drilling') and a highly effective technology transfer process for both parties. Over a 10-year period, during which three generations of prototype drills were utilized, this field drilling operation was used on approximately 200 separate fields on over 100 different properties, many of which were drilled for a number of successive years. Figure 19.17 shows the self-contained field operational machine.

While the primary purpose of this prototype drilling operation was to provide vital field performance information for the originating scientists and function as a technology transfer medium, the operation became the cornerstone for development and evaluation of new and innovative farm management techniques and strategies. And cooperating scientists and consultants used the opportunity as the means to introduce drought-tolerant pasture species into existing dryland grasslands by other scientists (Barr, 1986; Ritchie, 1986a, b; Milne and Fraser, 1990; Milne *et al.*, 1993).

Summary of Drill Development and Technology Transfer

1. There are few known or standardized experimental procedures for objectively evaluating no-tillage technologies.
2. The study of no-tillage drills, planters and openers requires developing knowledge about experimental procedures, mechanical performance and resulting plant growth.
3. Removing large soil blocks from the field in an undisturbed state to a climatically controlled environment is a useful method to control soil moisture, drill with openers to simulate field performance and control post-drilling climate.
4. Environmental requirements of seeds and seedlings within the seed slot involves studying such variables as: (i) soil moisture regime within the slot; (ii) soil-air humidity within the slot; (iii) soil oxygen within and around the slot; and (iv) soil temperature within the slot.
5. Soil disturbance by drill openers requires monitoring the parameters of: (i) soil strength; (ii) instantaneous soil pressure (stress); (iii) instantaneous and permanent soil displacement; (iv) soil bulk density; and (v) smearing.
6. Important aspects of seed position within the soil after drilling are: (i) seed

Fig. 19.17. A fully self-contained drilling machine for field testing and on-farm demonstrations.

spacing along the row; (ii) seed depth; and (iii) lateral position of the seed relative to the centre line of the slot.

7. The pathway seeds travel from metering to and through successful no-tillage openers is often more tortuous and less predictable than with simpler openers for tilled soils.

8. It is important to quantify the drag forces opposing rotation of disc openers to eliminate those that are unnecessary and minimize those that are useful.

9. Normal field wear of all drill components (blades, wings, discs, bearings, etc.) must be studied with continuous field drilling in undisturbed soil.

10. Adding components to openers for fertilizer placement may cause undesirable wear patterns or interfere with the ability of the opener to handle surface residues.

11. Field toolbars with multiple openers are useful to field-test prototype openers.

12. The ultimate objective of any seed drill development programme is to produce a field-capable machine that can prove itself in the normal commercial operation for which it is intended.

References

Abu-Hamdeh, N.H. (2003) Soil compaction and root distribution for okra as affected by tillage and vehicle parameters. *Soil and Tillage Research* 74, 25–35.

Addae, P.C., Collis-George, N. and Pearson, C.J. (1991) Over-riding effects of temperature and soil strength on wheat seedling under minimum and conventional tillage. *Field Crops Research* 28, 103–116.

Afzal, C.M. (1981) The requirements of dry fertilizer placement in direct drilled crops by an improved chisel coulter. Thesis, Massey University Library, New Zealand, 127 pp.

Allam, R.K. and Wiens, E.H. (1982) Air seeder testing. *Canadian Agricultural Engineering* 24, 91–101.

Allmaras, R.R., Schomberg, H.H., Douglas, C.L., Jr and Dao, T.H. (2000) Soil organic carbon sequestration potential of adopting conservation tillage in U.S. croplands. *Journal of Soil and Water Conservation* 55, 365–373.

Almeida, R.A. (1993) Adaptação da 'matraca' ao plantio direto em pequenas propriedades. In: *Encontro Latinoamerico de Planto Direitona Paquena Propriedade 1*, IAPAR, Ponta Grossa, Brazil, pp. 251–257.

Angers, D.A. and Simard, R.R. (1986) Relationships between organic matter content and soil bulk density. *Canadian Journal of Soil Science* 66, 743–746.

Ankeny, M.D., Kaspar, T.C. and Horton, R. (1990) Characterization of tillage and traffic effects on unconfined infiltration measurements. *Soil Science Society of America Journal* 54, 837–840.

Anon. (1993) *Sod Seeding Techniques.* Research Update 699, Prairie Agricultural Machinery Institute, Saskatoon, Saskatchewan, Canada.

Anon. (1995) North Otago sustainable land management guidelines [draft]. Otago Regional Council, Dunedin, New Zealand, 92 pp.

Anon. (2000) Study promotes soil benefits of no-till cropping. *Country Wide, New Zealand,* April, 22.

Anon. (2004) Mapping a route through precision drill maze. *Profi International* 6, June, 18–24.

Aquino, P. (1998) *The Adoption of Bed Planting of Wheat in the Yaqui Valley, Sonora, Mexico.* CIMMYT Wheat Special Report No. 17a, CIMMYT, Mexico, DF.

Araújo, A.G., Casão, R., Jr and Figueiredo, P.R.A. (1993) Recomendações para dimensionamento e construção do rolo-faca. In: *Encontro Latinoamericano sobre Plantio Direto na Pequena Propriedade, 1, 1993. Anais.* Instituto Agronômico do Paraná, Ponta Grossa, pp. 271–280.

Araújo, A.G. and Rodrigues, B.N. (2000) Manejo mecânico e químico da aveia preta e sua influencia sobre a taxa de decomposição e o controle de plantas daninhas em semeadura direta de milho. *Planta Daninha, Botucatu* 18(1), 151–160.

Avnimelech, Y. and Cohen, A. (1988) On the use of organic manures for amendment of compacted clay soils: effects of aerobic and anaerobic conditions. *Biological Wastes* 29, 331–339.

Bainer, R., Kepner, R.A. and Barger, E.L. (1955) *Principles of Farm Machinery.* John Wiley and Sons, New York, 571 pp.

Baker, C.J. (1969a) A tillage bin and tool testing apparatus for turf samples. *Journal of Agricultural Engineering Research* 14(4), 357–360.

Baker, C.J. (1969b) The present methods of pasture establishment. *Proceedings of the New Zealand Grassland Association* 31, 52–59.

Baker, C.J. (1970) A simple covering harrow for direct drilling. *New Zealand Farmer* 91, 62–63.

Baker, C.J. (1971) Filling up spaces by surface seeding. In: *Proceedings of the 24th New Zealand Weed and Pest Control Conference*, New Zealand Weed and Pest Control Society, pp. 75–79.

Baker, C.J. (1976a) An investigation into the techniques of direct drilling seeds into undisturbed sprayed pasture. Thesis, Massey University Library, New Zealand, 186 pp.

Baker, C.J. (1976b) Experiments relating to the techniques of direct drilling of seeds into untilled dead turf. *Journal of Agricultural Engineering Research* 21(2), 133–145.

Baker, C.J. (1976c) Some effects of cover, seed size, and soil moisture status on establishment of seedlings by direct drilling. *New Zealand Journal of Experimental Agriculture* 5, 47–53.

Baker, C.J. (1979) Equipment impact and recent developments. In: *Proceedings of the Conservation Tillage Technical Seminar, Christchurch, New Zealand*, Monsanto (NZ) Ltd, Johnsonville, New Zealand, pp. 5–12.

Baker, C.J. (1981a) How direct drilling equipment developments in Europe and North America relate to New Zealand. In: *Proceedings of the Conservation Tillage Technical Seminar, Christchurch, New Zealand*, Monsanto (NZ) Ltd, Johnsonville, New Zealand, pp. 105–121.

Baker, C.J. (1981b) Techniques of overdrilling for the introduction of improved pasture species in temperate grasslands. In: *Proceedings of the XIV International Grassland Congress*, International Grassland Congress, pp. 542–544.

Baker, C.J. (1993a) Inverted T drill openers for pasture establishment by conservation tillage. In: *Proceedings of the XVII International Grassland Congress, Palmerston North, New Zealand*, International Grassland Congress.

Baker, C.J. (1993b) The evolution of 'Cross Slot' no-tillage planters and drills. In: *Proceedings of the II National Congress of No Tillage, Cordoba, Argentina (Trabajos Presentados II Congreso Nacional de Siembra Directa)*, Asociación Argentina Productores en Siembra Directa, Rosario, Argentina, pp. 104–116.

Baker, C.J. (1993c) Machinery: drills. In: Pottinger, R.P., Lane, P.M. and Wilkins, J.A.R. (eds) *Pasture Renovation Manual*, 2nd edn. AgResearch, Hamilton, New Zealand, pp. 48–54.

Baker, C.J. (1994) The case for hi-technology no-tillage. In: *Proceedings of the Third National Congress of No-Tillage, Cordoba, Argentina*, Asociación Argentina Productores en Siembra Directa, Rosario, Argentina, pp. 145–157.

Baker, C.J. (1995) Is hi-tech no-tillage inevitable? In: *Proceedings of the Third National No-Tillage Conference, Indianapolis, Indiana, USA*, Lessiter Publications, Brookfield, Wisconsin, USA, pp. 153–164.

Baker, C.J. and Afzal, C.M. (1981) Some thoughts on fertilizer placement in direct drilling. In: *Proceedings of the Conservation Tillage Seminar, Christchurch, New Zealand*, Monsanto (NZ) Ltd, Johnsonville, New Zealand, pp. 343–354.

Baker, C.J. and Afzal, C.M. (1986) Dry fertilizer placement in conservation tillage: seed damage in direct drilling. *Soil and Tillage Research* 7, 241–250.

Baker, C.J. and Badger, E.M. (1979) Developments with seed drill coulters for direct drilling: II. Wear characteristics of an experimental chisel coulter. *New Zealand Journal of Experimental Agriculture* 7, 185–188.

Baker, C.J. and Choudhary, M.A. (1988) Seed placement and micro-management of residue in dryland no-till. In: *Proceedings of the International Conference on Dryland Farming, Amarillo, Texas, USA*, Texas Agricultural Experiment Station, Bushland, Amarillo, Texas, pp. 544–546.

Baker, C.J. and Mai, T.V. (1982a) Physical effects of direct drilling equipment on undisturbed soils: IV techniques for measuring soil compaction in the vicinity of drilling grooves. *New Zealand Journal of Agricultural Research* 25, 43–49.

Baker, C.J. and Mai, T.V. (1982b) Physical effects of direct drilling equipment on undisturbed soils. V. Groove compaction and seedling root development. *New Zealand Journal of Agricultural Research* 25, 51–60.

Baker, C.J. and Saxton, K.E. (1988) The Cross Slot conservation-tillage grain drill opener. In: *Winter Meeting of the American Society of Agricultural Engineers, Chicago, Illinois, USA*, American Society of Agricultural Engineers, St Joseph, Missouri, Paper no. 88–1568.

Baker, C.J., Badger, E.M., McDonald, J.H. and Rix, C.S. (1979a) Developments with seed drill coulters for direct drilling: I. Trash handling properties of coulters. *New Zealand Journal of Experimental Agriculture* 7, 175–184.

Baker, C.J., McDonald, J.H., Seebeck, K., Rix, C.S. and Griffiths, P.M. (1979b) Developments with seed drill coulters for direct drilling: III. An improved chisel coulter with trash handling and fertilizer placement capabilities. *New Zealand Journal of Experimental Agriculture* 7, 189–196.

Baker, C.J., Thom, E.R. and McKain, W.I. (1979c) Developments with seed drill coulters for direct drilling: IV. Band spraying for suppression of competition during over-drilling. *New Zealand Journal of Experimental Agriculture* 7, 411–416.

Baker, C.J., Chaudhry, A.D. and Springett, J.A. (1987) Barley seedling establishment and infiltration from direct drilling in a wet soil. *Proceedings of the Agronomy Society of New Zealand* 17, 59–66.

Baker, C.J., Chaudhry, A.D. and Springett, J.A. (1988) Barley seedling establishment by direct drilling in a wet soil. 3. Comparison of six sowing techniques. *Soil and Tillage Research* 11, 167–181.

Baker, C.J., Saxton, K.E. and Ritchie, W.R. (1996) *No-tillage Seeding: Science and Practice.* CAB International, Wallingford, UK, 258 pp.

Baker, C.J., Choudhary, M.A. and Collins, R.M. (2001) Factors affecting the uptake of no-tillage in Australia, Asia and New Zealand. In: *Proceedings 1 World Congress on Conservation Agriculture, Madrid, Spain,* Vol. 1, European Conservation Agriculture Federation (Brussels, Europe) and Food and Agriculture Organization of the United Nations, Rome, Italy, pp. 35–42.

Ball, B.C., Scott, A. and Parker, J.P. (1999) Field N_2O, CO_2 and CH_4 fluxes in relation to tillage, compaction and soil quality in Scotland. *Soil and Tillage Research* 53(1), 29–39.

Barr, S.J. (1980) Some factors affecting the success of pasture renovation using over-drilling with band spraying. Dissertation, Department of Agricultural Engineering Library, Massey University, New Zealand, 84 pp.

Barr, S.J. (1981) Specifying and evaluating the performance of overdrilling machines. In: *Proceedings of the Conservation Tillage Seminar, Christchurch, New Zealand,* Monsanto (NZ) Ltd, Johnsonville, New Zealand, pp. 151–163.

Barr, S.J. (1986) Direct drilling to increase pasture production. *New Zealand Dairy Farming Annual* 150–158.

Basker, A., MacGregor, A.N. and Kirkman, J.H. (1993) Exchangeable potassium and other cations in non-ingested soil and casts of two species of pasture earthworms. *Soil Biology and Biochemistry* 25, 1673–1677.

Baumer, C.R., Devito, C.R. and Gonzalez, N.C. (1994) *Sembradoras Directas de Granos Finos.* Boletin de Extension No. 9 (PAC-BAN No. 5), Instituto Nacional de Tecnologia Agropecuaria (INTA), Argentina, 25 pp.

Bayer, C., Martin-Neto, L., Mielniczuk, J. and Ceretta, C.A. (2000a) Effect of no-till cropping systems on soil organic matter in a sandy loam Acrisol from southern Brazil monitored by electron spin resonance and nuclear magnetic resonance. *Soil and Tillage Research* 53, 95–104.

Bayer, C., Mielniczuk, J., Amado, T.J.C., Martin-Neto, L. and Fernandes, S.V. (2000b) Organic matter storage in a sandy loam Acrisol affected by tillage and cropping systems in southern Brazil. *Soil and Tillage Research* 54, 101–109.

Benjamin, L.A. (1990) Variation in time of seedling emergence within populations: a feature that determines individual growth and development. *Advances in Agronomy* 44, 1–25.

Bernardi, R. and Lazaretti, A. (2004) Agricultura orgânica em sistema de plantio direto. In: *Encontro Nacional de Plantio Direto na Palha, 9°, Anais.* Federação Brasileira de Plantio Direto na Palha, Chapecó, Brazil, pp. 105–106.

Bevin, A.S. (1944) *The Awakening.* Wilson and Horton, Auckland, New Zealand, 78 pp.

Blackmore, L.W. (1955) The overdrilling of pastures. In: *Proceedings of the 7th New Zealand Grasslands Association Conference.* New Zealand Grasslands Association, Dunedin, pp. 139–148.

Blackwell, P.S., Ward, M.A., Lefevre, R.N. and Cowan, D.J. (1985) Compaction of a swelling clay soil by agricultural traffic, effects upon conditions for growth of winter cereals and evidence for some recovery of structure. *Journal of Soil Science* 36(4), 633–650.

Blevens, R.L. and Frye, W.W. (1993) Conservation tillage: an ecological approach to soil management. *Advances in Agronomy* 51, 37–78.

Bligh, K.J. (1991) Tined and disced no-till seeder comparison. In: *Proceedings of the Meeting of the American Society of Agricultural Engineers,* American Society of Agricultural Engineers, St Joseph, Missouri, Paper number 91–8005.

Borlaug, N.E. (1994) Feeding a human population that increasingly crowds a fragile planet. In: *Supplement to Transactions of the 15th World Congress of Soil Science, Acapulco, Mexico*, International Society of Soil Science. Supplement to 15th World Congress of Soil Science, Acapulco, Mexico.

Botta, G.F., Jorajuria, D., Balbuena, R. and Rosatto, H. (2004) Mechanical and cropping behavior of direct drilled soil under different traffic intensities: effect on soybean (*Glycine max* L.) yields. *Soil and Tillage Research* 78, 53–58.

Bouaziz, A., Souty, N. and Hicks, D. (1990) Emergence force exerted by wheat seedlings. *Soil and Tillage Research* 17, 211–219.

Braverman, M.P., Dusky, J.A., Locascio, S.J. and Hornsby, A.G. (1990) Sorption and degradation of thiobencarb in three Florida soils. *Weed Science* 38(6), 583–588.

Brougham, R.W. and Hodgson, J. (1992) Grasslands for the world. Foreword to *Prospectus for XVII International Grassland Congress*. International Grassland Congress, Palmerston North, New Zealand.

Brown, A.D. (1997) Soil physical conditions for optimum crop establishment. PhD thesis, Cranfield University, Silsoe, UK.

Brown, S.W. (1982) An investigation into wear characteristics of a direct drilling coulter (opener). Thesis, Massey University Library, Auckland, New Zealand, 228 pp.

Brown, S.W. and Baker, C.J. (1985) Wear characteristics of a direct-drilling opener. *Soil and Tillage Research* 6, 247–260.

Calegari, A. (1990) *Plantas para adubação verde de inverno no sudoeste do Paraná*. Boletim Técnico, 35, IAPAR, Londrina, 37 pp.

Cambardella, C.A. and Elliott, E.T. (1992) Particulate soil organic matter changes across a grassland cultivation sequence. *Soil Science Society of America Journal* 56, 777–783.

Campbell, A.J. (1985) An X-ray technique for determining seed placement in direct drilled soils. Thesis, Massey University Library, Auckland, New Zealand, 154 pp.

Campbell, A.J. and Baker, C.J. (1989) An X-ray technique for determining three-dimensional seed placement in soils. *Transactions of the American Society of Agricultural Engineers* 32(2), 379–384.

Campbell, B.D. (1981) Observations on early seedling establishment for overdrilled red clover. In: *Proceedings of the Conservation Tillage Seminar, Christchurch, New Zealand*. Monsanto (NZ) Ltd, Johnsonville, New Zealand, pp. 143–150.

Campbell, B.D., McDonald, J.H. and Baker, C.J. (1985) A mechanism to regulate winged coulter depth and overdrilled seed spacing in field experiments. *New Zealand Journal of Experimental Agriculture* 13, 175–179.

Campbell, C.A., Selles, F., Lafond, G.P. and Zentner, R.P. (2001) Adopting zero tillage management: impact on soil C and N under long-term crop rotations in a thin Black Chernozem. *Canadian Journal of Soil Science* 81, 139–148.

Campbell, D.J., Dickson, J.W., Ball, B.C. and Hunter, R. (1986) Controlled seedbed traffic after ploughing or direct drilling under winter barley in Scotland, 1980–1984. *Soil and Tillage Research* 8, 3–28.

Carpenter, A., Kain, W.M., Baker, C.J. and Sims, R.E.H. (1978) The effects of tillage technique on insect pests of seedling maize. In: *Proceedings of the 31st New Zealand Weed and Pest Control Conference*. New Zealand Weed and Pest Control Society, pp. 89–91.

Carter, L.M., Meek, B.D. and Rechel, E.A. (1991) Zone production system for cotton: soil response. *Transactions of the American Society of Agricultural Engineers* 34(2), 354–360.

Carter, M.R. (1994) *Conservation Tillage in Temperate Agroecosystems*. Lewis, Boca Raton, Florida, 390 pp.

Carter, P.A. (1986) Aspects of precision spacing of vegetable crop seeds and no-tillage. Thesis, Massey University Library, Auckland, New Zealand, 110 pp.

Casão, R., Jr and Yamaoka, R.S. (1990) Desenvolvimento de semeadora-adubadora direta'tração animal. In: *Congresso Brazileiro de Engenharia Agricola 1990, Piracicaba Anais*. SBEA, Piracicaba, pp. 766–777.

Chamen, W.C.T. and Longstaff, D.J. (1995) Traffic and tillage effects on soil conditions and crop growth on a swelling clay soil. *Soil Use and Management* 11, 168–176.

Chamen, W.C.T., Chittey, E.T., Leede, P.R., Goss, M.J. and Howse, K.R. (1990) The effect of tyre soil contact pressure and zero traffic on soil and crop responses when growing winter wheat. *Journal of Agricultural Engineering Research* 47, 1–21.

Chamen, W.C.T., Vermeulen, G.D., Campbell, D.J. and Sommer, C. (1992) Reduction of traffic-induced soil compaction: a synthesis. *Soil and Tillage Research* 24, 303–318.

Chaney, K. and Swift, R.S. (1984) The influence of organic matter on aggregate stability in some British soils. *Journal of Soil Science* 35, 223–230.

Chaudhry, A.D. (1985) Effects of direct drilling openers, surface residue and earthworms on seed and seedling performance in a wet soil. PhD thesis, Massey University Library, Auckland, New Zealand, 336 pp.

Chaudhry, A.D. and Baker, C.J. (1988) Barley seedling establishment by direct drilling in a wet soil: I. Effects of openers under simulated rainfall and high water-table conditions. *Soil and Tillage Research* 11, 443–461.

Choudhary, M.A. (1979) Interrelationships between performance of direct drilled seeds, soil micro-environment and drilling equipment. PhD thesis, Massey University Library, Auckland, New Zealand, 211 pp.

Choudhary, M.A. and Baker, C.J. (1980) Physical effects of direct drilling equipment on undisturbed soils: I. Wheat seedling emergence from a dry soil under controlled climates. *New Zealand Journal of Agricultural Research* 23, 489–496.

Choudhary, M.A. and Baker, C.J. (1981a) Physical effects of direct drilling equipment on undisturbed soils: II. Seed groove formation by a 'triple disc' coulter and seedling performance. *New Zealand Journal of Agricultural Research* 24, 183–187.

Choudhary, M.A. and Baker, C.J. (1981b) Physical effects of direct drilling equipment on undisturbed soils: III. Wheat seedling performance and in-groove micro-environment in a dry soil. *New Zealand Journal of Agricultural Research* 24, 189–195.

Choudhary, M.A. and Baker, C.J. (1982) Effects of drill coulter design and soil moisture status on emergence of wheat seedlings. *Soil and Tillage Research* 2, 131–142.

Choudhary, M.A. and Baker, C.J. (1994) In: Carter, M.R. (ed.) *Conservation Tillage in Temperate Agroecosystems.* Lewis Publishers, Boca Raton, Florida, pp. 183–207.

Choudhary, M.A., Guo, P.Y. and Baker, C.J. (1985) Seed placement effects on seedling establishment in direct drilled fields. *Soil and Tillage Research* 6, 79–93.

Choudhary, M.A., Baker, C.J. and Stiefel, W. (1988a) Dry fertilizer placement in direct drilling. *Soil and Tillage Research* 12, 213–221.

Choudhary, M.A., Baker, C.J., Currie, L.D. and Lynch, T.J. (1988b) Disposal of agricultural liquid waste by sub-soil injection. In: Bhamidimarri, R. (ed.) *Alternative Waste Treatment Systems.* Elsevier Applied Science, London.

Cockcroft, B. and Olsson, K.A. (2000) Degradation of soil structure due to coalescence of aggregates in no-till, no-traffic beds in irrigated crops. *Australian Journal of Soil Research* 38, 61–70.

Collins, R.M. (1970) A study of band spraying and direct drilling as a technique for increasing the winter production of pastures. Thesis, Massey University Library, Auckland, New Zealand, 142 pp.

Collis-George, N. and Lloyd, J.E. (1979) The basis for a procedure to specify soil physical properties of a seed bed for wheat. *Australian Journal of Agricultural Research* 30, 831–846.

Cook, J. and Veseth, R. (1993) *Wheat Health.* National Society of Plant Pathology.

Copp, L.G.L. (1961) A precision seeder operated by suction. *New Zealand Journal of Agricultural Research* 4, 441–443.

Cross, M.W. (1957) Overdrilling machinery and methods. In: *Massey Agricultural College Dairy Farmers' Conference, Massey University, New Zealand.* Massey University, Palmerston North, New Zealand, pp. 77–84.

Cross, M.W. (1959) Method of reducing pasture establishment costs. *Sheepfarming Annual, Massey University, New Zealand.* Massey University, Palmerston North, New Zealand, 107–116.

Crovetto, L.C. (1996) *Stubble over the soil: The Vital Role of Plant Residue in Soil Management to Improve Soil Quality.* American Society of Agronomy, Madison, Wisconsin, 245 pp.

Cullen, N. (1966) Pasture establishment on unploughable hill country in New Zealand. In: *Proceedings of the X International Grassland Congress*, International Grassland Congress, Palmerston North, New Zealand, pp. 851–855.

Dangol, B. (1968) A review of the factors affecting pasture establishment. Thesis, Massey University Library, Auckland, New Zealand, 88 pp.

Dick, W.A., Van Doren, D.A., Triplett, G.D., Jr and Henry, J.E. (1986a) *Influence of Long-term Tillage and Crop Rotation Combination on Crop Yields and Selected Soil Parameters. The Results Obtained for a Typic Fragiudalf Soil.* Research Bulletin 1180, OSU and OARDC, Wooster, Ohio.

Dick, W.A., Van Doren, D.A., Triplett, G.D., Jr and Henry, J.E. (1986b) *Influence of Long-term Tillage and Crop Rotation Combination on Crop Yields and Selected Soil Parameters. II. The Results Obtained for a Typical Fragiudalf Soil.* Research Bulletin 1181, OSU and OARDC, Wooster, Ohio.

Dickson, J.W. and Campbell, D.J. (1990) Soil and crop responses to zero- and conventional-traffic systems for winter barley in Scotland, 1982–1986. *Soil and Tillage Research* 18, 1–26.

Diekow, J., Mielniczuk, J., Knicker, H., Bayer, C., Dick, D.P. and Kögel-Knabner, I. (2004) Soil C and N stocks as affected by cropping systems and nitrogen fertilization in a southern Brazil Acrisol managed under no-tillage for 17 years. *Soil and Tillage Research* 81, 87–95.

Dixon, H.N. (1972) The effects of coulter design on soil compaction and root development of a cereal following direct drilling. Dissertation, Department of Agricultural Engineering Library, Massey University, Auckland, New Zealand, 51 pp.

Domínguez, J., Bohlen, P.J. and Parmelee, R.W. (2004) Earthworms increase nitrogen leaching to greater soil depths in row crop agroecosystems. *Ecosystems* 7, 672–685.

Doran, J.W. (1980) Microbial changes associated with residue management with reduced tillage. *Soil Science Society of America Journal* 44, 518–524.

Doran, J.W., Fraser, D.G., Culik, M.N. and Leibhardt, W.C. (1987) Influence of alternative and conventional agricultural management on soil microbial processes and N availability. *American Journal of Alternative Agriculture* 2, 99–106.

Dudek, D.J., Goffman, J. and Wade, S.M. (1997) Emissions trading in nonattainment areas: potential, requirements, and existing programs. In: Kosobud, R.F. and Zimmermann, J.M. (eds) *Market-based Approaches to Environmental Policy: Regulatory Innovations to the Fore.* Van Nostrand Reinhold, New York, pp. 151–185.

Dumanski, J., Desjardins, R.L., Tarnocai, C., Monreal, D., Gregorich, E.G., Kirkwood, V. and Campbell, C.A. (1998) Possibilities for future carbon sequestration in Canadian agriculture in relation to land-use changes. *Climate Change* 40, 81–103.

Duxbury, J.M., Harper, L.A. and Mosier, A.R. (1993) Contributions of agroecosystems to global climate change. In: Harper, L. *et al.* (eds) *Agroecosystem Effects on Radiatively Important Trace Gases and Global Climate Change.* ASA Publication 55, ASA, Madison, Wisconsin, pp. 1–18.

Edwards, J.H., Wood, C.W., Thurlow, D.L. and Ruf, M.E. (1992) Tillage and crop rotation effects on fertility status of a Hapludalt soil. *Soil Science Society of America Journal* 56, 1577–1582.

Edwards, W.M., Shipitalo, M.J. and Norton, L.D. (1988) Contribution of macroporosity to infiltration into a continuous corn no-tilled watershed: implications for contaminant movement. *Journal of Contaminant Hydrology* 3, 193–205.

Ellert, B.H. and Janzen, H.H. (1999) Short-term influence of tillage on CO_2 fluxes from a semi-arid soil on the Canadian prairies. *Soil and Tillage Research* 50, 21–32.

El-Swaify, S.A. (ed.) (1999) *Sustaining the Global Farm – Strategic Issues, Principles, and Approaches.* International Soil Conservation Organization (ISCO) and Department of Agronomy and Soil Science, University of Hawaii, Manoa, Hawaii, 60 pp.

Faulkner, E. (1943) *Ploughman's Folly.* Michael Joseph, London, 142 pp.

Fick, C. (2000) On the money. *Farm Journal* April, 12–13.

Flerchinger, G.N. and Saxton, K.E. (1989a) Simultaneous heat and water model of a freezing snow–residue–soil system: I. Theory and development. *Transactions of the American Society of Agricultural Engineers* 32(2), 565–571.

Flerchinger, G.N. and Saxton, K.E. (1989b) Simultaneous heat and water model of a freezing snow–residue–soil system: II. Field verification. *Transactions of the American Society of Agricultural Engineers* 32(2), 573–578.

Flowers, M.D. and Lal, R. (1998) Axle load and tillage effects on soil physical properties and soybean grain yield on a Mollic Ochraqualf in northwest Ohio. *Soil and Tillage Research* 48, 21–35.

Follas, G.B. (1981) The effects of coulter design on slug damage in direct drilling. In: *Proceedings of the Conservation Tillage Seminar, Christchurch, New Zealand.* Monsanto (NZ) Ltd, Johnsonville, New Zealand, pp. 201–216.

Follas, G.B. (1982) A study of the effects of ground cover on overwintering slug populations and the effect of coulter design on slug incidence in direct drilling. Thesis, Massey University Library, Auckland, New Zealand, 121 pp.

Follett, R.F. (2001) Soil management concepts and carbon sequestration in cropland soils. *Soil and Tillage Research* 61, 77–92.

Franzluebbers, A.J. and Arshad, M.A. (1996) Soil organic matter pools during early adoption of conservation tillage in northwestern Canada. *Soil Science Society of America Journal* 60, 1422–1427.

Franzluebbers, A.J., Hons, F.M. and Zuberer, D.A. (1995a) Tillage and crop effects on seasonal dynamics of soil CO_2 evolution, water content, temperature and bulk density. *Soil Science Society of America Journal* 2, 95–109.

Franzluebbers, A.J., Hons, F.M. and Zuberer, D.A. (1995b) Tillage-induced seasonal changes in soil physical properties affecting soil CO_2 evolution under intensive cropping. *Soil and Tillage Research* 34, 41–60.

Frye, W.W. (1984) Energy requirements in no tillage. In: Phillips, R.E. and Phillips, S.H. (eds) *No-tillage Agricultural Principles and Practices.* VanNostrand Reinhold, New York, pp. 127–151.

Fundação ABC (1993) Custos de Mecanizacao Agricola Castro. *Revista Batavo* 21, 23.

Gaffney, J. and Wilson, A. (2003) The economics of zero tillage and controlled traffic farming for Western Downs farms. In: *Proceedings of the International Soil Tillage Research Organisation 16th Triennial Conference, Brisbane, 13–18 July.* Elsevier, Netherlands, pp. 458–464.

Gamero, C.A., Siqueira, R., Levien, R. and Silva, S.L. (1997) Decomposição da aveia preta (*Avena strigosa* Schreb.) manejada com rolo-faca e triturador de palhas. In: *Congresso Brasileiro de Engenharia Agrícola, 1997, Campina Grande. Anais do Congresso Brasileiro de Engenharia Agrícola, 1997,* v. cd-rom, pp. 1–13.

Garcia-Torres, L., Benites, J. and Martinez-Vilela, A. (2001) Conservation agriculture: a worldwide challenge. In: *Proceedings of the 1st World Congress on Conservation Agriculture, Madrid, Spain, 1–5 October 2001.* XUL, Cordova, Spain, p. 387.

Gerik, T.J., Morrison, J.E., Jr and Chichester, F.W. (1987) Effects of controlled-traffic on soil physical properties and crop rooting. *Agronomy Journal* 79, 434–438.

Giles, D.M. (1994) Crop residue decomposition under no-tillage in Manawatu. Dissertation, Department of Agricultural Engineering Library, Massey University, Auckland, New Zealand, 67 pp.

Gilley, J.E. and Doran, J.W. (1997) Tillage effects on soil erosion potential and soil quality of a former conservation reserve program site. *Journal of Soil and Water Conservation* 52, 184–188.

Goh, K.M. (2004) Carbon sequestration and stabilization in soils: implications for soil productivity and climate change. *Soil Science and Plant Nutrition* 50, 467–476.

Govaerts, B., Sayre, K.D. and Deckers, J. (2004) Stable high yields with zero-tillage and permanent bed planting. *Field Crops Research.* Article in press but available online at www.sciencedirect.com

Grovers, G., Vandaele, K., Desmet, P.J.J., Poesen, J. and Bunte, K. (1994) The role of tillage in soil redistribution on hillslopes. *European Journal of Soil Science* 45, 469–478.

Grabski, A.S., Schafer, B.M. and Desborough, P.J. (1995) A comparison of the impact of 14 years of conventional and no-till cultivation on physical properties and crop yields of a loam soil at Grafton, NSW. In: *Proceedings of the National Controlled Traffic Conference, Rockhampton, Australia*, pp. 97–102.

Grant, F.R. (1997) Changes in soil organic matter under different tillage and rotations: Mathematical modeling in ecosystems. *Soil Science Society of America Journal* 61, 1159–1175.

Graves, D.W. (1994) No tillage: soil inversion in perspective. In: *Proceedings of the Federation of Organic Agriculture Movements, Canterbury, New Zealand* (in press).

Green, M. and Eliason, M. (1999) *Equipment Issues in Crop Residue Management for Direct Seeding.* Factsheets in Direct Seeding, November, Alberta Agriculture, Food and Development.

Hadfield, A.B. (1993) Seeding depth effects on the performance of wheat and lupin seedlings under no-tillage. Thesis, Massey University Library, Auckland, New Zealand, 115 pp.

Haggar, R.J. (1977) Herbicides and low-cost grassland establishment, with special reference to clean seedbeds and one-pass seeding. In: *Proceedings of the International Conference on Energy Conservation in Crop Production, Palmerston North, New Zealand.* Massey University, Palmerston North, New Zealand, 13 pp. 31–38.

Hall, D.G.M., Reeve, M.J., Thomasson, A.J. and Wright, V.F. (1977) *Water Retention, Porosity and Density of Field Soils.* Soil Survey Technical Monograph No. 9, Rothamsted Experimental Station, Harpenden, UK, 75 pp.

Hamilton-Manns, M. (1994) Effects of date of sowing on pasture establishment by no-tillage. Thesis, Massey University Library, Auckland, New Zealand, 138 pp.

Havlin, J.L., Kissel, D.E., Maddux, L.D., Classen, M.M. and Long, J.H. (1990) Crop rotation and tillage effects on soil organic carbon and nitrogen. *Soil Science Society of America Journal* 54, 448–452.

Heege, H.J. (1993) Seeding methods performance for cereals, rape and beans. *Transactions of the American Society of Agricultural Engineers* 36(3), 653–661.

Heywood, A.L. (1977) Selected plant and soil responses to three methods of seedbed preparation. Dissertation, Department of Agricultural Engineering Library, Massey University, Auckland, New Zealand, 87 pp.

Hobbs, P.R. and Gupta, R.K. (2003) Resource-conserving technologies for wheat in rice–wheat systems. In: Ladha, J.K., Hill, J., Gupta, R.K., Duxbury, J. and Buresh, R.J. (eds) *Improving the Productivity and Sustainability of Rice–Wheat Systems: Issues and Impact.* ASA Special Publication 65, Agronomy Society of America, Madison, Wisconsin, pp. 149–171.

Hobbs, P.R. and Gupta, R.K. (2004) Problems and challenges of no-till farming for the rice–wheat systems of the Indo-Gangetic plains in South Asia. In: Lal, R., Hobbs, P., Uphoff, N. and Hansen, D.O. (eds) *Sustainable Agriculture and the Rice–Wheat System.* Ohio State University, Columbus, Ohio, pp. 101–119; Marcel Dekker, New York, pp. 101–121.

Hollbrook, J.R. (1995) Direct drilling in the UK and Europe and subsequent research into the effect of light on development of wheat. In: *Proceedings of the 2nd EC Workshop of the Concerted Action on Experiences with the Applicability of No-tillage Crop Production in the West European Countries, Silsoe, 15–17 May 1995.* Justus Liebig University, Giessen, Germany.

Houghton, R.A., Hobbie, J.A., Melillo, J.M., More, B., Peterson, B.J., Shaver, G.R. and Woodwell, G.M. (1983) Changes in the carbon content of terrestrial biota in soils between 1860 and 1980: a net release of CO_2 to the atmosphere. *Ecological Monographs* 53, 235–262.

Hudson, B.D. (1994) Soil organic matter and available water capacity. *Journal of Soil and Water Conservation* 49(2), 189–194.

Hughes, K.A. (1975) A study of tillage mechanisms in relation to soil properties and crop growth. Thesis, Massey University Library, Auckland, New Zealand, 139 pp.

Hughes, K.A. and Baker, C.J. (1977) The effects of tillage and zero-tillage systems on soil aggregates in a silt loam. *Journal of Agricultural Engineering Research* 22, 291–301.

Hyde, G.M., Johnson, G.E., Simpson, J.B. and Payton, D.M. (1979) Grain drill design concepts for Pacific Northwest conservation farming. In: *Winter Meeting of the American Society of Agricultural Engineers, St Joseph, Missouri, USA.* American Society of Agricultural Engineers, St Joseph, Missouri, Paper no. 79–1525.

Hyde, G.M., Wilkins, D.E., Saxton, K.E., Hammel, J., Swanson, G., Hermanson, R., Dowding, E.A., Simpson, J.B. and Peterson, C.L.M. (1987) Reduced tillage seeding equipment developments. In: Elliot, L.F. (ed.) *STEEP – Conservation Concepts and Accomplishments.* Washington State University Press, Pullman, Washington.

Hyde, G.M., George, J.E., Saxton, K.E. and Simpson, J.B. (1989) Slot mulch insertion machine: design and performance. In: *American Society of Agricultural Engineers Winter Conference, Chicago, Illinois, USA*, Paper no. 89–1502.

Inwood, S.B. (1990) The effects on seedling performance of different establishment techniques and drilling patterns in pasture renewal. Dissertation, Department of Agricultural Engineering Library, Massey University, Auckland, New Zealand, 61 pp.

IPCC (2000) *Land Use, Land-use Change, and Forestry.* Special Report of the Intergovernmental Panel on Climate Change, Cambridge University Press, Cambridge, UK, 377 pp.

Javed, T. (1992) Determination of drag torque of the rotating disc of the 'Bioblade' opener. Thesis, Massey University Library, Auckland, New Zealand, 162 pp.

Justice, S. and Biggs, S. (2004a) Socially responsible rural mechanization processes and policies in Nepal. *Agricultural Research and Extension Network Newsletter* 50, July.

Justice, S. and Biggs, S. (2004b) Socially equitable mechanisation in Nepal. *Appropriate Technology* 31, March, 1.

Justice, S., Haque, E., Meisner, C., Hossain, I., Sah, G., Tripathi, J., Rashid, M.H. and Amin, M.R. (2004) Giving South Asia farmers a choice: a single drill for reduced and strip till crops for 2-wheel tractors. Paper presented at the 2004 International Commission on Agricultural Engineering (CIGR) International Conference, Beijing, China, 11–14 October 2004.

Kanchanasut, P., Scotter, D.R. and Tillman, R.W. (1978) Preferential solute movement through larger soil voids, 11 experiments with saturated soil. *Australian Journal of Soil Research* 16, 257–267.

Karonka, P. (1973) Machinery development for direct drilling. *Outlook on Agriculture* 7(4), 190–195.

Kern, J.S. and Johnson, M.G. (1993) Conservation tillage impacts on national soil and atmospheric carbon levels. *Soil Science Society of America Journal* 57, 200–210.

Kim, D.A. (1971) The effects of sowing and renovation techniques on the establishment of, and competition between, pasture species. Thesis, Massey University Library, Auckland, New Zealand, 152 pp.

Kimble, J.M., Lal, R. and Follett, R.R. (2002) Agricultural practices and policy options for carbon sequestration: what we know and where we need to go. In: Kimble, J.M. *et al.* (eds) *Agricultural Practices and Policies for Carbon Sequestration in Soil.* CRC Press, Boca Raton, Florida, pp. 495–501.

King, J.A., Bradley, R.I., Harrison, R. and Carter, A.D. (2004) Carbon sequestration and saving potential associated with changes to the management of agricultural soils in England. *Soil Use and Management* 20, 394–402.

Kirby, E.J.M. (1993) Effect of sowing depth on seedling emergence, growth and development in barley and wheat. *Field Crops Research* 35, 101–111.

Kurachi, S.A.H., Costa, J.A.S., Bernardi, J.A., Silveira, G.M. and Coelho, J.L.D. (1993) *Avaliação Tecnológica: resultados de ensaios de mecanismos dosadores de sementes de semeadoras-adubadoras de precisão.* Boletim Técnico 28, Instituto Agronômico, Campinas.

Lafond, G.P., Derksen, D.A., Loeppky, H.A. and Struthers, D. (1994) An agronomic evaluation of conservation tillage systems and continuous cropping in East Central Saskatchewan. *Journal of Soil and Water Conservation* 49, 387–393.

Lal, R. (1984) Soil erosion from tropical arable lands and its control. *Advances in Agronomy* 37, 183–248.

Lal, R. (1989) Conservation tillage for sustainable agriculture: tropics vs temperate environments. *Advances in Agronomy* 42, 1073–1082.

Lal, R. (1997) Residue management, conservation tillage and soil restoration for mitigating greenhouse effect by CO_2 enrichment. *Soil and Tillage Research* 43, 81–107.

Lal, R. (1999) Global carbon pools and fluxes and the impact of agricultural intensification and judicious land use. In: *Prevention of Land Degradation, Enhancement of Carbon Sequestration and Conservation of Biodiversity through Land Use Change and Sustainable Land Management with a Focus on Latin America and the Caribbean.* World Soil Resources Report 86, FAO, Rome, pp. 45–52.

Lal, R. (2000) World cropland soils as a source or sink for atmospheric carbon. *Advances in Agronomy* 71, 145–191.

Lal, R. (2002) Why carbon sequestration in agricultural soils? In: Kimble, J.M. *et al.* (eds) *Agricultural Practices and Policies for Carbon Sequestration in Soil.* CRC Press, Boca Raton, Florida, pp. 21–30.

Lal, R. (2004) Carbon emission from farm operations. *Environment International* 30, 981–990.

Lal, R., Logan, T.J., Eckert, D.J., Dick, W.A. and Shipitalo, M.J. (1994a) Conservation tillage in the corn belt of the United States. In: Carter, M.R. (ed.) *Conservation Tillage in Temperate Agroecosystems.* CRC Press, Boca Raton, Florida, pp. 73–114.

Lal, R., Mahboubi, A.A. and Fausey, N.R. (1994b) Long-term tillage affects on Mollic Ochraqualf in northwestern Ohio. III. Soil nutrient profile. *Soil and Tillage Research* 15, 371–382.

Lal, R., Kimble, J.M., Follet, R.F. and Cole, V. (1998) *Potential of U.S. Cropland for Carbon Sequestration and Greenhouse Effect Mitigation.* Ann Arbor Press, Chelsea, Michigan, 128 pp.

Lamarca, C.C. (1998) *Stubble over the Soil.* American Society of Agronomy, Madison, Wisconsin, 245 pp.

Lamers, J.G., Perdok, U.D., Lumkes, L.M. and Klooster, J.J. (1986) Controlled traffic farming systems in the Netherlands. *Soil and Tillage Research* 8, 65–76.

Lane, P.M.S., Pottinger, R.P. and Kaligariff, P.J. (1993) Pasture renovation: costs and benefits. In: Pottinger, R.P., Lane, P.M.S. and Wilkins, J.R. (eds) *Pasture Renovation Manual,* 2nd edn. AgResearch, Hamilton, New Zealand, pp. 115–120.

Le Bissonnais, Y. (1990) Experimental study and modelling of soil surface crusting processes. In: Bryan, R.B. (ed.) *Soil Erosion: Experiments and Models.* Catena Verlag, Cremlingen-Destedt, pp. 13–28.

Lee, K.E. (1985) *Earthworms: Their Ecology and Relationship with Soils and Land Use.* Academic Press, New York.

Li, Y.X., Tullberg, J.N. and Freebairn, D.M. (2001) Traffic and residue cover effects on infiltration. *Australian Journal of Soil Research* 39, 239–247.

Lindstrom, M.J., Nelson, W.W. and Schumacher, T.E. (1992) Quantifying tillage erosion rates due to moldboard plowing. *Soil and Tillage Research* 24, 243–255.

Little, C.E. (1987) *Green Fields Forever: The Conservation Tillage Revolution in America.* Island Press, Washington, DC, 192 pp.

Lobb, D.A. and Kachanoski, R.G. (1999) Modelling tillage translocation using steppe, near plateau, and exponential functions. *Soil and Tillage Research* 51, 261–277.

Lobb, D.A., Kachanoski, R.J. and Miller, M.H. (1995) Tillage translocation and tillage erosion on shoulder slope landscape positions measured using 137Cesium as a tracer. *Canadian Journal of Soil Science* 75, 211–218.

Lobb, D.A., Lindstrom, M.J., Quine, T.A. and Govers, G. (2000) Tillage at the threshold of the 21st century: new directions in response to tillage translocation and tillage erosion. In: *Proceedings of the 15th Conference of the International Soil and Tillage Research Organization, Elsevier, Netherlands, Fort Worth, Texas, 2–6 July 2000.*

Logan, T.J., Lal, R. and Dick, W.A. (1991) Tillage systems and soil properties in North America. *Soil and Tillage Research* 20, 241–270.

Lücke, W. and von Hörsten, D. (2004) Design and development of a new combine straw chopper with huge spreading range. In: *EurAgEng 2004: Proceedings of the AgEng 2004 Conference, Belgium.*

Lynch, J.M. (1977) Phytotoxicity of acetic acid produced in an anaerobic decomposition of wheat straw. *Journal of Applied Bacteriology* 10, 131–135.

Lynch, J.M. (1978) Production of a phototoxicity of acetic acid in anaerobic soils containing plant residues. *Journal of Soil Biology* 10, 131–135.

Lynch, J.M., Ellis, F.A.B., Harper, S.H.T. and Christian, D.G. (1980) The effect of straw on the establishment and growth of winter cereals. Agricultural Research Council report. *Agriculture and the Environment* 5, 321–328.

Mai, T.V. (1978) The effects of drill coulter designs on soil physical properties and plant responses in untilled seedbeds. Thesis, Massey University Library, Auckland, New Zealand, 220 pp.

Martin, D.L. and Thrailkill, D.J. (1993) Moisture and humidity requirements for germination of surface seeded corn. *Applied Engineering in Agriculture* 9(1), 43–48.

Mason, R.M., Page, J.R., Tullberg, J.N. and Buttsworth, R.K. (1995) The economics of controlled traffic: South Burnett case study. In: *Proceedings of the National Controlled Traffic Conference, Rockhampton, Queensland, 13–14 September*, pp. 109–114.

McKenzie, A.F., Fan, M.X. and Cardin, F. (1998) Nitrous oxide emission in three years as affected by tillage, corn–soybeans–alfalfa rotations and nitrogen fertilization. *Journal of Environmental Quality* 27, 698–703.

McQueen, D.J. and Shepherd, T.G. (2002) Physical changes and compaction sensitivity of a fine-textured, poorly drained soil (Typic Endoaquept) under varying durations of cropping, Manawatu Region, New Zealand. *Soil and Tillage Research* 63(3–4), 93–107.

Meek, B.D., DeTar, W.R., Rolph, D., Rechel, E.R. and Carter, L.M. (1990) Infiltration rate as affected by an alfalfa and no-till cropping system. *Soil Science Society of America Journal* 54(2), 505–508.

Meisner, C.A., Bodruzzaman, M., Amin, M.R., Baksh, E., Hossain, A.B.S., Ahmed, M. and Sadat, M.A. (2003) Conservation tillage options for the poor, small landholders in South Asia. In: *Conservation Agriculture, 2003.* Kluwer Academic Publishers, The Netherlands.

Milne, G.D. and Fraser, T. (1990) Establishment of 1600 hectares of dryland species around Oamaru/Timaru. *Proceedings of the New Zealand Grasslands Association* 52, 133–137.

Milne, G.D., Moloney, S.C. and Smith, D.R. (1993) Demonstration of dryland species on 90 east coast North Island farms. *Proceedings of the New Zealand Grassland Association* 55, 39–44.

Mitchell, J. (1983) Direct drilling project. Dissertation, Department of Agricultural Engineering Library, Massey University, Auckland, New Zealand, 20 pp.

Mitchell, J.P., Munk, D.S., Prys, B., Klonsky, K.K., Wroble, J.F. and De Moura, R.L. (2004a) Conservation tillage cotton production systems in the San Joaquin Valley. In: *Proceedings of the Western States Conservation Tillage Conference, 8–9 September 2004, Five Points, California. Kearney Agricultural Center*, University of California Agriculture and Natural Resources, Parlier, California, pp. 57–64.

Mitchell, J.P., Jackson, L. and Miyao, G. (2004b) *Minimum Tillage Vegetable Crop Production in California.* Publication 8132, University of California Agriculture and Natural Resources, Oakland, California, 9 pp.

Moens, R. (1989) Factors affecting slug damage and control measure decisions. In: Henderson, I. (ed.) *Slugs and Snails in World Agriculture.* Monograph 41, The British Crop Protection Council, Thornton Heath, pp. 227–236.

Monegat, C. (1991) *Plantas de Cobertura do Solo. Características e manejo em pequenas propriedades.* Chapecó, Brazil, 337 pp.

Morrison, J.E. (1988a) Interactive planter depth control and pneumatic downpressure system. *Transactions of the American Society of Agricultural Engineers* 31(1), 14–18.

Morrison, J.E. (1988b) Hydraulic downpressure system performance for conservation planting machines. *Transactions of the American Society of Agricultural Engineers* 31(2), 19–23.

Morrison, J.E., Allen, R.R., Jr, Wilkins, D.E., Powell, G.M., Grisso, R.D., Erbach, D.C., Herdon, L.P., Murray, D.L., Formanek, G.E., Pfost, D.L., Herron, M.M. and Baumert, D.J. (1988) Conservation planter, drill and air-type seeder selection guidelines. *Applied Engineering in Agriculture* 4(4), 300–309.

Mosier, A., Shimel, D., Valentine, D., Bronson, K. and Parton, W. (1991) Methane and nitrous oxide fluxes in native, fertilized and cultivated grasslands. *Nature* 350, 330–332.

Mosier, A., Kroez, C., Nevison, C., Oenema, O., Seitzinger, S. and Van Cleemput, O. (1998) Closing the global N_2O budget: nitrous oxide emissions through the agricultural nitrogen cycle. *Nutrient Cycling in Agroecosystems* 52, 225–248.

Nichols, K.A. and Wright, S.F. (2004) Contributions of soil fungi to organic matter in agricultural soils. In: Magdoff, F. and Weil, R. (eds) *Functions and Management of Soil Organic Matter in Agroecosystems.* CRC Press, Boca Raton, Florida, pp. 179–198.

Nix, J. (2001) *Farm Management Pocketbook,* 31st edn. Imperial College at Wye, Ashford, Kent, UK, 256 pp.

Pangnakorn, U., George, D.L., Tullberg, J.N. and Gupta, M.L. (2003) Effect of tillage and traffic on earthworm populations in a vertisol in South-East Queensland. In: *Proceedings of the International Soil Tillage Research Organisation 16th Triennial Conference, Brisbane, 13–18 July,* Elsevier, Netherlands, pp. 881–885.

Papendick, R.I., McCool, D.K. and Krauss, H.A. (1983) Soil conservation: Pacific Northwest. In: Dregne, H.E. and Willis, W.O. (eds) *Dryland Agriculture.* Agronomy Monograph 23, ASA, CSSA, and SSSA, Madison, Wisconsin.

Paul, E., Paustian, K., Elliott, E.T. and Cole, C.V. (eds) (1997) *Soil Organic Matter in Temperate Agroecosystems: Long-term Experiments in North America.* CRC Press, Boca Raton, Florida.

Paustian, K., Andren, O., Janzen, H.H., Lal, R., Smith, P., Tian, G., Tiessen, H., VanNoordwijk, M. and Woomer, P.L. (1997a) Agricultural soils as a sink to mitigate CO_2 emissions. *Soil Use Management* 13, 230–244.

Paustian, K., Collins, H.P. and Paul, E.A. (1997b) Management controls on soil carbon. In: Paul, E.A. *et al.* (eds) *Soil Organic Matter in Temperate Agroecosystems: Long-term Experiments in North America.* CRC Press, Boca Raton, Florida, pp. 15–49.

Pellow, R.W. (1992) The potential for transplanting seedlings by no-tillage. Thesis, Massey University Library, Auckland, New Zealand, 87 pp.

Pidgeon, J.D. (1981) A preliminary study of minimum tillage systems (including broadcasting) for spring barley in Scotland. *Soil and Tillage Research* 1, 139–151.

Poesen, J., Wesenael, B., Govers, G., Martinez-Fernadez, J., Desmet, B., Vandaele, K., Quine, T. and Degraer, G. (1997) Patterns of rock fragment covered generated by tillage erosion. *Geomorphology* 18, 193–197.

Pottinger, R.P., Lane, P.M.S. and Wilkins, J.R. (eds) (1993) *Pasture Renovation Manual.* AgResearch. Hamilton, New Zealand, 151 pp.

Praat, J.P. (1988) A study of seeding depth with direct drilled peas. Dissertation, Department of Agricultural Engineering Library, Massey University, Auckland, New Zealand, 78 pp.

Praat, J.P. (1995) Row spacing and seedling rate interactions in perennial ryegrass and tall fescue swards established by direct drilling. PhD thesis, Massey University Library, Auckland, New Zealand, 213 pp.

Pradhan, G., Khan, S.U., Justice, S. and Gami, V.K. (1997) Farmer participatory research on the Chinese hand tractor. In: Hobbs, P. and Rajbandari, N. (eds) *Proceedings of the Rice–Wheat Research End-of-Project Workshop, 1–3 October 1997, Hotel Shangrala, Kathmandu, Nepal.* Nepal Agricultural Research Council (NARC), International Maize and Wheat Improvement Centre (CIMMYT) and Rice–Wheat Research Consortium, Mexico DF.

Prior, S.A., Runion, G.B., Torbert, H.A., Rogers, H.H. and Reeves, D.W. (2003) Effects of elevated atmospheric CO_2 on biomass production and C sequestration: conventional and conservation cropping systems. In: *Proceedings of the International Soil Tillage Research Organisation 16th Triennial Conference, Brisbane, 13–18 July.* Elsevier, Netherlands, pp. 943–947.

Radford, B.J., Davis, R.J., McGarry, D., Pillai, U.P., Rickman, J.F., Walsh, P.A. and Yule, D.F. (2000) Changes in the properties of a Vertisol and responses of wheat after compaction with harvester traffic. *Soil and Tillage Research* 54, 155–170.

Radford, B.J., Wilson-Rummenie, A.C., Simpson, G.B., Bell, K.L. and Ferguson, M.A. (2001) Compacted soil affects soil macrofauna populations in a semi-arid environment in central Queensland. *Soil Biology and Biochemistry* 33(12–13), 1869–1872.

Radford, R. and Yule, D. (2003) Compaction persists for five years after cessation of wheel traffic on a Vertisol in Central Queensland. In: *Proceedings of the International Soil Tillage Research Organisation 16th Triennial Conference, Brisbane, 13–18 July.* Elsevier, Netherlands, pp. 949–954.

Ralisch R. *et al.* (1998) In: *Encontro Latinoameriacno de Plantio Direto NA Pequena Propriedade, III.* IAPAR, Pato Branco.

Rasmussen, P.E. and Rohde, C.R. (1988) Long-term tillage and nitrogen fertilization affects on organic N and C in a semi-arid soil. *Soil Science Society of America Journal* 44, 596–600.

Rawson, H.M. (2004) *Report of the Team Leader, FAO Second Mission: March 9 to April 13, 2004. Project Titled 'Intensification of Sustainable Production of Wheat and Rice–Wheat Systems'.* Food and Agriculture Organization of the United Nations, Rome, Italy.

Reicosky, D.C. (1990) Canopy gas exchange in the field: closed chambers. *Remote Sensing Reviews* 5(1), 163–177.

Reicosky, D.C. (1997) Tillage-induced CO_2 emission from soil. *Nutrient Cycling in Agroecosystems* 49, 273–285.

Reicosky, D.C. (1998) Strip tillage methods: impact on soil and air quality. In: Mulvey, P.J. (ed.) *Environmental Benefits of Soil Management. Proceedings of the ASSSI National Soils Conference, Brisbane, Australia*, pp. 56–60.

Reicosky, D.C. (2001a) Conservation agriculture: global environmental benefits of soil carbon management. In: Garcia-Torres, L. *et al.* (eds) *Conservation Agriculture: A Worldwide Challenge.* XUL, Cordoba, Spain, pp. 3–12.

Reicosky, D.C. (2001b) Tillage-induced CO_2 emissions and carbon sequestration: effect of secondary tillage and compaction. In: Garcia-Torres, L. *et al.* (eds) *Conservation Agriculture: A Worldwide Challenge.* XUL, Cordoba, Spain, pp. 265–274.

Reicosky, D.C. and Lindstrom, M.J. (1993) Fall tillage method: effect on short-term carbon dioxide flux from soil. *Agronomy Journal* 85, 1237–1243.

Reicosky, D.C. and Lindstrom, M.J. (1995) Impact of fall tillage on short-term carbon dioxide flux. In: Lal, R. *et al.* (eds) *Soils and Global Change.* CRC Press, Boca Raton, Florida, pp. 177–187.

Reicosky, D.C., Wagner, S.W. and Devine, O.J. (1990) Methods of calculating carbon dioxide exchange rates for maize and soybean using a portable field chamber. *Photosynthetica* 24(1), 22–38.

Reicosky, D.C., Kemper, W.D., Langdale, G.W., Douglas, C.W., Jr and Rasmussen, P.E. (1995) Soil organic matter changes resulting from tillage and biomass production. *Journal of Soil and Water Conservation* 50, 253–261.

Reicosky, D.C., Kemper, W.D., Langdale, G.W., Douglas, C.W. Jr., Rasmussen, P.E. (1996) Soil organic matter changes resulting from tillage and biomass production. *Proceedings National No-Tillage Conference, St Louis, Missouri, USA.* 07–105. Lessiter Publications, Brookfield, Wisconsin.

Reicosky, D.C., Evans, S.D., Cambardella, C.A., Allmaras, R.R., Wilts, A.R. and Huggins, D.R. (2002) Continuous corn with moldboard tillage: residue and fertility effects on soil carbon. *Journal of Soil and Water Conservation* 57(5), 277–284.

Ribeiro, M.F.S. (2004) *Farmer's Perception, Impacts and Dissemination Issues of Conservation Agriculture/Labour Saving Technologies. Report of IFAD/FAO/DMC. Case Study in Tanzania*, 61 pp.

Ribeiro, M.F.S., *et al.* (1993) Implicacoes do plantia direto em pequenas propriedades no Centro-Sul do Parana. In: *Encontro Latinamericano sobre Plantio Direto na Pequena Propriedade 1.* Instituto Agronomico do Parana, Ponta Grossa, Brazil, pp. 271–280.

Ribeiro, M.F.S., Benassi, D.A., Siqueira, E.M., Milleo, R.D.S., Hennerich, J.E., Richter, S. and Krieck, R. (1998) Avaliação de semeadoras de plantio direto a tração animal, safra 1999/2000. In: *Encontro Latinoamericano de Plantio Direto NA Pequena Propriedade, III, 1998.* IAPAR, Pato Branco.

Ritchie, W.R. (1982) Aspects of seed transfer within a direct drilling coulter (opener). Thesis, Massey University Library, Auckland, New Zealand, 148 pp.

Ritchie, W.R. (1986a) Pasture renovation by overdrilling. *Proceedings of the New Zealand Grasslands Association* 47, 159–164.

Ritchie, W.R. (1986b) Pasture renovation or pasture renewal: *New Zealand Dairy Farming Annual* 136–143.

Ritchie, W.R. (1988) Turf renovation – an equipment viewpoint. *New Zealand Turf Management Journal* 2(1), 20–22.

Ritchie, W.R. and Baker, C.J. (1987) A field programme for evaluating direct drilling technology. *Proceedings of the Agronomy Society of New Zealand* 17, 77–82.

Ritchie, W.R. and Cox, T.I. (1981) Towards precision sowing in uncultivated ground. In: *Proceedings of the Conservation Tillage Technical Seminar, Christchurch, New Zealand.* Monsanto (NZ) Ltd, Johnsonville, New Zealand, pp. 123–132.

Ritchie, W.R., Wrigley, M.P. and Hill, K. (1994) The seed-rate debate – the last word? *New Zealand Turf Management Journal* August, 26–27.

Ritchie, W.R., Baker, C.J. and Hamilton-Manns, M. (2000) *Successful No-tillage in Crop and Pasture Establishment.* Caxton Press, New Zealand, 96 pp.

Rizzardi, M., Boller, W. and Dalloglio, R. (1994) Distribuição de plantas de milho, na linha de semeadura, e seus efeitos nos componentes de produção. *Pesquisa Agropecuária Brasileira* 29(8), pp.1231–1236.

Robert, M. (1996) Aluminum toxicity a major stress for microbes in the environment. In: Huang, P.M. *et al.* (eds) *Environmental Impacts,* Vol. 2, *Soil Component Interactions.* CRC Press, Boca Raton, Florida, pp. 227–242.

Robertson, G.P., Paul, E. and Harwood, R.R. (2000) Greenhouse gases in intensive agriculture: contribution of individual guesses to the radiate of forcing of the atmosphere. *Science* 289, 1922–1925.

Robinson, G.S. (1957) What pastures should be overdrilled? In: *Massey Agricultural College Dairy Farmers' Conference, New Zealand*, pp. 71–76.

Robinson, G.S. and Cross, M.W. (1957) Overdrilling pumice pastures. *New Zealand Journal of Agriculture* 95, 283–288.

Ross, C. (2001) *A Tick for No-till.* Soil Horizons Issue 5, Manaaki Whenua – Landcare Research, Lincoln, New Zealand.

Ross, C. (2002) The structure and function of untilled soils. In: *Proceedings of the Monsanto Conservation Agriculture Seminars, February, Taupo and Ashburton.* Monsanto (NZ) Ltd, Johnsonville, New Zealand, pp. 27–37.

Ross, C., Saggar, S., Yeates, G.W., Dando, J. and Shepherd, T.G. (2000) Soil quality under long-term cropping by no-tillage and conventional cultivation, and permanent pasture in the Manawatu. In: *Extended Abstracts, Proceedings of the NZSSS/ASSSI Soil 2000 Conference, 3–8 December, Lincoln University.* New Zealand Society of Soil Science and Australian Society of Soil Science.

Ross, C., Dando, J., Saggar, S., Yeates, G. and Shepherd, T.G. (2002a) Soil quality under long-term cropping by no-tillage compared with conventional cultivation, and with permanent pasture. In: *Proceedings Soil Quality and Sustainable Land Management Conference, NZ Association Resource Management, 3–5 April 2002.* Massey University, Palmerston North, New Zealand, pp. 41–45.

Ross, C., Saggar, S., Yeates, G., Dando, J. and Shepherd, T.G. (2002b) Soil quality under long-term cropping by no-tillage compared with conventional cultivation and permanent pasture. In: Currie, L.D. and Loganathan, P. (eds) *Dairy Farm Soil Management.* Occasional Report No. 15, Fertilizer and Lime Research Centre, Massey University, Palmerston North, New Zealand, pp. 119–126.

Row, C., Sampson, R.N. and Hair, D. (1996) Environmental and land-use changes from increasing forest area and timber growth. In: Sampson, R.N. and Hair, D. (eds) *Forest and Global Change,* Vol. 2: *Forest Management Opportunities for Mitigating Carbon Emissions.* American Forests, Washington, DC.

Rice–Wheat Consortium (RWC) Website. Available at: http://www.rwc-prism.cgiar.org/rwc

Ryan, D.L., Hay, R.J.M. and Baker, C.J. (1979) Response of some ryegrass cultivars to over-drilling in Southland. *Proceedings of the New Zealand Grassland Association* 40, 136–144.

Sa, J.C.D., Cerri, C.C., Dick, W.A., Lal, R., Venske, S.P., Piccolo, M.C. and Feigl, B.E. (2001) Organic matter dynamics and carbon sequestration rates for a tillage chronosequence in a Brazilian Oxisol. *Soil Science Society of America Journal* 65, 1486–1499.

Sah, R.P., Tripathi, J., Sah, G. and Bhattarai, S. (2004) National Coordinators Report for the Regional Technical Coordination Committee, RWC. In: *Proceedings of RSC/RTCC Meeting, Islamabad, Pakistan, January 2004.*

Salmon, R. (2005) Brassica seed treatment trials. In: *Proceedings of Cross Slot No-Tillage Systems Annual Conference,* Methven, New Zealand. Baker No-Tillage Ltd, Feilding, New Zealand.

Sangoi, L. (1990) Arranjo de plantas e características de genótipos de miilho em sois níveis de fertilidade. *Pesquisa Agropecuária Brasileira* 25(7), 945–953.

Saxton, K.E. (1988) Conservation of soil water by residue and tillage. In: *Proceedings of the International Commission on Irrigation and Drainage Symposium.* Vol. 4, *Regional and Local Water Management Systems.* Dubrovnik, Yugoslavia, pp. 291–298.

Saxton, K.E. (1990) Criteria for conservation-tillage and the cross-slot opener. In: *Proceedings of the 12th Annual Manitoba North Dakota Zero-Tillage Workshop, January 17–19, Bismark, North Dakota, USA,* pp. 69–80.

Saxton, K.E. and Baker, C.J. (1990) The Cross Slot drill opener for conservation tillage. In: *Proceedings of the Great Plains Conservation Tillage Symposium, Bismark, North Dakota, USA*, Great Plains Agricultural Council Bulletin No. 131, Bismark, North Dakota, pp. 65–72.

Saxton, K.E., Bristow, K.L., Omer, M.A. and Flerchinger, G.N. (1988a) Tillage and crop residue management for water conservation. In: *Proceedings International Conference on Dryland Farming, Amarillo, Texas, USA.* Texas Agricultural Experiment Station, Bushland, Amarillo, Texas, pp. 493–497.

Saxton, K.E., McCool, D.K. and Kenny, J.F. (1988b) Tillage and residue impacts on infiltration. In: *Conference Proceedings of International Conference on Infiltration Development and Application, Honolulu, Hawaii, USA*, pp. 508–516.

Sayre, K.D. and Hobbs, P.R. (2004) The raised-bed system of cultivation for irrigated production conditions. In: Lal, R., Hobbs, P., Uphoff, N. and Hansen, D.O. (eds) *Sustainable Agriculture and the Rice–Wheat System.* Ohio State University, Columbus, Ohio, pp. 337–355.

Schimitz, H., Sommer, M. and Walter, S. (1991) *Animal Traction in Rainfed Agriculture in Africa and South America: Determinants and Experiences.* GATE/GTZ, Wieveg, pp. 102–103.

Schlesinger, W.H. (1985) Changes in soil carbon storage and associated properties with disturbance and recovery. In: Trabalha, J.R. *et al.* (eds) *The Changing Carbon Cycle: A Global Analysis.* Springer Verlag, New York, pp. 194–220.

Schumacher, T.E., Lindstrom, M.J., Schumacher, J.A. and Lemme, G.D. (1999) Modelling spatial variation and productivity due to tillage and water erosion. *Soil and Tillage Research* 51, 331–339.

Scotter, D.R. (1976) Liquid and vapour phase transport in soil. *Australian Journal of Soil Research* 14, 33–41.

Shepherd, T.G., Parshotam, A. and Newman, R.H. (2006) Dynamics of organic carbon fractions and physical properties of a clayey soil continuously cropped and then returned to pasture. *European Journal of Soil Science* (in press).

Siqueira, R. and Araújo, A.G. (1999) Máquinas para o manejo mecânico das planta e dos resíduos culturais. In: Castro Filho, C. and Muzilli, O. (eds) *Uso e Manejo dos Solos de Baixa Aptidão Agrícola.* IAPAR Circular Técnica, Instituto Agronômico do Paraná, Londrina, Brazil, pp. 124–138.

Siqueira, R. and Casão, R., Jr (2004) *Trabalhador no cultivo de grãos e oleaginosas: máquinas para manejo de coberturas e semeadura no sistema plantio direto.* SENAR, Curitiba, Brazil, 87 pp.

Sisti, C.P.J., dos Santos, H.P., Kohhann, R., Alves, B.J.R., Urquiaga, S. and Boddey, R.M. (2004) Changing carbon and nitrogen stocks in soil under 13 years of conventional or zero tillage in southern Brazil. *Soil and Tillage Research* 76, 39–58.

Six, J., Conant, R.T., Paul, E.A. and Paustian, K. (2002) Stabilization mechanisms of soil organic matter: implications for C-saturation of soils. *Plant Soil* 241, 155–176.

Six, J., Ogle, S.M., Breidt, F.J., Conant, R.T., Mosier, A.R. and Paustian, K. (2004) The potential to mitigate global warming with no-tillage management is only realized when practiced in the long-term. *Global Change Biology* 10, 155–160.

Skidmore, E.L., Kumar, M. and Larson, W.E. (1979) Crop residue management for wind erosion control in the Great Plains. *Journal of Soil and Water Conservation* 34, 90–94.

Smith, O.H., Petersen, G.W. and Needelman, B.A. (2000) Environmental indicators of agroecosystems. *Advances in Agronomy* 69, 75–97.

Smith, P., Powlson, D.S., Glendining, M.J. and Smith, J.U. (1998) Preliminary estimates of the potential for carbon mitigation in European soils through no-till farming. *Global Change Biology* 4, 679–685.

Soane, B.D. (1990) The role of organic matter in the soil compactibility: a review of some practical aspects. *Soil and Tillage Research* 16, 179–202.

Spoor, G. (1994) Application of mole drainage in the solution of subsoil management problems. In: Jayawardane, N.S. and Stewart, B.A. (eds) *Advances in Soil Science: Subsoil Management Techniques.* Lewis Publishers, Chelsea, Michigan, pp. 67–107.

Spoor, G., Tijink, F.G.J. and Weisskopf, P. (2003) Subsoil compaction: risk, avoidance, identification and alleviation. *Soil and Tillage Research* 73, 175–182.

Stibbe, E., Terpstra, R. and Kouwenhoven, J.K. (1980) Effect of spring tillage on seedbed characteristics. Plant establishment and yield of silage corn on a light sandy soil. *Soil and Tillage Research* 1, 47–57.

Stockfisch, N., Forstreuter, T. and Ehlers, W. (1999) Ploughing effects on soil organic matter after 20 years of conservation tillage in Lower Saxony, Germany. *Soil and Tillage Research* 52, 91–101.

Thom, E.R., Sheath, G.W., Bryant, A.M. and Cox, N.R. (1986) Renovation of pastures containing paspalum: 1. Persistence of overdrilled ryegrass and prairie grass and effect on seasonal pasture production. *New Zealand Journal of Agricultural Research* 29, 575–585.

Thom, E.R. and Ritchie, W.R. (1993) Banded versus blanket spraying and direct drilling. In: Pottinger, R.P., Lane, P.M.S. and Wilkins, J.R. (eds) *Pasture Renovation Manual.* AgResearch, Ruakura, Hamilton, New Zealand, pp. 55–58.

Thompson, C. (1993) The interactions between machine design and soil conditions influencing in-groove vertical seed placement by the Bioblade direct drill. Thesis, Department of Agricultural Engineering Library, Massey University, Auckland, New Zealand.

Tisdall, J.M. (1996) Formation of soil aggregates in the accumulation of soil organic matter. In: Carter, M.R. and Stewart, B.A. (eds) *Structure and Soil Organic Matter Storage in Agricultural Soils.* Advances in Soil Science, CRC Press, Boca Raton, Florida, pp. 57–96.

Torbert, H.A. and Reeves, D.W. (1995) Traffic and residue management systems: effects on fate of fertilizer N in corn. *Soil and Tillage Research* 33(3–4), 197–213.

Tracy, P.W., Westfall, D.G., Elliott, E.T., Peterson, G.A. and Cole, C.V. (1990) Carbon, nitrogen, phosphorus and sulfur mineralization in plow and no-till cultivation. *Soil Science Society of America Journal* 54, 457–461.

Unger, P.W. (1996) Soil bulk density, penetration resistance, and hydraulic conductivity under controlled traffic conditions. *Soil and Tillage Research* 37(1), 67–75.

United Nations Framework Convention on Climate Change Secretariat (1997) *Kyoto Protocol to the United Nations Framework Convention on Climate Change.* United Nations Framework Convention on Climate Change Secretariat, Bonn, Germany.

Uri, D.M. (2000) An evaluation of the economic benefits and costs of conservation tillage. *Environmental Geology* 39, 238–248.

Van Raij, B., Silva, N.M., Bataglia, O.C., Quaggio, J., Hiroce, R., Cantarella, H., Belinazzi, J.R., Jr, Dechen, A.R. and Trani, P.E. (1985) *Recomendações de adubação e calagem para o Estado de São Paulo.* Boletim Técnico 100, Instituto Agronômico, Campinas, Brazil, 107 pp.

Veseth, R., Saxton, K.E. and McCool, I.K. (1993) Tillage and residue management strategies for variable cropland. In: *Proceedings. Precision Farming Variable Cropland for Profit and Conservation, Pullman, Washington, USA*, pp. 33–45.

Vinther, F.P. (1992) Measured in simulated denitrification activity in a cropped sandy and loamy soil. *Biology and Fertility of Soils* 14, 43–48.

Wagger, M.G. and Denton, H.P. (1989) Influence of cover crop and wheel traffic on soil physical properties in continuous no-till corn. *Soil Science Society of America Journal* 53(4), 120–121.

Wakelin, A.M., Lorraine-Colwill, D.F. and Preston, C. (2004) Glyphosate resistance in four different populations of *Lolium rigidum* is associated with reduced translocation of glyphosate to meristematic zones. *Weed Research* 44, 453–459.

Walsh, M.J. (2002) Growing the market: recent developments in agricultural sector carbon trading. In: Kimble, J.M. *et al.* (eds) *Agricultural Practices and Policies for Carbon Sequestration in Soil.* CRC Press, Boca Raton, Florida, pp. 375–385.

Wander, M.M., Bidar, M.G. and Aref, S. (1998) Tillage impacts on depth distribution of total and particular organic matter in three Illinois soils. *Soil Science Society of America Journal* 62, 1704–1711.

Wang, X., Gao, H.W. and Li, H.W. (2003) Runoff and soil erosion from farmland under conservation tillage. In: *Proceedings of the International Soil Tillage Research Organisation 16th Triennial Conference, Brisbane, 13–18 July*, Elsevier, Netherlands, pp. 1347–1353.

Webb, B., Blackwell, P., Riethmuller, G. and Lemon, J. (2000) *Tramline Farming Systems Technical Manual.* Bulletin 4607, Department of Agriculture, Western Australia, Perth, 88 pp.

West, T.O. and Marland, G. (2002) A synthesis of carbon sequestration, carbon emissions and net carbon flux in agriculture: comparing tillage practices in the United States. *Agriculture, Ecosystems & Environment* 91, 217–232.

West, T.O. and Post, W.M. (2002) Soil organic carbon sequestration rates by tillage and crop rotation: a global data analysis. *Soil Science Society of America Journal* 66, 1930–1946.

Wilkins, D.E., Bolton, F. and Saxton, K.E. (1992) Evaluating seeders for conservation tillage production of peas. *Applied Engineering in Agriculture* 8(2), 165–170.

Wolkowski, R.P. (1990) Relationship between wheel-traffic-induced soil compaction, nutrient availability and crop growth: a review. *Journal of Production Agriculture* 3(4), 460–469.

Wolkowski, R.P. (1991) Corn growth response to K fertilization on three compacted soils. *Soil and Tillage Research* 21, 287–298.

Wuest, S.B. (2002) Water transfer from soil to seed: the role of vapor transport. *Soil Science Society of America Journal* 66(6), 1760–1763.

Yano, E.H. and Mello, L.M.M. (2000) Tamanho do fragmento de guandú (Cajanus cajan) manejado com triturador de palhas, rolo-faca e roçadora. In: *Congresso Brasileiro de Engenharia Agrícola, 29, Fortaleza, Anais. Sociedade Brasileira de Engenharia Agrícola.*

Yavuzcan, H.G. (2000) Wheel traffic impact on soil conditions as influenced by tillage system in Central Anatolia. *Soil and Tillage Research* 54(3–4), 129–138.

Young, D.L., Kwon, T.J. and Young, F.L. (1994) Profit and risk for integrated conservation farming systems in the Palouse. *Journal of Soil and Water Conservation* 49(6), 601–606

Index

Page numbers in **bold** refer to figures in the text; those in *italics* refer to tables or boxes

Printed and bound by CPI Group (UK) Ltd, Croydon, CR0 4YY
16/09/2026
14963056-0006